Lecture Notes in Mathematics

Edited by A. Dold and B. Eckmann

847

Ulrich Koschorke

Vector Fields and Other Vector Bundle Morphisms – A Singularity Approach

Springer-Verlag
Berlin Heidelberg New York 1981

Author

Ulrich Koschorke
The Institute for
Advanced Study
School of Mathematics
Princeton, NJ 08540/USA
and Mathematik V
Universität – Gesamthochschule Siegen
5900 Siegen 21
Federal Republic of Germany

AMS Subject Classifications (1980): 55 N 22, 55 S 35, 57 R XX

ISBN 3-540-10572-7 Springer-Verlag Berlin Heidelberg New York
ISBN 0-387-10572-7 Springer-Verlag New York Heidelberg Berlin

Printed in Germany

Printing and binding: Beltz Offsetdruck, Hemsbach/Bergstr.
2141/3140-543210

Math
Sep

TABLE OF CONTENTS

CHAPTER III - FRAMEFIELDS

INTRODUCTION.

Let M be a closed connected smooth n-dimensional manifold. A vectorfield v on M assigns a tangent vector v(x) to every point x in M.

Problem. When is there a continuous vectorfield without zeroes on M?

To answer this question, we pick a vectorfield v with isolated zeroes (this presents no difficulty since there are very many such vectorfields). Analyzing the behaviour of v around each zero, we obtain "local obstructions" whose sum is denoted by

$$\mathrm{Index}(v) \quad \in \quad \pi_{n-1}(S^{n-1}) \;=\; \mathbb{Z} .$$

Theorem of Poincaré-Hopf (see e.g. [28] (1926) or [49])

(1) There exists a nowhere vanishing vector field on M if and only if Index(v) = 0.

(2) Index(v) is the Euler number $\chi(M)$ of M.

Now consider an arbitrary integer $k \geq 1$. A k-field v on M assigns a whole system $v_1(x), v_2(x), \ldots, v_k(x) \in T_x M$ of tangent vectors to every point $x \in M$. The singularity of v is the locus

$$S \;=\; \{\, x \in M \mid v_1(x),\ldots,v_k(x) \text{ linearly dependent } \} .$$

A k-field with empty singularity is called a k-framefield. *)

We may generalize our original question as follows.

Basic problem. When is there a k-framefield on M?

The answer for k = 1, given above, is classical and one of the

*) This differs from the notations of E.Thomas [65] who already assumes k-fields to be nonsingular.

very central results in differential topology. For $k > 1$, the situation is drastically more complicated (e.g. even in the simple case when $k = 2$, and M is nonorientable and evendimensional, the solution was given only in the last decade). However, there are several powerful methods to attack the problem. For instance, one may start from a k-field with finite singularities and study its index which now lies in the homotopy group $\pi_{n-1}(V_{n,k})$ of the Stiefel manifold $V_{n,k}$ of k-frames in $\mathbb{R}^n$. There are drawbacks, though: Index(v) is not always defined (since a k-field with finite singularities need not exist for $k > 1$), and, if defined, the index will in general depend on the choice of v and not on M alone. Nevertheless, this approach has proved to be very fruitful. It has led to an analogon of the theorem of Poincaré-Hopf for orientable M and $k \leq 3$ (see the work of Thomas [65], Frank and especially Atiyah and Dupont [6]).

In this monograph we will adopt a different (and, in fact, quite opposite) approach.

Assume $n > 2k$. Consider a k-field v which is nondegenerate (as a (k-1)-morphism, see definition 1.4). This is a transversality condition which can be easily fulfilled. E.g. when $k = 1$, it just means that the vectorfield v is transverse to the zero section in the tangent bundle TM.

We will often identify v with the vector bundle homomorphism

$$v : M \times \mathbb{R}^k \longrightarrow TM$$

which maps (x, i-th standart basis vector) to $v_i(x) \in T_x M$. The nondegeneracy of v then implies the following facts.

(i) The singularity S of v forms a closed smooth (k-1)-dimensional submanifold of M. (So we are dealing with isolated points only when $k = 1$; for $k = 2,3,..$ S will be a union of circles, or a surface, etc. Clearly, v is injective over $M - S$).

(ii) At all points of S the kernel of v is one-dimensional.

In order to see whether the singularity can be removed altogether, we need to study the behaviour of v around S. The following data turn out to contain the relevant information:

(i) the smooth manifold S;

(ii) the continuous map

$$g = (\ker, \text{inclusion}) \; : \; S \longrightarrow P^{k-1} \times M ,$$

where the "kernel map" ker assigns to every $x \in S$ the kernel of $v_x : \mathbb{R}^k \longrightarrow T_x M$ (considered as a point in real projective (k-1)-space);

(iii) a vector bundle isomorphism $\bar{g}$ which describes the stable normal bundle of S in terms of the pullback (under g) of a certain (virtual) vectorbundle ϕ_M over $P^{k-1} \times M$.

If we start from another nondegenerate k-field v' on M, we can find a suitable nondegenerate vector bundle homomorphism V over $M \times I$ which restricts to v and v' over $M \times \{0\}$ and $M \times \{1\}$ respectively. Then the singularity data of V provide a bordism from the triple $(S, g, \bar{g})$ to the corresponding triple of singularity data of v'. So clearly we obtain a welldefined obstruction

$$\omega_k(M) = [S, g, \bar{g}] \in \Omega_{k-1}(P^{k-1} \times M; \phi_M)$$

where the expression to the right denotes the "normal bordism" group of triples such as $(S, g, \bar{g})$.

Theorem 13.3*. *There exists a k-framefield on M if and only if $\omega_k(M) = 0$.*

While $\omega_k(M)$ may be considered as a "characteristic class" lying in a group which varies with M, we can also define the "characteristic number"

$$\omega_k'(M) = [S, (\ker, \tau|S), \bar{g}] \in \Omega_{k-1}(P^{k-1} \times B(S)O(n); \phi)$$

where τ is a classifying map of TM (going into BSO(n) or BO(n) according to whether M is oriented or not), and ϕ is a suitable virtual vector bundle over $P^{k-1} \times B(S)O(n)$. Actually, $\omega_k'(M)$ does not really depend on M anymore, but only on the bordism class of M in the "fine" (or vector field) bordism group $\mathfrak{N}_n(0,0)$ or $\Omega_n(0,0)$ (here bordisms are required to carry nonvanishing vector fields pointing inward on one side and outward on the other, see also Reinhart [55]).

Theorem 13.4*. M is bordant (in $\mathfrak{N}_n(0,0)$ or $\Omega_n(0,0)$, respectively) to a manifold M' which allows a k-framefield if and only if $\omega_k'(M) = 0$.

Furthermore, $\omega_k'(M)$ depends only on the following classical invariants of M: in the unoriented theory on the Stiefel-Whitney numbers and the Euler number of M, and in the oriented theory in addition on the Pontrjagin numbers, and for $n \equiv 1(4)$, on the (real) Kervaire semicharacteristic $k(M)$ of M.

Let us check to what extend the previous results contain a Poincaré-Hopf type solution of our "basic problem". Theorem 13.3* generalizes fully the first part of the Poincaré-Hopf theorem (in the metastable range $n > 2k$). An analogon of the second part would amount to computing our obstruction $\omega_k(M)$ in terms of classical invariants of M. This can be split up into two problems: to get some control over the transition from $\omega_k(M)$ to $\omega_k'(M)$ (which is often feasible for low k), and to calculate $\omega_k'(M)$; here theorem 13.4* indicates which invariants of M are possibly involved.

At this point it is natural to contemplate the following important notion. The pair (n,k) is called $\mathfrak{N}$-pleasant (or Ω-pleasant resp.)

if any class in Thom's bordism group $\mathfrak{N}_n$ (resp. in Ω_n) contains a manifold with a k-framefield as soon as the classical necessary conditions in terms of the Stiefel-Whitney numbers (and of the signature, taken modulo suitable powers of 2) are fulfilled (see definitions 12.12 and 16.27). As an example, we show that (n,k) is both $\mathfrak{N}$- and Ω-pleasant when $k \leq 4$ (see theorems 12.13 and 19.37). Now, this notion is closely connected with the full computation of $\omega_k'(M)$. E.g. if $k \geq 1$ and (n,k) is $\mathfrak{N}$-pleasant, then $\omega_k'(M) \in \Omega_{k-1}(P^{k-1} \times BO(n);\phi)$ is zero if and only if the Euler number and all Stiefel-Whitney numbers of M involving $w_i(M)$, $i > n-k$, vanish.

The approach sketched above is one of the lines along which we reach existence results for 2-, 3- and 4-framefields (in sections 14 and 15). Here are a few samples.

Theorem 14.12. Let M be a connected closed smooth manifold (orientable or not) of dimension $n > 6$, $n \equiv 2(4)$.

Then M allows a 3-framefield if and only if the Euler number $\chi(M)$ and the Stiefel-Whitney class $w_{n-2}(M)$ vanish.

This generalizes a result of Atiyah and Dupont ([6], [18]) who treated the case when M is orientable.

Theorem 14.16*. Let M be a closed smooth manifold of dimension $n \geq 7$, $n \equiv 3(4)$, such that the homomorphism

$$w_1(M)\cdot \ : \ H^1(M;\mathbb{Z}_2) \longrightarrow H^2(M;\mathbb{Z}_2)$$

is injective.

Then M has a 3-framefield if and only if the (integer, twisted) Stiefel-Whitney class $W_{n-2}(M)$ vanishes.

The strong nonorientability condition here is characteristic for some of the assumptions which we make in our existence theorems: it gives us control over the transition from $\omega_k(M)$ to $\omega'_k(M)$. For this reason, the computational techniques of the singularity approach seem to be more suited to the unoriented setting, thus complementing the methods e.g. of Atiyah and Dupont [6] very nicely.

The situation is very similar when we compare spin manifolds with manifolds having a (highly) nontrivial second Whitney class. Note the striking analogy between the following result and work of E. Thomas (cf. [65], theorem 6), who showed for an odd-dimensional manifold M satisfying $w_1(M)^2 \neq 0$ that M carries a 2-framefield if and only if $w_{n-1}(M) = 0$.

Corollary 15.13*. Let M be a closed connected orientable manifold of dimension $n \geqq 11$, $n \equiv 3(4)$, such that $w_2(M)^2 \neq 0$.

Then: M carries a 4-framefield if and only if $w_{n-3}(M) = 0$.

This overlaps with work of D. Randall ([53], [54]).

The following extends the result of E. Thomas mentioned above. The cases $(n,k) = (2q,2)$ or $(4q,3)$ were also treated independently by D. Frank [23].

Theorem (see 14.3, 14.12, 14.13, 14.14 and 14.16). Assume $k = 2$ or 3, $n > 2k$. Let M be a closed connected nonorientable n-manifold; if n is odd, $k = 2$, assume even that $w_1(M)^2 \neq 0$; if $n \equiv 1(4)$, $k = 3$, assume that $w_1(M)^2$ is not of the form $w_1(M) \cdot y + y^2$ for any $y \in H^1(M;\mathbb{Z}_2)$. Suppose M has a k-field with finite singularities.

Then M has a k-framefield if and only if $\chi(M) = 0$.

As this last result suggests, there is, of course, a very close connection between our invariants ω_k, ω'_k and the index of a k-field u with finite singularities if it exists (see diagram 13.7 and proposition 13.8). In particular, we can define a homomorphism

$$\Theta^{(or)} : \pi_{n-1}(V_{n,k}) \longrightarrow \Omega_{k-1}(P^{k-1} \times B(S)O(n);\phi)$$

which always makes the index independent of u and dependent only on the underlying manifold M (since it maps Index(u) to $\omega_k'(M)$). $\Theta^{(or)}$ turns out to be universal among such homomorphisms: they all must factor through $\Theta^{(or)}(\pi_{n-1}(V_{n,k}))$ (see theorem 16.11). An example is the homomorphism

$$\theta^S : \pi_{n-1}(V_{n,k}) \longrightarrow KR^S(P^{k+s-1}, P^{s-1})$$

introduced, and studied in detail, by Atiyah and Dupont [6]; in particular, they conjecture the above universal property for θ^S. In view of the bordism theoretic meaning of ω_k' and of $\Theta^{(or)}$, it is not surprising that this conjecture can be linked to the bordism problem which imposed itself above: if (n,k) is Ω-pleasant, the conjecture is true in some suitable form (see proposition 16.28).

The group $\pi_{n-1}(V_{n,k})$ also occurs in the exact homotopy sequence of the fibration

$$\begin{array}{ccc} S^{n-k-1} & \subset & V_{n,k+1} \\ & & \big\downarrow \text{proj} \\ & & V_{n,k} \,. \end{array}$$

The image $proj_*(\pi_{n-1}(V_{n,k+1}))$ turns out to be very closely related to the group $\pi(n,k)$ of all elements in $\pi_{n-1}(V_{n,k})$ which can occur as the index of some k-field with finite singularities on any closed n-manifold. (In fact, $\pi(n,k)$ and $proj_*(\pi_{n-1}(V_{n,k}))$ may always coincide in the metastable range). In section 16 we study these and other subgroups of $\pi_{n-1}(V_{n,k})$, such as the kernel of $\Theta^{(or)}$. Also we calculate them for $k \leq 4$ (see the tables in 16.15; note in particular the amazing regularity of $\ker \Theta^{or}$). Sometimes the knowledge of these subgroups leads directly to existence results.

Theorem 16.21. Assume that $n \equiv 2(4)$, $n > 2k$ and $1 \leq k \leq 6$. Let M be a closed connected n-manifold allowing a k-field u with finite singularities.

Then M carries a k-framefield if and only if $\chi(M) = 0$.

Proof. For $k \neq 2$, the index of u lies in the group $\pi(n,k) \cong \mathbb{Z}$ (see § 16) which can accomodate only the Euler number $\chi(M)$. If $k = 2$, the result follows from the work of E. Thomas (when M is orientable, see [65]), or else from the last theorem.∎

Our "basic problem" can be modified as follows. Assume we have already a k-framefield v on M, when can we extend it to a (k+1)-framefield? To answer this question, we define framed bordism classes

$$\chi(M,v) \in \Omega_k^{fr}(M) \quad \text{and} \quad \chi'(M,v) \in \Omega_k^{fr}(B(S)O(n-k))$$

which play much the same rôle that $\omega_k(M)$ and $\omega'_k(M)$ played before (see theorem 19.3). However, there is a new aspect: on one hand $\chi(M,v)$ and $\chi'(M,v)$ are the obstructions to existence problems; but on the other hand, they are able, to some extend, to distinguish between different k-framefields. Even if we weaken our invariants via the obvious augmentation to define

$$\chi''(M,v) \in \Omega_k^{fr} \cong \pi_k^S ,$$

this classification aspect remains remarkable. Here is a sample result.

Theorem 19.13*. If $n > 2k+2$, the invariant $\chi''(M,-)$ detects at least as many different regular homotopy (actually, even fine bordism) classes of k-framefields on M as there are elements in the kernel of

$$\text{inclusion}_* : \pi_{n-1}(S^{n-k-1}) \longrightarrow \pi_{n-1}(V_{n,k+1}) .$$

For $n \equiv k(2)$, this kernel contains the odd torsion of $\pi_{n-1}(S^{n-k-1}) \cong \pi_k^S$.

In particular, if n is odd (and as large as required above), then M has at least twelve 3-framefields (and at least fifteen 7-framefields, and at least sixty-three 11-framefields resp.) if any, detected by χ''.

In section 19 we study the invariants χ, χ' and χ'' in some detail. For low k we calculate their value sets explicitly (see 19.17, 19.24 and 19.26). As an immediate consequence of our tables we obtain e.g. the following result which sharpens work of E. Thomas and of Atiyah and Dupont somewhat.

Theorem 19.27*. Let M be a closed orientable manifold of dimension $n \geq 6$.

If $n \equiv 2(4)$, every nowhere vanishing vectorfield on M extends to a 2-framefield.

If $n \equiv 3(4)$, every 2-framefield on M extends to a 3-framefield.

Various important aspects of χ' can best be illustrated by the long exact sequence (cf. theorem 19.5)

$$.. \longrightarrow \Omega_n(k+1,\ k+1) \xrightarrow{\ f\ } \Omega_n(k,k) \xrightarrow{\ \chi'\ } \Omega_k^{fr}(BSO(n-k)) \xrightarrow{\ \partial\ } \Omega_{n-1}(k,k) \rightarrow ..$$

and its unoriented analogue. Here $\Omega_n(k,k)$ denotes the "fine" bordism group of oriented n-manifolds with k-framefields, and f is the obvious forgetful homomorphism. The system of all these unbounded exact sequences gives rise to a spectral sequence (see 19.28) whose E^∞-term is closely related to the notion of Ω-pleasantness.

In section 19 we also compare singularity data of a morphism on a compact n-manifold M and of its restriction to the boundary. Theorem 19.6 gives a far-reaching generalization e.g. of the wellknown identity

$$\chi(\partial M) = (1 - (-1)^n)\ \chi(M)$$

for Euler numbers. One of the many corollaries is a calculation of the first differential in our spectral sequence. This can be used to compute the groups $\Omega_n(k,k)$ almost completely for $k \leq 3$ (see theorems 19.39 - 41; cf. also § 12). As a side product of the proofs, we can determine the "index groups" $\pi^{or}(8q + 4, k)$ for $k = 3$ and 4, thus completing our tables in section 16. (This is one further example showing how natural and powerful the bordism theoretic approach can be).

The singularity approach provides also the machinery for a fulfledged homotopy classification of k-framefields (in the range $n > 2k+1$). As an illustration, we determine the number of homotopy classes of k-fields on a closed manifold (orientable or not) when $k = 1$ or $k = 2$, and, in one instance, even when $k = 3$ (see 18.2, 18.6 and 18.10).

As a final application of our theory we compare the span of a closed n-manifold M (i.e. the maximum number k such that TM contains a trivial k-dimensional subbundle) to the stable span (i.e. the maximum k such that $TM \oplus (M \times \mathbb{R})$ contains a trivial (k+1)-dimensional subbundle). As the example of the spheres shows, these two numbers can differ dramatically. In § 20 we introduce an integer invariant $s(M)$ satisfying

$$\text{span}\,(S^n) \leq s(M) \leq n .$$

Among other results (which yield e.g. a simple geometric proof of much of the work of Bredon and Kosinski [10] on the span of π-manifolds) we establish the following.

<u>Corollary 20.5*.</u> <u>Let M be a closed connected n-manifold such that $2 \cdot s(M) \leq n$. Then the span of M is equal to $s(M)$ or to the stable span of M.</u>

The distinction between the two cases involves the invariant $\omega_{s(M)+1}(M)$ which equals the Euler number $\chi(M)$ for even n and which is $\mathbb{Z}_2$-valued in the interesting odd-dimensional situations.

Now $s(M)$ can sometimes be computed (see propositions 20.6, 20.8, 20.11 and 20.26). E.g. $s(M) = \text{span}(S^n)$ if n is even, or $n \equiv 1(4)$ and $w_1(M)^2 = 0$, or $n \equiv 3(8)$ and $w_1(M)^4 + w_1(M)w_3(M) + w_2(M)^2 = 0$. Combined with results such as corollary 20.5^* these and other calculations have motivated the conjecture that span(M) always equals $\text{span}(S^n)$ or stable span(M) (see remark 20.21). In 20.18, 20.27, 20.28 and 20.29 we give many counterexamples of the type $M = N \times S^q$ where both q and the Euler number of N are odd. It is a curious fact that often the span of such a product is also different from both the span and the stable span of S^q.

A survey of all our various singularity invariants for k-fields, and of the connections between them, is given in theorem 13.11.

All of the above suggests just one possible area of application of the singularity approach. A similar program has been carried out for immersions in the recent thesis of C. Ølk [50]. Our techniques can also be applied to planefields and foliations (see [32], [33], [38] for some related work), and, very generally, to k-morphisms over (not necessarily compact) smooth manifolds M (i.e. to vector bundle homomorphisms whose rank is at least k at all points of M). Thus general vector bundle mono- and epimorphisms are included in the discussion, and so are the k-mersions of S.D. Feit [21].

In chapter I we study k-morphisms in a generic situation ("nondegeneracy"), and we analyze their behaviour around the singularity (i.e. the locus of all points in M where the rank is precisely k). The relevant singularity data give rise to an obstruction which must vanish if the minimum rank can be increased to $k + 1$ (after a suitable deformation). As a striking fact, the main version of this invariant turns out to coincide with an obstruction to sectioning a certain sphere bundle (proposition 2.15); hence it can always be defined, even if there is no k-morphism whose singularity could be used.

The central result of the first chapter is the existence theorem 3.1 which states that under certain dimension conditions the vanishing of our singularity invariant is not only necessary but also sufficient for the existence of a (k+1)-morphism. In the case of (stable) monomorphisms over a compact base versions of this result were previously proved by Dax ([15], II. 6.2, and [16]), Hatcher-Quinn ([25], 4.3) and Salomonsen ([57], 8.1).

Our existence theorem is phrased as the solution of an extension problem, and therefore it leads also to the classification of (k+1)-morphisms up to homotopy or concordance (in section 4). As an interesting consequence we can in particular reduce our existence and classification questions to problems of sections of sphere bundles (corollaries 3.5 and 4.11). Also, we obtain e.g. that in many situations there are only finitely many concordance classes of monomorphisms (see theorem 4.14).

In chapter II we consider a cruder classification, and we study "fine" bordism groups involving k-morphisms. A typical example is the group $\mathfrak{N}_n(m,k)$ built from closed smooth n-dimensional manifolds M, together with a k-morphism of the form $\bar{g} : TM \longrightarrow M \times \mathbb{R}^m$. This covers (unoriented) bordism of manifolds with framefields, or of immersions, or, more generally, of k-mersions.

In the development of our techniques we will use two basic ingredients. The first one, more technical in nature, is the stability theorem 6.6. It implies e.g. that the stable homotopy of spheres can be interpreted as parallelized (not just stably parallelized!) bordism of parallelized manifolds (see corollary 7.20). Another consequence is that every closed manifold M whose stable span is at least as large as a given number r (i.e. $TM \oplus \mathbb{R}$ allows $r+1$ independent sections), is bordant to a manifold with an r-framefield (see corollary 7.22). In general, the stability theorem assures that when we modify a morphism by generalized surgery, we can make it suitably tangent to newly arising boundaries (except maybe in some rare, lowdimensional cases).

The second, more important ingredient is again the singularity approach. It leads to a very powerful tool, namely the long exact sequence 7.12. Here the third term, which measures singularities, is a normal bordism group. Since such groups are included in our notion of fine bordism, we can iterate our method. Thus, when we rebuild bordism groups from exact sequences, we can distinguish contributions of different orders.

An invariant ρ which would classify as of second order in this sense, is defined in (7.14) and (7.14'). It satisfies an identity (see proposition 7.15) which we can sometimes use to split our exact sequence up to torsion, or to detect elements of order 2^s, $s > 1$.

As a first application of this machinery, we give a complete description of the rational bordism groups $\mathfrak{N}_n(m,k) \otimes \mathbb{Q}$ and of their oriented analogues $\Omega_n(m,k) \otimes \mathbb{Q}$ (see theorems 8.1 and 8.26). This solves a problem raised in J. Maffei [43]; it also overlaps with some work of R. Wells [68] who computed the immersion groups $\mathfrak{N}_n(m,n) \otimes \mathbb{Q}$. One interesting feature in the calculation of $\Omega_n(m,k) \otimes \mathbb{Q}$ is the use of the universal model for singularities: this is a finite codimensional submanifold of the space of Fredholm operators (on an infinite dimensional real Hilbert space). Actually, I know of no way to avoid this excursion into infinite dimensional topology.

Using the exact sequence 7.12, we can extend another result of R. Wells and compute the unoriented bordism groups of immersions $M^n \looparrowright \mathbb{R}^{2n-k}$ for $k \leq 2$ (see theorem 10.8); in particular, it is amusing to see how, in our framework, the bordism group $\mathfrak{N}_2(3,2) \cong \mathbb{Z}_8$

(of surfaces immersed into $\mathbb{R}^3$) is built up from three copies of $\mathbb{Z}_2$, of first, second, and third "order" respectively. In such calculations, it is important to know how 7.12 can be cut up into short exact sequences. This amounts mainly to a search for relations between Stiefel-Whitney numbers, or for counterexamples to such relations.

As a further example, we determine various bordism groups of manifolds with framefields (see section 12). However, here it would be more efficient to use the techniques of § 19 rather than 7.12.

The main application of the singularity sequence 7.12, however, is theorem 9.3. Together with the tool kit assembled in the rest of section 9, it gives us crucial information about low-dimensional normal bordism groups. Thus we can apply the singularity approach not only to develop our obstruction theory, but also to calculate our invariants in terms of classical ones.

Our approach is very geometric. Almost all of the arguments in the development of our theory involve only elementary differential topology, the central concept being transversality. In particular, no homotopy theory is needed. Actually, in § 5 we show how several classical homotopy theoretical results of Freudenthal, James and Haefliger-Hirsch follow as easy consequences of our classification theory. However, when we apply our general results to specific situations, and when explicit calculations in terms of classical invariants are needed, we make strong use of characteristic classes (see e.g. Milnor [46]) and of the fundamental facts of bordism theory (see e.g. the first two chapters in Conner-Floyd [13]).

This work, developped mainly in 1973 and early 1974, was supported in part by the National Science Foundation, and later by the Sonderforschungsbereich Theoretische Mathematik in Bonn. My thanks go also to Mrs. Sprenger for her very dedicated typing, and to Wilfried Köhler, Bernd Lübcke and Christof Olk for their help in proofreading. In addition, several mathematical improvements by Christof Olk are gratefully acknowledged.

Notation

γ, γ_q	universal (q-) plane bundle
λ, (γ_1)	canonical line bundles over
P^∞, P^k	real projective spaces of indicated dimensions
<u>Möbius</u>	nontrivial line bundle over the 1-sphere S^1
$\underline{\mathbb{R}}^k$	trivial k-plane bundle
proj, π	obvious projection
π_1, π_2	projection on indicated factor

Vector bundles and their classifying maps will be confused sometimes.

Obvious pullbacks will often not be indicated.

CHAPTER I. EXISTENCE AND HOMOTOPY CLASSIFICATION.

§1. Non-degenerate singularities and their structure.

Throughout this paper "manifold" will mean smooth manifold (the topology of which satisfies the Hausdorff axiom and has a countable base), possibly with boundary. We will fix an n-manifold M, and over it two smooth real vector bundles α and β of dimensions a and b (whose fibers at a point $x \in M$ will be denoted by α_x and β_x), and we will also fix an integer k such that $0 \leq k \leq \min(a,b)$. From § 2 on we will also assume that a vector bundle morphism $u^{\partial}: \alpha | \partial M \longrightarrow \beta | \partial M$ whose rank is nowhere less than $k+1$, has been given over the boundary ∂M of M.

First we discuss briefly some important fiber bundles associated with these data. Let

$$\tilde{p} : \underset{\sim}{\mathrm{Hom}} (\alpha,\beta) \longrightarrow M$$

denote the homomorphism bundle from α to β. Define subfibrations by

$$A^k(\alpha,\beta) = \bigcup_{x \in M} A^k (\alpha_x,\beta_x) \qquad \text{and}$$

$$W^k(\alpha,\beta) = \bigcup_{x \in M} W^k (\alpha_x,\beta_x) \quad ,$$

where $A^k(\alpha_x,\beta_x)$, resp $W^k(\alpha_x,\beta_x)$, stands for the set of all linear maps from α_x to β_x of rank $=k$, resp $\geq k$. We will need the following facts about these spaces (cf. [31], where they are denoted by $A_{a-k,b-k}(\alpha,\beta)$ and $V_{a-k,b-k}(\alpha,\beta)$; see especially pp. 99f).

(i) $W^k(\alpha,\beta)$ is open in $\underset{\sim}{\mathrm{Hom}} (\alpha,\beta)$.

(ii) $A^k(\alpha,\beta)$ is a closed smooth submanifold of $W^k(\alpha,\beta)$.

Furthermore, there are canonical smooth vector bundles $\widetilde{\mathrm{Ker}}$, $\widetilde{\mathrm{Coker}}$ and $\widetilde{\mathrm{Im}}$ over $A^k(\alpha,\beta)$, where e.g. $\widetilde{\mathrm{Ker}}$ denotes the subbundle

$$\widetilde{\mathrm{Ker}} = \bigcup_{h \in A^k(\alpha,\beta)} h \times \ker h$$

of the pullback $\tilde{p}^*(\alpha)$.

(iii) The normal bundle of $A^k(\alpha,\beta)$ in $W^k(\alpha,\beta)$ (and hence in $\underset{\sim}{Hom}(\alpha,\beta)$) is described by the canonical smooth vector bundle isomorphism

$$(1.1) \qquad \tilde{j} : \nu(A^k(\alpha,\beta),\ W^k(\alpha,\beta)) \xrightarrow{\cong} \underset{\sim}{Hom}(\widetilde{\underset{\sim}{Ker}},\ \widetilde{\underset{\sim}{Coker}})$$

defined as follows. Given $h \in A^k(\alpha,\beta)$ with $\tilde{p}(h) = x \in M$, a normal vector $[\tilde{h}] \in T_h(\underset{\sim}{Hom}(\alpha,\beta))/T_h(A^k(\alpha,\beta))$ at h can be represented by some $\tilde{h}$ tangent to the fiber $\underset{\sim}{Hom}(\alpha,\beta)_x$, or, in other words, by a linear map $\tilde{h} : \alpha_x \longrightarrow \beta_x$. Composing $\tilde{h}$ with the inclusion $\ker h \subset \alpha_x$ on one side, and with the projection $\beta_x \longrightarrow \operatorname{Coker} h$ on the other, we obtain the desired element $\tilde{j}[\tilde{h}] \in \underset{\sim}{Hom}(\widetilde{\underset{\sim}{Ker}},\ \widetilde{\underset{\sim}{Coker}})$.

Next consider the Grassmannian bundle

$$p : G_{a-k}(\alpha) \longrightarrow M$$

of (a-k)-planes in α. By definition an element $g \in G_{a-k}(\alpha)$ is a linear subspace of $\alpha_{p(g)} = (p^*(\alpha))_g$; thus we have, in a natural fashion, canonical smooth vector bundles

$$\gamma \subset p^*(\alpha) \quad \text{and} \quad \gamma^{\perp} = p^*(\alpha)/\gamma$$

of dimension a-k, resp. k, over $G_{a-k}(\alpha)$. Note that the vector bundle tangent to the fibers of p is determined by the canonical smooth vector bundle isomorphism

$$(1.2) \qquad j : TF(G_{a-k}(\alpha)) \xrightarrow{\cong} \underset{\sim}{Hom}(\gamma,\gamma^{\perp})$$

defined as follows (compare [17], or [31], p. 97). Let $g : (-\varepsilon,\varepsilon) \longrightarrow G_{a-k}(\alpha_x)$ be a smooth path representing the tangent vector $v = \dot{g}(0)$ at $g(0)$. If $g(0)^{\perp}$ is a complement of the subspace $g(0)$ of α_x, then, for t close to 0, we can consider the space $g(t)$ as the graph of a linear map $\ell(t) : g(0) \longrightarrow g(0)^{\perp}$.

Define $j(v)$ to be the derivative $\dot{\ell}(0)$, considered as a map from $g(0)$ to $\alpha_x/g(0)$.

The fiber maps $\tilde{p}$ and p are related by the commutative diagram

(1.3)
$$\begin{array}{ccc} A^k(\alpha,\beta) & \xrightarrow{\tilde{p}_{ker}} & G_{a-k}(\alpha) \\ & {\scriptstyle\tilde{p}}\searrow \quad \swarrow{\scriptstyle p} & \\ & M & \end{array},$$

where $\tilde{p}_{ker}$ associates to each linear map $h : \alpha_x \longrightarrow \beta_x$ of rank k its $(a-k)$-dimensional kernel. Actually, $\tilde{p}_{ker}$ is itself a fibration, since $A^k(\alpha,\beta)$ can be identified smoothly with the total space of the monomorphism bundle $A^k(\gamma^{\perp}, p^*(\beta))$ over $G_{a-k}(\alpha)$.

We are now ready to start our discussion of non-degenerate singularities.

Definition 1.4. Let $u : \alpha \longrightarrow \beta$ be a smooth vector bundle homomorphism, and let s_u denote the corresponding section of $\underset{\sim}{\mathrm{Hom}}(\alpha,\beta)$.

We call u a <u>k-morphism</u> if for all $x \in M$ the rank of $u_x : \alpha_x \longrightarrow \beta_x$ is at least k (or, equivalently, if s_u goes into $W^k(\alpha,\beta)$).

We call u a <u>non-degenerate k-morphism</u> if in addition s_u is transverse to $A^k(\alpha,\beta)$.

Classical transversality results (see e.g. [47], theorem 1.35) imply the following density property. Given a k-morphism $u : \alpha \longrightarrow \beta$, a closed subspace $L \subset M$ where s_u is already transversal to $A^k(\alpha,\beta)$, and a neighborhood V of $s_u(M)$ in $W^k(\alpha,\beta)$, there is a non-degenerate k-morphism u' with $u'|L = u|L$ and $s_{u'}(M) \subset V$, and such that u and u' are (linearly) homotopic through k-morphisms.

Now consider a fixed non-degenerate k-morphism u from α to β; to simplify matters assume that $u|\partial M$ is a $(k+1)$-morphism. Then the singularity

$$S = \{x \in M \mid \mathrm{rank}(u_x) = k\} \quad = \quad s_u^{-1}(A^k(\alpha,\beta))$$

is a closed submanifold of dimension $n-(a-k)(b-k)$ of the interior of M. It is naturally equipped with two important data. First, we have the vector bundle homomorphism

$$(1.5) \qquad u|S \; : \; \alpha|S \longrightarrow \beta|S$$

of constant rank k. This leads in particular to the canonical smooth vector bundles Ker, Coker and Im (of dimensions $a-k$, $b-k$ and k) over S, where e.g. Ker is the subbundle of $\alpha|S$ given by $\mathrm{Ker} = \bigcup_{x\in S} \mathrm{Ker}\, u_x$. Second, if $\nu(S,M)$ denotes the normal bundle of S in M, we have the canonical isomorphism

$$(1.6) \qquad \tilde{j}\cdot s_{u*} : \nu(S,M) \xrightarrow{\cong} \mathrm{Hom}(\mathrm{Ker}, \mathrm{Coker})$$

deduced from (1.1) and the tangent map of s_u.

We will see that these data essentially characterize the behavior of u in a neighborhood of S.

Let us first use (1.5) to construct a model for u. Choose a complement Coim of Ker in $\alpha|S$, and identify a complement of Im in $\beta|S$ in the obvious way with Coker to obtain splittings

$$(1.7) \qquad \alpha|S = \mathrm{Ker} \oplus \mathrm{Coim} \quad \text{and} \quad \beta|S = \mathrm{Coker} \oplus \mathrm{Im} \quad ;$$

clearly $u|S$ restricts to an isomorphism

$$(1.8) \qquad u| \quad : \quad \mathrm{Coim} \xrightarrow{\cong} \mathrm{Im}$$

Next define

$$\hat{M} = \mathrm{Hom}\,(\mathrm{Ker}, \mathrm{Coker})$$

and identify the zero section with S.

Moreover, let

$$\hat{\alpha} = \widehat{\underset{\sim}{\text{Ker}}} \oplus \widehat{\underset{\sim}{\text{Coim}}} \quad \text{and} \quad \hat{\beta} = \widehat{\underset{\sim}{\text{Coker}}} \oplus \widehat{\underset{\sim}{\text{Im}}}$$

be the obvious pullbacks of the bundles in (1.7) by the bundle projection $\hat{\pi} : \hat{M} \to S$. By definition, a point $y \in \hat{M}$ is just a linear map from $\underset{\sim}{\text{Ker}}_{\hat{\pi}(y)} = \widehat{\underset{\sim}{\text{Ker}}}_y$ to $\underset{\sim}{\text{Coker}}_{\hat{\pi}(y)} = \widehat{\underset{\sim}{\text{Coker}}}_y$. Thus we have the "footpoint homomorphism"

$$\hat{u}_1 : \widehat{\underset{\sim}{\text{Ker}}} \longrightarrow \widehat{\underset{\sim}{\text{Coker}}}.$$

Furthermore, the isomorphism $u|$ (cf. 1.8) pulls back to an isomorphism

$$\hat{u}_2 : \widehat{\underset{\sim}{\text{Coim}}} \longrightarrow \widehat{\underset{\sim}{\text{Im}}}.$$

Putting these two together, we obtain the desired model homomorphism

(1.9) $$\hat{u} = \hat{u}_1 \oplus \hat{u}_2 \; : \; \hat{\alpha} \longrightarrow \hat{\beta}$$

over $\hat{M}$. Clearly $\hat{u}$ is a non-degenerate k-morphism, and at its singularity S it coincides with u; moreover $\tilde{j} \cdot s_{\hat{u}*} = \text{Id}$.

The following shows that our model reflects the essential features of the behavior of u around S.

Structure lemma 1.10. There are open neighborhoods $\mathcal{V}$ of S in M and $\hat{\mathcal{V}}$ of S in $\hat{M}$, and there is a diffeomorphism $f : \mathcal{V} \longrightarrow \hat{\mathcal{V}}$, covered by isomorphisms $\bar{f}_\alpha : \alpha|\mathcal{V} \longrightarrow \hat{\alpha}|\hat{\mathcal{V}}$ and $\bar{f}_\beta : \beta|\mathcal{V} \longrightarrow \hat{\beta}|\hat{\mathcal{V}}$, such that

(i) f leaves S pointwise fixed,

(ii) $\bar{f}_\alpha$ and $\bar{f}_\beta$ restrict to the identity over S, and

(iii) the diagram

$$\begin{array}{ccc} \alpha|\mathcal{V} & \xrightarrow{u|\mathcal{V}} & \beta|\mathcal{V} \\ \downarrow \bar{f}_\alpha & & \downarrow \bar{f}_\beta \\ \hat{\alpha}|\hat{\mathcal{V}} & \xrightarrow{\hat{u}|\hat{\mathcal{V}}} & \hat{\beta}|\hat{\mathcal{V}} \end{array}$$

commutes.

Proof. We start with an open tubular neighborhood $\mathcal{V}$ of S in $\mathring{M}$, with projection $\pi: \mathcal{V} \longrightarrow S$; when necessary in the following proof, we will feel free to replace $\mathcal{V}$ by a smaller tubular neighborhood. In this sense the splittings in (1.7) extend to splittings

$$(1.7') \qquad \alpha|\mathcal{V} = \alpha_1 \oplus \alpha_2 \quad \text{and} \quad \beta|\mathcal{V} = \beta_1 \oplus \beta_2$$

such that $u|\alpha_2$ is injective, $\beta_2 = \operatorname{Im}(u|\alpha_2)$, and $\alpha_1 = \operatorname{Ker}(\mathrm{proj}_2 \circ u|\mathcal{V})$, where proj_2 denotes the projection of β onto β_2. Choose isomorphisms

$$\bar{f}'_\alpha : \alpha|\mathcal{V} \longrightarrow \pi^*(\alpha|S) \quad \text{and} \quad \bar{f}'_\beta : \beta|\mathcal{V} \longrightarrow \pi^*(\beta|S)$$

which are compatible with the splittings (1.7) and (1.7'), cover the identity map of $\mathcal{V}$ and restrict to the identity over S. This allows us to identify $u|\mathcal{V}$ with a homomorphism of the form

$$u_1 \oplus u_2 : \pi^*(\underset{\sim}{\mathrm{Ker}}) \oplus \pi^*(\underset{\sim}{\mathrm{Coim}}) \longrightarrow \pi^*(\underset{\sim}{\mathrm{Coker}}) \oplus \pi^*(\underset{\sim}{\mathrm{Im}}).$$

We may assume that $u_2 = \pi^*(u|)$; otherwise we need only replace $\bar{f}'_\beta$ by $(\mathrm{Id} \oplus \pi^*(u|)\circ u_2^{-1})\circ \bar{f}'_\beta$. Moreover, it follows from the transversality condition on u that u_1, considered as a map from $\mathcal{V}$ to $\underset{\sim}{\mathrm{Hom}}(\underset{\sim}{\mathrm{Ker}}, \underset{\sim}{\mathrm{Coker}}) = \hat{M}$, is a diffeomorphism around S, since it induces an isomorphic tangent map. Now put $f = u_1$ and $\hat{\mathcal{V}} = f(\mathcal{V})$. Also compose $\bar{f}'_\alpha$ with the isomorphism

$$\pi^*(\alpha|S) = (\hat{\pi}\circ f)^* (\alpha|S) = f^*(\hat{\alpha}),$$

and modify $\bar{f}'_\beta$ similarly, to obtain the required isomorphisms $\bar{f}_\alpha$ and $\bar{f}_\beta$.

Remark 1.11. Note that the isomorphism of normal bundles

$$f_* : \nu(S,M) \longrightarrow \nu(S,\hat{M}) = \underset{\sim}{\mathrm{Hom}}(\underset{\sim}{\mathrm{Ker}}, \underset{\sim}{\mathrm{Coker}}),$$

induced by the tangent map of any diffeomorphism f as in the statement of (1.10), satisfies the condition

(iii') f_* coincides with the isomorphism $\tilde{j}\cdot s_{u*}$ given by (1.6).

Now consider any set of data f, $\bar{f}_\alpha$, $\bar{f}_\beta$, ϑ, $\hat{\vartheta}$ as in the statement of lemma 1.10, but with condition (iii) replaced by (iii'). Such data are determined up to isotopy by $\tilde{j}\cdot s_{u*}$; this follows essentially from the isotopy theorem for tubular neighborhoods ([41],p.77). Thus, while (1.5) allows us to construct the model homomorphism $\hat{u}$, (1.6) provides the data to carry $\hat{u}$ back to a neighborhood of S. The smooth homotopy class (through non-degenerate k-morphisms which are defined around S and which all have their singularity precisely at S) of the resulting k-morphism

$$\bar{f}_\beta^{-1} \circ \hat{u}|\hat{\vartheta} \circ \bar{f}_\alpha$$

is independent of all choices involved in its construction and depends only on the homotopy class (in the obvious sense) of the singularity data given by (1.5) and (1.6). It follows from lemma 1.10 that this construction establishes a one-to-one onto correspondence between the two types of homotopy classes.∎

Exercise 1.12. Consider the obvious vector bundle map

$$u : \mathbb{R}^a \subset \mathbb{R}^{n+1} = TS^n \oplus \mathbb{R} \longrightarrow TS^n$$

over the sphere $S^n (a \leq n+1)$. Show that u is a nondegenerate $(a-1)$-morphism and study its singularity data.∎

In a certain range of dimensions (sometimes called the "stable range") we need not bother to analyse singularities at all: their high codimension will guarantee that they are automatically empty. So, starting from a nondegenerate 0-morphism (which will avoid $A^0(\alpha,\beta)$ for dimension reasons), we make it transverse to $A^1(\alpha,\beta)$ (which it will avoid again) and so forth, until we obtain a mono- (or epi-)morphism. This procedure is the key to what follows.

<u>Exercise 1.13.</u> Prove the following claims.

(i) If $n + a \leq b$, then there exists a monomorphism $\alpha \hookrightarrow \beta$.

(ii) If $n + a < b$, then any two such monomorphisms are homotopic through monomorphisms.

(iii) If ξ and η are smooth vector bundles over a paracompact manifold M such that $\dim \xi = \dim \eta > \dim M$, then any isomorphism $h : \xi \oplus \underset{\sim}{\mathbb{R}}^a \cong \eta \oplus \underset{\sim}{\mathbb{R}}^a$, $a \geq 0$, can be deformed into an isomorphism of the form is $\oplus\, id_{\underset{\sim}{\mathbb{R}}^a}$. (Deform first $h|\underset{\sim}{\mathbb{R}}^a$ into the inclusion $\underset{\sim}{\mathbb{R}}^a \subset \eta \oplus \underset{\sim}{\mathbb{R}}^a$).

(iv) The obvious homomorphism

$$i_* : \pi_1(SO(q)) \longrightarrow \pi_1(SO(q+1))$$

is bijective for $q \geq 3$ and onto for $q = 2$. (Actually, $\pi_1(SO(2)) = \pi_1(S^1) \cong \mathbb{Z}$ and $\pi_1(SO(3)) \cong \pi_1(P^3) = \mathbb{Z}_2$). ■

§2. Singularity invariants

In this section we introduce and study the basic invariants of our obstruction theory.

We start with a brief discussion of normal bordism functors (see also [15], [57], [67]), since they provide the groups in which our obstructions lie.

Let X be a topological space, equipped with a continuous map $p : X \longrightarrow Y$ into some Hausdorff space Y. Also, let ϕ be a virtual real vector bundle over X, i.e. an ordered pair (ϕ^+, ϕ^-) of vector bundles written $\phi = \phi^+ - \phi^-$. Fix an integer r.

Now consider tripels of the form $(S, g, \bar{g})$ where

(i) S is an r-manifold without boundary (but not necessarily compact),

(ii) $g : S \longrightarrow X$ is a continuous map such that $p \circ g$ is proper (i.e. the inverse image of every compact subset of Y is compact), and

(iii) $\bar{g} : \mathbb{R}^s \oplus TS \oplus g^*(\phi^+) \longrightarrow \mathbb{R}^t \oplus g^*(\phi^-)$ is a vector bundle isomorphism, s and t being suitable integers.

Two such tripels $(S_0, g_0, \bar{g}_0)$ and $(S_1, g_1, \bar{g}_1)$ are called bordant if there is

(i) a manifold $\mathcal{S}$ whose boundary is the disjoint union of its closed submanifolds S_0 and S_1,

(ii) a continuous map $G : \mathcal{S} \longrightarrow X$ extending $g_0 \cup g_1$ and such that $p \circ G$ is proper; and

(iii) an isomorphism $\bar{G} : \mathbb{R}^{\tilde{s}} \oplus T\mathcal{S} \oplus G^*(\phi^+) \longrightarrow \mathbb{R}^{\tilde{t}} \oplus G^*(\phi^-)$ over $\mathcal{S}$ such that $\bar{G}|\partial\mathcal{S} = Id_0 \oplus \bar{g}_0 \cup Id_1 \oplus \bar{g}_1$; here the symbol Id stands for the identity map of trivial bundles of suitable dimensions, and $T\mathcal{S}|\partial\mathcal{S}$ is identified with $\mathbb{R} \oplus T(\partial\mathcal{S})$ via a vector field normal to $\partial\mathcal{S}$, pointing inward along S_0 and outward along S_1.

Definition 2.1. The set of bordism classes $[S,g,\bar{g}]$ of tripels $(S,g,\bar{g})$ as above, with the group structure given by disjoint union, is called the rth normal bordism group of (X,p) with coefficients in ϕ, and is denoted by $\Omega_r(X,p;\phi)$.

If $f : X' \longrightarrow X$ is a continuous map, the induced homomorphism

$$f_* : \Omega_r(X', p\circ f; f^*(\phi)) \longrightarrow \Omega_r(X,p;\phi)$$

is defined by $f_* [S,g',\bar{g}'] = [S,f\circ g',\bar{g}']$.

Note that a bordism class $[S,g,\bar{g}] \in \Omega_r(X,p;\phi)$ depends only on S, g, and the (stable) homotopy class of the isomorphism $\bar{g}$. Consequently, if $\phi_1 = \phi_1^+ - \phi_1^-$ is another virtual bundle over X and if isomorphisms $\phi^+ \cong \phi_1^+$, $\phi^- \cong \phi_1^-$ are given canonically up to homotopy, then there is a canonical isomorphism $\Omega_r(X,p;\phi) = \Omega_r(X,p;\phi_1)$. Thus for our purposes we can consider ϕ and ϕ_1 as equal. E.g. if a homomorphism bundle Hom(ξ,η) occurs as a summand in the description of ϕ, we can safely replace it by the tensor product $\xi \otimes \eta$. Similarly, for any vector bundle ψ over X, we will identify the virtual bundle $(\phi^+ \oplus \psi) - (\phi^- \oplus \psi)$ with ϕ, since the corresponding normal bordism groups are canonically isomorphic. Furthermore we can choose natural isomorphisms between the groups corresponding to the coefficient bundles $(\phi^+ \oplus \mathbb{R}) - \phi^-$, ϕ, and $\phi^+ - (\phi^- \oplus \mathbb{R})$, respectively.

Our bordism group remains also unchanged if we replace the map p by $p'\circ p$, where $p' : Y \longrightarrow Y'$ is any proper map into a Hausdorff space Y'. It follows that for compact X the group $\Omega_r(X,p;\phi)$ is entirely independent of p. This, however, need not hold in general. E.g. if X is a non-compact connected manifold $\neq \emptyset$, and if p_o, respectively Id_X, denotes a constant map, respectively the identity map, on X, then $\Omega_o(X,p_o;\phi)$ is $\mathbb{Z}$ or $\mathbb{Z}_2$ (according to whether or not the orientation line bundles of ϕ^+ and ϕ^- are isomorphic), but $\Omega_o(X, Id_X; \phi) = 0$.

We can easily extend (2.1) to define a "generalized homology theory" on an appropriate category of (pairs of) spaces, equipped with a collection of "p-compacta" and with a virtual vector bundle, and of maps compatible with the first structure and covered by strict homomorphism between corresponding parts of the virtual bundles. Furthermore, if ϕ is expressed in the form $\phi = \phi^+$ - trivial bundle (e.g. if X is compact or has the homotopy type of a finite-dimensional CW-complex), and if X is paracompact and Y can be chosen to be locally compact and to have a countable basis (in which case we may replace p by a map into $\mathbb{R}$), then the Pontryagin-Thom construction leads to a canonical isomorphism

(2.2) $$\Omega_r(X,p;\phi) \cong \pi^S_{r+\ell}(T(\phi^+,p)).$$

Here $T(\phi^+,p)$ is the disjoint union of (the total space of) ϕ^+ and a point ∞, with the topology given by the basis consisting of all open subspaces of ϕ^+ and of all complements of unit disk bundles of the form $D_\mu(\phi^+)|p^{-1}(K)$ where μ is any euclidean inner product of ϕ^+ and $K \subset Y$ is any compactum; ℓ denotes the dimension of ϕ^+.

In the special case when p is a trivial map taking only one value, we will drop it from our notations. Thus $\Omega_r(X;\phi)$ denotes the bordism group of compact singular r-manifolds in X, equipped with a description of their stable tangent (or normal) bundle in terms of ϕ. This case has been studied in detail by Dax [15] and Salomonsen [57] (note however, the opposite sign convention for ϕ in [57]).

Now let M, α, β and k be again as in section 1. From now on we will also fix a (k+1)-morphism

$$u^\partial : \alpha|\partial M \longrightarrow \beta|\partial M.$$

<u>Definition 2.3.</u> Two k-morphisms $u_0, u_1 : \alpha \longrightarrow \beta$ are called <u>k-homotopic mod ∂M</u> if they are homotopic through k-morphisms which all restrict to the same morphism over ∂M.

Let $[u]$ be a mod ∂M k-homotopy class of k-morphisms from α to β which extend u^∂ over all of M. We may pick a non-degenerate k-morphism u as a representative. In the notation of §1, the singularity S of u is then equipped with a continuous map

(2.4) $$\tilde{g} = s_u|S \quad : \quad S \longrightarrow A^k(\alpha,\beta)$$

and with an isomorphism

(2.5) $$\bar{\tilde{g}} : TS \oplus (\underset{\sim}{\mathrm{Ker}} \otimes \underset{\sim}{\mathrm{Coker}}) \xrightarrow{\mathrm{Id} \oplus (\tilde{j} \cdot s_{u*})^{-1}} TS \oplus \nu(S,M) = TM|S$$

(compare with (1.5) and (1.6)). Observe that $\tilde{g}$, when composed with the fiber map

$$\tilde{p} : A^k(\alpha,\beta) \longrightarrow M,$$

gives the proper inclusion map $S \subset M$. Also

$$\underset{\sim}{\mathrm{Ker}} \otimes \underset{\sim}{\mathrm{Coker}} = \tilde{g}^*(\widetilde{\underset{\sim}{\mathrm{Ker}}} \otimes \widetilde{\underset{\sim}{\mathrm{Coker}}}) \text{ and } TM|S = \tilde{g}^*(\tilde{p}^*(TM)).$$

Thus the bordism class of $(S,\tilde{g},\bar{\tilde{g}})$ defines an invariant

(2.6) $$\tilde{\omega}_{k+1}(\alpha,\beta, u^\partial,[u]) \in \Omega_{n-(a-k)(b-k)}(A^k(\alpha,\beta),\tilde{p};\tilde{\phi}).$$

Here

(2.7) $$\tilde{\phi} = \widetilde{\underset{\sim}{\mathrm{Ker}}} \otimes \widetilde{\underset{\sim}{\mathrm{Coker}}} - \tilde{p}^*(TM);$$

adding $\widetilde{\underset{\sim}{\mathrm{Ker}}} \otimes \widetilde{\underset{\sim}{\mathrm{Im}}} \cong \widetilde{\underset{\sim}{\mathrm{Ker}}} \otimes (\tilde{p}^*(\alpha)/\widetilde{\underset{\sim}{\mathrm{Ker}}})$ to both sides of the minus sign, resp. adding the bundle tangent to the fibers of $\tilde{p}$, we can also write

(2.7') $$\tilde{\phi} = \widetilde{\underset{\sim}{\mathrm{Ker}}} \otimes \tilde{p}^*(\beta) - (\tilde{p}^*(TM) \oplus (\widetilde{\underset{\sim}{\mathrm{Ker}}} \otimes (\tilde{p}^*(\alpha)/\widetilde{\underset{\sim}{\mathrm{Ker}}})))$$

and

(2.7") $\tilde{\phi} = \tilde{p}^* (\underset{\sim}{\mathrm{Hom}}(\alpha,\beta)) - T(A^k(\alpha,\beta))$.

Lemma 2.8. The invariant $\tilde{\omega}_{k+1}(\alpha,\beta,u^\partial,[u])$ is well-defined, i.e. independent of the choice of u within the class $[u]$.

Proof. Pick any other non-degenerate k-morphism $u' \in [u]$. Then a homotopy between u and u' gives rise to a k-morphism u_I from the obvious pullback of α over $M \times I$ to that of β. We may assume that in a whole neighborhood of $M \times \{0\}$, resp. $M \times \{1\}$, u_I is given by u_o, resp. u_1, and that u_I is non-degenerate. The singularity of u_I and its data provide the necessary bordism between $(S,\tilde{g},\bar{\tilde{g}})$ and the corresponding tripel for u'.

Lemma 2.9. If $k=0$ or $n+1<(a-k+1)(b-k+1)$, then there is precisely one mod ∂M k-homotopy class $[u]$ of k-morphisms extending u^∂. Hence in this case we may drop $[u]$ from our notations.

Proof. When we try to increase the minimum rank of any homomorphism from α to β (or of any homotopy between k-morphisms) step by step to k, all occurring non-degenerate singularities will have a codimension larger than $n+1$ and therefore will be empty.

Next we weaken our invariant somewhat to obtain a version which is simpler and always independent of $[u]$.

In the discussion leading to (2.6) the kernel bundle $\underset{\sim}{\mathrm{Ker}} \subset \alpha|S$ of $u|S$ determines a (continuous, proper) "kernel map"

(2.10) $g : S \longrightarrow G_{a-k}(\alpha)$.

Furthermore we have the isomorphism

(2.11)

$$\bar{g} : TS \oplus (\underset{\sim}{\mathrm{Ker}} \otimes \beta|S) = TS \oplus (\underset{\sim}{\mathrm{Ker}} \otimes \underset{\sim}{\mathrm{Coker}}) \oplus (\underset{\sim}{\mathrm{Ker}} \otimes \underset{\sim}{\mathrm{Im}}) \longrightarrow TM|S \oplus (\underset{\sim}{\mathrm{Ker}} \otimes ((\alpha|S)/\underset{\sim}{\mathrm{Ker}}))$$

derived from (2.5) and (1.8). Clearly $\underset{\sim}{\mathrm{Ker}} = g^*(\gamma)$; hence $\underset{\sim}{\mathrm{Ker}} \otimes (\beta | S) = g^*(\gamma \otimes p^*(\beta))$, and the right hand term in (2.11) equals $g^*(p^*(TM) \oplus (\gamma \otimes \gamma^{\perp}))$ which in turn can be identified with $g^*(T(G_{a-k}(\alpha)))$ (see 1.2)). Thus the tripel $(S, g, \bar{g})$ gives rise to a bordism class

$$(2.12) \qquad \omega_{k+1}(\alpha, \beta, u^{\partial}) \in \Omega_{n-(a-k)(b-k)}(G_{a-k}(\alpha), \mathrm{Id}; \phi),$$

where

$$(2.13) \qquad \phi = \gamma \otimes p^*(\beta) - T(G_{a-k}(\alpha))$$

$$(2.13') \qquad = \gamma \otimes p^*(\beta) - (p^*(TM) \oplus (\gamma \otimes \gamma^{\perp}))$$

$$(2.13'') \qquad = (\gamma \otimes p^*(\beta) \oplus \gamma \otimes \gamma) - (p^*(TM) \oplus (\gamma \otimes p^*(\alpha))).$$

Now observe the striking resemblance between these data and the corresponding data for 0-morphisms from γ to $p^*(\beta)$ over $G_{a-k}(\alpha)$; note especially the form of ϕ in (2.13) and the fact that

$$\dim(G_{a-k}(\alpha)) - \dim(\gamma) \cdot \dim(p^*(\beta)) = n - (a-k)(b-k).$$

Therefore it seems worthwhile to study the operator p^{**} which associates to any morphism $\dot{u}: \alpha \to \beta$ over some subspace L of M the morphism

$$(2.14) \qquad p^{**}(\dot{u}) : \gamma \subset p^*(\alpha) \xrightarrow{p^*(\dot{u})} p^*(\beta)$$

over the corresponding subspace $p^{-1}(L)$ of $G_{a-k}(\alpha)$. Clearly p^{**} lowers the minumum rank of $\dot{u}$ by no more than k, the codimension of γ in $p^*(\alpha)$. In particular

$$p^{**}(u^{\partial}) \quad : \quad \gamma | \partial G_{a-k}(\alpha) \longrightarrow p^*(\beta) | \partial G_{a-k}(\alpha)$$

is a 1-morphism.

Proposition 2.15. *We have the identity*

$$\omega_{k+1}(\alpha,\beta,u^{\partial}) = \omega_1(\gamma,p^*(\beta),p^{**}(u^{\partial})).$$

Consequently, the invariant $\omega_{k+1}(\alpha,\beta,u^{\partial})$ *is independent of the k-morphism* u *originally involved in its definition. Moreover, the identity above can be used to define* $\omega_{k+1}(\alpha,\beta,u^{\partial})$ *even if no appropriate* u *exists.*

Proof. In order to describe $\omega_1(\gamma,p^*(\beta),p^{**}(u^{\partial}))$, consider the 0-morphism $p^{**}(u)$ over $G_{a-k}(\alpha)$. It vanishes precisely at the (a-k)-planes contained in the kernel of u. Hence the singularity of $p^{**}(u)$ lies at $g(S)$.

It remains to determine the normal bundle map

$$\bar{j} : \nu(g(S),G_{a-k}(\alpha)) = (p^*(\nu(S,M)) \oplus TF(G_{a-k}(\alpha)))|g(S)$$
$$\longrightarrow \underset{\sim}{\mathrm{Hom}}(\gamma,p^*(\beta))|g(S) = (\underset{\sim}{\mathrm{Hom}}(\gamma,p^*(\underset{\sim}{\mathrm{Coker}})) \oplus \underset{\sim}{\mathrm{Hom}}(\gamma,p^*(\underset{\sim}{\mathrm{Im}})))|g(S)$$

induced by $p^{**}(u)$. Here we have identified $p^*(\nu(S,M))|g(S)$ with the normal bundle of $g(S)$ in $G(\mathfrak{V})$, where G is the section of $G_{a-k}(\alpha)$, defined in a neighborhood $\mathfrak{V}$ of S by an extended kernel bundle $\alpha_1 = \bar{f}_{\alpha}^{-1}(\widehat{\underset{\sim}{\mathrm{Ker}}})$ as in lemma 1.10. It is not hard to see that on the first component bundle $\bar{j}$ coincides with the pullback of the isomorphism (1.6). Also note that if $x' \in S$, then $p^{**}(u)$ is given over $G_{a-k}(\alpha_x)$ by restricting the fixed linear map $u_x : \alpha_x \longrightarrow \beta_x$ to the various (a-k)-planes in α_x. Moreover, the isomorphism (1.2) at $g(x)$ is the tangent map of an obvious chart from a neighborhood of $g(x)$ in $G_{a-k}(\alpha_x)$ onto $\mathrm{Hom}(\mathrm{Ker}_x, \mathrm{Coim}_x)$, which can be covered in a standard way by a trivialization of γ. This can be used to show that $\bar{j}$ restricts also to the standard isomorphism between the second component bundles.

In particular the 0-morphism $p^{**}(u)$ is non-degenerate. The identity in our proposition follows immediately. Furthermore, lemmas

2.8 and 2.9 imply that

$$\omega_1(\gamma,p^*(\beta),p^{**}(u^\partial)) = \tilde{\omega}_1(\gamma,p^*(\beta),p^{**}(u^\partial),[p^{**}(u)])$$

depends only on α, β, u^∂ and k, and therefore so does $\omega_{k+1}(\alpha,\beta,u^\partial)$.

Next we compare the two versions of our invariant. Note that the map $\tilde{p}_{ker} : A^k(\alpha,\beta) \longrightarrow G_{a-k}(\alpha)$ (cf. 1.3) pulls ϕ back to $\tilde{\phi}$ (use (2.7') and (2.13')), and hence induces a homomorphism

$$\tilde{p}_{ker*} : \Omega_r(A^k(\alpha,\beta),\tilde{p};\tilde{\phi}) \longrightarrow \Omega_r(G_{a-k}(\alpha),Id;\phi).$$

Clearly

(2.16) $\quad \tilde{p}_{ker*}(\tilde{\omega}_{k+1}(\alpha,\beta,u^\partial,[u])) = \omega_{k+1}(\alpha,\beta,u^\partial).$

The following result implies that for

$$n < (a-k+1)(b-k)$$

we lose no information if we consider $\omega_{k+1}(\alpha,\beta,u^\partial)$ instead of $\omega_{k+1}(\alpha,\beta,u^\partial,[u])$.

<u>Lemma 2.17.</u> <u>If</u> $r<b-k$, <u>then</u> $\tilde{p}_{ker*}$ <u>is an isomorphism</u>.

<u>Proof.</u> Let T be an r-dimensional manifold or a bordism of such manifolds. Factoring a map $g : T \longrightarrow G_{a-k}(\alpha)$ through $\tilde{p}_{ker}$ amounts to finding a k-morphism from $g^*(\gamma^\perp)$ to $g^*(p^*(\beta))$ (cf. the observation following (1.3)). But such a k-morphism, with or without prescribed boundary behavior, can be constructed as in the proof of lemma 2.9.

The same technique can also be used to prove the following result which sometimes allows us to express our obstruction groups in a more familiar way.

Lemma 2.18. If $r<k$, then a classifying map of γ gives rise to the canonical isomorphism

$$\Omega_r(G_{a-k}(\alpha),\mathrm{Id};\phi) \cong \Omega_r(M\times BO(a-k),\ \text{first projection};\phi_\infty),$$

where ϕ_∞ is defined by the obvious modification of (2.13").

Finally note that $\tilde{\omega}_{k+1}(\alpha,\beta,u^\partial,[u])$ vanishes necessarily if $[u]$ contains a (k+1)-morphism, and that $\omega_{k+1}(\alpha,\beta,u^\partial) = 0$ if u^∂ extends to any (k+1)-morphism over M at all. ∎

Example 2.19. Consider the case where M is connected, $\alpha = \underset{\sim}{\mathbb{R}}$, $\dim \beta = n = \dim M$ and $k = 0$. We can identify the vector bundle $\underset{\sim}{\mathrm{Hom}}(\alpha,\beta)$ with β so that a nondegenerate k-morphism from α to β is basically just a section s of β which is transverse to the zero-section; its singularity consists of isolated zeroes.

We may also identify $A^k(\alpha,\beta)$ and $G_1(\alpha)$ with M via $\tilde{p}$ and p. So $\tilde{\omega}_1(\alpha,\beta,s^\partial) = \omega_1(\alpha,\beta,s^\partial)$ lies in

$$\Omega_0(M,\ \mathrm{Id};\ \beta - TM) \cong \begin{cases} 0 & \text{if } M \text{ noncompact,} \\ \mathbb{Z}_2 & \text{if } M \text{ compact, } \Lambda^n\beta \not\cong \Lambda^n TM, \\ \mathbb{Z} & \text{if } M \text{ compact, } \Lambda^n\beta \cong \Lambda^n TM. \end{cases}$$

The points in $S = s^{-1}(0)$ are counted zero, resp. mod 2, resp. with signs $+1$ or -1 : if the orientations of β and M correspond to one another in a bijective and continuous fashion, pick such a correspondance $h : \Lambda^n\beta \cong \Lambda^n TM$ and count a zero z of s positively or negatively according to whether or not the (n-th exterior power of the) tangent map s_* agrees with h^{-1} at z.

We want to determine the number $\overset{(\sim)}{\omega}_1(\alpha,\beta,s^\partial)$ in a few cases.

If M is closed (i.e. compact and without boundary), it follows easily from the definition of the mod 2 Euler class $w_n(\beta)$ (cf.[46]) that $\overset{(\sim)}{\omega}_1(\alpha,\beta)$, taken mod 2 if necessary, equals $w_n(\beta)[M]$ (where $[M]$

denotes the fundamental class of M).

If M is closed and odd-dimensional, and $\Lambda^n\beta \cong \Lambda^n TM$, then $\overset{(\sim)}{\omega}_1(\alpha,\beta) = 0$; indeed, note that the section $-s$ of β has the same zeroes as s, but counted with opposite signs, so that $\overset{(\sim)}{\omega}_1(\alpha,\beta) = -\overset{(\sim)}{\omega}_1(\alpha,\beta)$ in $\mathbb{Z}$ (use lemma 2.8).

Our last case is very classical. If M is compact, $\beta = TM$, and the tangential vectorfield s points outward at all points of the boundary ∂M, then $\overset{(\sim)}{\omega}_1(\mathbb{R},TM,s|\partial M)$ is the number of zeroes of s, counted correctly. According to lemma 2.8, this number is an invariant of M alone. Actually, it is just the Euler number $\chi(M)$ of M, by the theorem of Poincaré and Hopf (see [49]). (So the preceding paragraph implies also the wellknown fact that $\chi(M) = 0$ for closed odd-dimensional M).∎

§3. Existence of (k+1)-morphisms.

We are now ready to state and exploit the main result of this chapter.

Existence theorem 3.1. Let α and β be vector bundles of dimensions a and b over an n-manifold M (which need not be compact). Let $u^{\partial} : \alpha|\partial M \longrightarrow \beta|\partial M$ be a (k+1)-morphism (and $u : \alpha \longrightarrow \beta$ a k-morphism extending u^{∂}). Moreover assume that

$$(3.2) \qquad n+2 < 2(a-k)(b-k) \quad .$$

Then u^{∂} can be extended to a (k+1)-morphism over all of M (which is rel ∂M k-homotopic to u) if and only if the invariant $\tilde{\omega}_{k+1}(\alpha,\beta,u^{\partial};[u])$ vanishes.

The proof will be given below. Note that we can drop the reference to u, provided the additional condition

$$(3.3) \qquad n+1 < (a-k+1)(b-k+1)$$

is satisfied or $k = 0$ (see 2.9). If even

$$(3.4) \qquad n < (a-k+1)(b-k)$$

holds, we may also replace $\tilde{\omega}_{k+1}(\alpha,\beta,u^{\partial})$ by $\omega_{k+1}(\alpha,\beta,u^{\partial})$ (see 2.17). Furthermore, in this case we can use proposition 2.15 to reduce the whole question to an extension problem for sections of sphere bundles:

Corollary 3.5. Under the assumption of the theorem and under the additional condition (3.4), u^{∂} can be extended to a (k+1)-morphism over all of M if and only if the section corresponding to $p^{**}(u^{\partial})$ can be extended to a nowhere vanishing section of the vector bundle $\mathrm{Hom}(\gamma,p^*(\beta))$ over all of $G_{a-k}(\alpha)$.

Next suppose that ∂M is the disjoint union of the closed subspaces $\partial_1 M$ and $\partial_2 M$, and that the boundary condition u^∂ is only given over $\partial_1 M$ while no restrictions are imposed at $\partial_2 M$. We can easily reduce this seemingly more general situation to the setting of the theorem. Indeed, u can be deformed (through k-morphisms which all restrict to u^∂) into a (k+1)-morphism if and only if there is such a deformation for $u|M-\partial_2 M$. This can be shown by "pushing" $\partial_2 M$ into the interior of M along some inward pointing vector field.

Recall that a smooth map is called a (k+1)-mersion if the rank of its tangent map is nowhere less than k+1. We can now deduce the following result from the work of S. Feit [21].

<u>Corollary 3.6.</u> <u>Let</u> M_1 <u>and</u> M_2 <u>be manifolds of dimensions</u> n_1 <u>and</u> n_2, <u>possibly with boundary, and let</u> $f : M_1 \longrightarrow M_2$ <u>be a continuous map. Assume that</u> M_1 <u>is not closed or that</u> $k+1 < n_2$, <u>and assume furthermore that</u>

$$n_1+1 < \min\{2(n_1-k)(n_2-k)-1,\ (n_1-k+1)\cdot(n_2-k+1)\} \quad .$$

<u>Then</u> f <u>is homotopic to a (k+1)-mersion if and only if</u>

$$\tilde{\omega}_{k+1}\ (T\mathring{M}_1,\ f^*(TM_2)|\mathring{M}_1,\ \emptyset) = 0\ ,$$

<u>where</u> $\mathring{M}_1 = M_1 - \partial M_1$. ∎

Our existence theorem has particular importance when $k+1$ is the maximum possible rank. Since problems concerning epimorphisms can be translated, via dualizing, into questions about monomorphisms, we need to consider only the latter. Then $a-k=1$ and hence γ is just the canonical line bundle over the projectification $RP(\alpha) = G_1(\alpha)$ of α.

Theorem 3.7. *Let α and β be vector bundles of dimensions a and b over the n-manifold M. Let $u^{\partial} : \alpha|\partial M \hookrightarrow \beta|\partial M$ be a monomorphism. Assume that $n < 2(b-a)$.*

Then u^{∂} can be extended to a monomorphism over all of M if and only if the invariant

$$\omega_a(\alpha,\beta,u^{\partial}) \in \Omega_{n-b+a-1}(RP(\alpha),Id;\gamma\otimes p^*(\beta)-(\gamma\otimes p^*(\alpha)\oplus p^*(TM)))$$

vanishes.

Proof of Theorem 3.1+. Assume that $\tilde{\omega}_{k+1}(\alpha,\beta,u^{\partial};[u])$ is zero and that the k-morphism u is non-degenerate. Thus, in the discussion leading to (2.6), the triple $(S,\tilde{g},\bar{\tilde{g}})$ admits a zero bordism $(\mathcal{S},\tilde{G},\bar{G})$. Now consider the pullbacks $pr^*(\alpha)$ and $pr^*(\beta)$ of α and β by the first projection map $pr : M\times I \longrightarrow M$. We will use the bordism above to construct a k-morphism

$$u_I : pr^*(\alpha) \longrightarrow pr^*(\beta)$$

over $M\times I$ which restricts to u over $M\times\{0\}$, to the obvious pullback u_I^{∂} over $(\partial M)\times I$, and to a $(k+1)$-morphism over $M\times\{1\}$. This will provide the required homotopy.

First observe that we can find an embedding

$$i = (i_1,i_2) : \mathcal{S} \hookrightarrow M\times I$$

which realizes $\mathcal{S}$ as a closed submanifold of $M\times I$ such that $\mathcal{S} \cap \partial(M\times I) = \partial\mathcal{S} = S \subset M\times\{0\}$. Indeed, since condition (3.2) implies the inequality

$$2\dim(\mathcal{S}) = 2n + 2 - 2(a-k)(b-k) < n ,$$

+ After I had written this proof, Bruce Williams drew my attention to the work of Hatcher and Quinn [25] which contains similar arguments.

we can deform the proper map

$$\tilde{p} \circ \tilde{G} : \mathcal{S} \longrightarrow M$$

into a proper embedding i_1. This follows from classical embedding theorems, e.g. by globalizing the results 1.28 and 1.29 in [47]. To complete the construction of i, let i_2 be a smooth function on $\mathcal{S}$ which, on some collar $S\times[0,\frac{1}{2}]$ around S, is essentially given by the second projection, and which takes the constant value $\frac{1}{2}$ outside of this collar.

Without changing anything at $S=\partial\mathcal{S}$, we can also deform the data $\tilde{G}$ and $\tilde{\tilde{G}}$ appropriately. Then G defines a vector bundle morphism

$$u_I|\mathcal{S} : pr^*(\alpha)|\mathcal{S} \longrightarrow pr^*(\beta)|\mathcal{S}$$

of constant rank k. Moreover, $\tilde{\tilde{G}}$ gives an isomorphism

$$T\mathcal{S} \oplus \underset{\sim}{Hom}(\underset{\sim}{Ker}(u_I|\mathcal{S}), \underset{\sim}{Coker}(u_I|\mathcal{S})) \longrightarrow T(M\times I)|\mathcal{S}$$

which, in view of (3.2), can be deformed until it maps $T\mathcal{S}$ identically into $T\mathcal{S} \subset T(M\times I)|\mathcal{S}$. Therefore we obtain an isomorphism

$$j_I : \nu(\mathcal{S}, M\times I) \xrightarrow{\cong} \underset{\sim}{Hom}(\underset{\sim}{Ker}(u_I|\mathcal{S}), \underset{\sim}{Coker}(u_I|\mathcal{S})).$$

Now we can apply the constructions of section 1 (see in particular 1.11) to $\mathcal{S}$, $u_I|\mathcal{S}$ and j_I. The resulting k-morphism $u_I|T$, defined over some closed tubular neighborhood T of $\mathcal{S}$ in $M\times I$, can be made to coincide with u over $(M\times\{0\})\cap T$.

Thus u_I is now defined over $R_0 = (\partial M)\times I \cup M\times\{0\} \cup T$. Furthermore, without introducing new singularities where the rank drops below $k+1$, we can extend u_I also over the manifold

$$R_1 = \{(pr(x),t) \mid x\in\mathcal{S},\ t\in[0,i_2(x))\},$$

and even over a neighborhood of $R_0 \cup R_1$. Here the inequality (3.2) is crucial; it implies that the dimension of R_1 is smaller than the codimensions of possibly occurring non-degenerate singularities.

Now clearly there is a smooth function $h : M \to (0,1]$ such that the domain of u_I contains the subspace $R = \{(x,t) \in M \times I \mid t \leq h(x)\}$ of $M \times I$ which, in turn, contains $\mathcal{S} \cup (\partial M) \times I$ in its interior. In particular, u_I is a (k+1)-morphism over the graph of h. To finish the proof, use the obvious diffeomorphism between $M \times I$ and R.

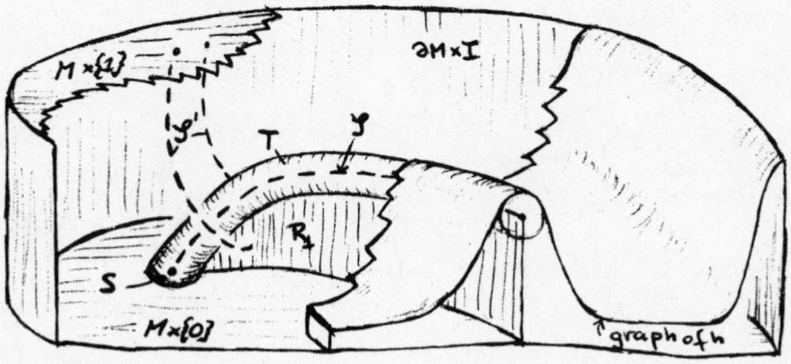

Figure 3.8. The construction of the homotopy u_I. Since (3.2) is not satisfied here, $\mathcal{S}$ may possibly link with a new singularity $\mathcal{S}'$.

Remark 3.9. We have used the dimension condition (3.2) three times in the proof of 3.1. For the first two times, namely when we embed $\mathcal{S}$ into $M \times I$ and when we "destabilize" the description of its normal bundle given by $\tilde{\tilde{G}}$, we could also allow equality in (3.2). Actually, there would not be any difficulty here at all if we had defined our obstruction in a suitable finer bordism set. However, when we extend u_I from R_0 over all of $M \times I$, we have at least to make sure that the singularity at $\mathcal{S}$ does not link with possible new singularities, and at this point the condition (3.2) seems to be crucial. ∎

The discussion in example 2.19 allows us to give a first (simple but quite useful) application of our existence theorem (in the form of 3.7).

Proposition 3.10. Let β be an n-plane bundle over a closed connected n-manifold M, $n \geq 3$.

(i) if $w_1(\beta) \neq w_1(M)$, we have:

β has a nowhere vanishing section precisely if its top Stiefel-Whitney class $w_n(\beta) \in H^n(M;\mathbb{Z}_2)$ is zero.

(ii) If $w_1(\beta) = w_1(M)$ and n is odd, then β has a nowhere vanishing section.

The whole proposition still holds for all $n \geq 1$ (see 13.19).

In particular, any odd-dimensional closed manifold has a nowhere zero vectorfield; moreover, its orientation bundle $\Lambda^n TM$ can be described as a pullback of the canonical line bundle over projective space P^{n-1} (apply 3.10 (ii) to $\beta = \underset{\sim}{\mathrm{Hom}}(\Lambda^n TM, \underset{\sim}{\mathbb{R}}^n)$ to obtain a monomorphism $\Lambda^n TM \hookrightarrow \underset{\sim}{\mathbb{R}}^n$).

Exercise 3.11. Let β be an n-plane bundle on a connected noncompact n-manifold. Show that β has a nowhere vanishing section (at least if $n \geq 3$).

§4. Homotopy and concordance classification of (k+1)-morphisms.

In this section we will spell out the classification theorems corresponding to the existence results of §3.

Let

$$u_0,\ u_1 : \alpha \longrightarrow \beta$$

be two (k+1)-morphisms extending u^∂ over all of M. In order to decide whether u_0 and u_1 are (k+1)-homotopic rel ∂M, we will apply the construction of section 2 and 3 to the vector bundles $pr_I^*(\alpha)$ and $pr_I^*(\beta)$ over $M\times I$ and to the boundary condition given by the (k+1)-morphism $u_0 \cup pr_I^*(u^\partial) \cup u_1$. Here pr_1 denotes the first projection of $M\times I$, $pr_I^*(u^\partial)$ is the obvious pullback of u^∂ over $(\partial M)\times I$, and we consider u_i, $i = 0,1$, as morphisms over $M\times\{i\} \subset M\times I$. The fact that the manifold $M\times I$ may have corners does not cause any difficulties.

If $k = 0$ or $n+2<(a-k+1)(b-k+1)$ (compare 2.9), we obtain the welldefined invariant

$$(4.1)\qquad \tilde{d}_{k+1}(u_0,u_1) := \tilde{\omega}_{k+1}(pr_I^*(\alpha),pr_I^*(\beta),\ u_0 \cup pr_I^*(u^\partial) \cup u_1)$$

which, after an obvious identification, lies in the group $\Omega_{n+1-(a-k)(b-k)}(A^k(\alpha,\beta),\tilde{p};\tilde{\phi})$. No dimension restriction is needed (cf. 2.15) to define the slightly weaker version

$$(4.2)\qquad d_{k+1}(u_0,u_1) := \omega_{k+1}(pr_I^*(\alpha),pr_I^*(\beta),u_0 \cup pr_I^*(u^\partial) \cup u_1)$$

which can be considered as an element in $\Omega_{n+1-(a-k)(b-k)}(G_{a-k}(\alpha),Id;\phi)$.

We have

$$\tilde{p}_{ker*}(\tilde{d}_{k+1}(u_0,u_1)) = d_{k+1}(u_0,u_1)$$

(cf. 2.16). Clearly both versions of this difference invariant depend on u_0 and u_1 only up to (k+1)-homotopy rel ∂M. Also, for three (k+1)-morphisms u_0, u_1, u_2 we have obviously

(4.3) $$d_{k+1}(u_0,u_2) = d_{k+1}(u_0,u_1) + d_{k+1}(u_1,u_2)$$

and the corresponding formula for $\tilde{d}_{k+1}(u_0,u_2)$.

Next we introduce another equivalence relation between (k+1)-morphisms which is interesting e.g. in the setting of planefields and of foliations.

Definition 4.4. Two (k+1)-morphisms $u_o, u_1 : \alpha \longrightarrow \beta$ are called (k+1)-concordant rel ∂M if the (k+1)-morphisms $(u_0,0),(u_1,0) : \alpha \longrightarrow \beta \oplus \mathbb{R}$ are (k+1)-homotopic rel ∂M.

To relate the difference invariants for homotopy to the ones for concordance, interprete $A^k(\alpha,\beta)$ as a subbundle of $A^k(\alpha,\beta \oplus \mathbb{R})$ and extend $\tilde{p}$ and $\widetilde{Ker} \oplus \tilde{\phi}$ in the obvious way over $A^k(\alpha,\beta \oplus \mathbb{R})$ to obtain $\tilde{p}_1$ and $\tilde{\phi}_1$ etc.. Consider the homomorphism

(4.5) $$\tilde{\Delta} : \Omega_r(A^k(\alpha,\beta),\tilde{p},\tilde{\phi}) \longrightarrow \Omega_{r-(a-k)}(A^k(\alpha,\beta \oplus \mathbb{R}),\tilde{p}_1;\tilde{\phi}_1)$$

which maps a bordism class $[S,\tilde{g},\bar{\tilde{g}}]$ to the class $[S_1,\tilde{g}_1,\bar{\tilde{g}}_1]$ where

(i) $S_1 \subset S$ is the zero set of a non-degenerate section of the (a-k)-dimensional vector bundle $\tilde{g}^*(\widetilde{Ker})$ over S,

(ii) $\tilde{g}_1$ = inclusion $\tilde{g} | : S_1 \longrightarrow A^k(\alpha,\beta) \subset A^k(\alpha,\beta \oplus \mathbb{R})$, and

(iii) $\bar{\tilde{g}}_1$ is obtained from $\bar{\tilde{g}}|S_1$ by using the isomorphism $TS_1 \oplus \tilde{g}^*(\widetilde{Ker})|S_1 \cong TS|S_1$.

Similarly, if we use non-degenerate sections of pullbacks of the canonical vector bundle γ, we obtain a homomorphism

(4.6) $$\Delta : \Omega_r(G_{a-k}(\alpha),Id;\phi) \longrightarrow \Omega_{r-(a-k)}(G_{a-k}(\alpha),Id;\gamma \oplus \phi) .$$

These homomorphisms are related by the identity $\Delta \circ \tilde{p}_{ker*} = \tilde{p}_{ker_1 *} \circ \tilde{\Delta}$.

Lemma 4.7. We have

$$\tilde{d}_{k+1}((u_0,0),(u_1,0)) = \tilde{\Delta}(\tilde{d}_{k+1}(u_0,u_1))$$

(whenever the right hand term is defined), and

$$d_{k+1}((u_0,0),(u_1,0)) = \Delta(d_{k+1}(u_0,u_1)).$$

Proof. Let S be the singularity of a non-degenerate k-morphism $u : pr_I^*(\alpha) \longrightarrow pr_I^*(\beta)$ extending $u_0 \cup pr_I^*(u^\partial) \cup u_1$. Choose a suitable homomorphism $v : pr_I^*(\alpha) \longrightarrow \underset{\sim}{\mathbb{R}}$ which vanishes over $\partial(M \times I)$ and which restricts to a non-degenerate 0-morphism on $\underset{\sim}{Ker}(u)$ over S. Then the non-degenerate k-morphism $(u,v) : pr_I^*(\alpha) \longrightarrow pr_I^*(\beta) \oplus \underset{\sim}{\mathbb{R}}$ has its singularity at the zero set S_1 of $v \mid \underset{\sim}{Ker}(u)$. An analysis of the singularity data completes the proof.

Classification theorem 4.8. Let α and β be vector bundles of dimensions a and b over an n-manifold M, and let $u^\partial : \alpha|\partial M \to \beta|\partial M$ be a (k+1)-morphism. Assume that $n+3<2(a-k)(b-k)$, and assume in addition that $n+2 < (a-k+1)(b-k+1)$ or $k = 0$.

Then two (k+1)-morphisms $u_0, u_1 : \alpha \to \beta$, which restrict to u^∂ over ∂M, are (k+1)-homotopic (resp. (k+1)-concordant) rel ∂M if and only if $\tilde{d}_{k+1}(u_0,u_1)$ (resp. $\tilde{\Delta}(\tilde{d}_{k+1}(u_0,u_1))$) vanishes. Furthermore, given a fixed (k+1)-morphism u_0 extending u^∂ over all of M, the invariant $\tilde{d}_{k+1}(u_0,-)$ (resp. $\tilde{\Delta}(\tilde{d}_{k+1}(u_0,-))$) defines a one-to-one onto correspondence between the set of rel ∂M (k+1)-homotopy (resp. (k+1)-concordance) classes of (k+1)-morphisms from α to β, extending u^∂, on one hand, and the group $\Omega_{n+1-(a-k)(b-k)}(A^k(\alpha,\beta),\tilde{p};\tilde{\varphi})$ (resp. its image under $\tilde{\Delta}$) on the other hand.

If in addition

(4.9) $\qquad n+1 < (a-k+1)(b-k)$,

then in all the above we can replace $A^k(\alpha,\beta)$ by $G_{a-k}(\alpha)$ and drop the tildas.

Proof. We need to consider only the case of homotopy classification. The first statement of the theorem follows from 3.1., applied to the bundles $pr_I^*(\alpha)$ and $pr_I^*(\beta)$ over $M\times I$ and to the boundary condition $u_0 \cup pr_I^*(u^\partial) \cup u_1$. As a consequence, (4.3) implies that the correspondence $\tilde{d}_{k+1}(u_0,-)$ is injective. To see that it is also onto, represent an arbitrary element $\tilde{d}$ of $\Omega_{n+1-(a-k)(b-k)}(A^k(\alpha,\beta),\tilde{p};\tilde{\phi})$ by a tripel $(\mathcal{S},\tilde{G},\tilde{\tilde{G}})$ and apply the construction in the proof of 3.1 to $u = u_0$ and to the zero bordism provided by this tripel; if u_1 is the (k+1)-morphism at the end of the resulting k-homotopy, clearly $\tilde{d}_{k+1}(u_0,u_1) = \tilde{d}$.

Remark 4.10. In the case of concordance classification the results of theorem 4.8 still hold under weaker dimension assumptions, e.g. at least when in the inequalities also the equal sign is allowed.

Corollary 4.11. Under the assumptions of the theorem and the additional condition 4.9 , the operation p^{**} (cf. 2.14) establishes a one-to-one onto correspondence between rel ∂M (k+1)-homotopy (resp. -concordance) classes of (k+1)-morphisms from α to β, extending u^∂, on one hand, and equivalence classes of nowhere vanishing sections of the vector bundle $\mathrm{Hom}(\gamma,p^*(\beta))$ over $G_{a-k}(\alpha)$, extending $p^{**}(u^\partial)$, on the other hand. Here such sections of $\mathrm{Hom}(\gamma,p^*(\beta))$ are considered equivalent if they are homotopic through nowhere vanishing sections of $\mathrm{Hom}(\gamma,p^*(\beta))$ (respectively $\mathrm{Hom}(\gamma,p^*(\beta)\oplus \mathbb{R})$) which all extend $p^{**}(u^\partial)$.

Remark 4.12. The corollary remains valid when M is an arbitrary CW-complex of dimension n and the (k+1)-morphism u^∂ is prescribed over any closed subcomplex. To see this, apply 4.11 or 3.5 with increasing q to the q-cells (whose models, after all, are smooth manifolds).∎

Often theorem 4.8 can be used to classify also homomorphisms from a vector bundle α over a manifold M_1 to a vector bundle β over a possibly different manifold M_2.

Indeed, equip the space $C^o(M_1,M_2)$ of continuous maps with the compact-open topology and with a base point f. Also, let $\pi_o(\mathrm{Aut}(f^*(\beta)))$ denote the group of homotopy classes of automorphisms of the pullback of β under the continous map f. Then there is a canonical group homomorphism

$$\pi_1(C^o(M_1,M_2),f) \longrightarrow \pi_o(\mathrm{Aut}(f^*(\beta)))$$

defined as follows. Represent an element of $\pi_1(C^o(M_1,M_2),f)$ by a continuous map

$$F : M_1 \times I \longrightarrow M_2$$

such that $F_0 = F_1 = f$. Moreover, extend the identity map of $f^*(\beta)$, given over $M_1 = M_1 \times \{0\}$, to an isomorphism $\bar{i} : pr_I^*(f^*(\beta)) \xrightarrow{\cong} F^*(\beta)$ over all of $M \times I$. When restricted to $M_1 \times \{1\}$, $\bar{i}$ gives the desired class of automorphisms of $f^*(\beta)$.

If the action of $\pi_1(C^o(M_1,M_2),f)$ on $\pi_0(\mathrm{Aut}(f^*(\beta)))$ is trivial (e.g. when the bundle β is trivial over the (n_1+1)-skeleton of some CW-decomposition of M_2), then vector bundle homomorphisms from α to β which cover maps homotopic to f and which have at least rank k+1 everywhere, correspond, up to suitable homotopies, to (k+1)-morphisms from α to $f^*(\beta)$; and we can apply the classification theorem.

<u>Corollary 4.13.</u> <u>Let</u> M_1 <u>and</u> M_2 <u>be manifolds of dimensions</u> n_1 <u>and</u> n_2, <u>and let</u> $f : M_1 \longrightarrow M_2$ <u>be a (k+1)-mersion such that the action of</u> $\pi_1(C^o(M_1,M_2),f)$ <u>on</u> $\pi_o(\mathrm{Aut}(f^*(TM_2))$ <u>is trivial. Assume that</u> M_1 <u>is not closed or that</u> $k+1 < n_2$, <u>and assume furthermore</u>

that $\partial M_2 = \emptyset$ *and* $n_1+2 < \min\{2(n_1-k)(n_2-k)-1, (n_1-k+1)(n_2-k+1)\}$.

Then the (k+1)-regular homotopy classes (in the sense of [21]) of all (k+1)-mersions from M_1 *to* M_2, *which are homotopic (as continuous maps) to* f, *are in one-to-one onto correspondence with the elements of*

$$\Omega_{n_1+1+(n_1-k)(n_2-k)}(A^k(T\mathring{M}_1, f^*(TM_2)|\mathring{M}_1), \tilde{p}; \widetilde{Ker}\otimes\widetilde{\mathring{C}oker}-\tilde{p}^*(T\mathring{M}_1)). \blacksquare$$

Finally we apply our classification theorem to the particularly important case when $a = k+1$. Again we denote the projective space bundle of α by $RP(\alpha)$ and the canonical line bundle over it by γ.

Theorem 4.14. *Let* α *and* β *be vector bundles of dimensions* a *and* b *over the n-dimensional manifold* M, *and fix a monomorphism* $u^\partial : \alpha|\partial M \hookrightarrow \beta|\partial M$. *Moreover, assume that* $n+1 < 2(b-a)$ (*resp.* $n < 2(b-a)$).

Then, given a monomorphism $u_o : \alpha \hookrightarrow \beta$ *extending* u^∂ *over all of* M, *the difference invariant* $d_a(u_o,-)$ *defines a one-to-one onto correspondence between the rel* ∂M *homotopy (resp. concordance) classes of all monomorphisms from* α *to* β *which extend* u^∂, *and the elements of the group*

$$\Omega_{n-b+a}(RP(\alpha), Id; \gamma\otimes p^*(\beta) - ((\gamma\otimes p^*(\alpha)) \oplus p^*(TM)))$$

(resp. of the quotient of this group by its subgroup $Ker\Delta$ *).*

Furthermore, for every element d *of this last quotient we have*

$$(4.15) \qquad 2^{n-b+a} \cdot d = 0 ;$$

hence, if M *is compact, there are only finitely many rel* ∂M *concordance classes of monomorphisms from* α *to* β *which extend* u^∂.

Proof. Formula (4.15) follows from a spectral sequence argument as in [57], theorem 10.1. For the last statement observe that $\Omega_{n-b+a}(RP(\alpha), Id; \phi)$ is a stable homotopy group of a Thomspace which admits the structure of a finite CW-complex when M is compact.

Example 4.16. (Vector fields on spheres). For $k \geq 3$ the formula

$$u_0(x_1,y_1;\ x_2,y_2;\dots;\ x_k,y_k) = (-y_1,x_1;\ -y_2,x_2;\dots;\ -y_k,x_k)$$
$$u_1(x_1,y_1;\ x_2,y_2;\dots;\ x_k,y_k) = (y_1,-x_1;\ -y_2,x_2;\dots;\ -y_k,x_k)$$

defines vectorfields u_o and u_1 on the sphere S^{2k-1} which get carried over into one another by the reflection $-\mathrm{Id} \times \mathrm{Id}$ on $\mathbb{R} \times \mathbb{R}^{2k-1} = \mathbb{R}^{2k}$. Then u_o and u_1 represent different (and, in fact, all) homotopy classes of non-vanishing vectorfields on S^{2k-1}.

To see this, interprete u_o and u_1 as monomorphisms from $\alpha = S^{2k-1} \times \mathbb{R}$ to $\beta = TS^{2k-1}$. Then, in theorem 4.14, $RP(\alpha) = S^{2k-1}$, γ is trivial, and $d_1(u_o,u_1)$ lies in $\Omega_1(S^{2k-1};\text{trivial})$; by transversality, this group is isomorphic to $\Omega_1(S^{2k-1} - *;\text{trivial}) \cong \Omega_1(\text{point};\text{trivial})$. Now note that the linear homotopy between u_o and u_1 defines a nondegenerate 0-morphism over $S^{2k-1} \times I$ whose singularity lies at the circle $\{(x_1,y_1;\ 0,\dots,0) | x_1^2 + y_1^2 = 1\} \times \{\frac{1}{2}\}$. An analysis of the singularity data then shows that $d_1(u_o,u_1)$ is represented by a parallelized circle. Hence $d_1(u_o,u_1) \neq 0$. Indeed, if the parallelized circle had a zero bordism N, it would come with a monomorphism $u : \underset{\sim}{\mathbb{R}} \oplus \underset{\sim}{\mathbb{R}}^q \hookrightarrow TN \oplus \underset{\sim}{\mathbb{R}}^q$ which, over ∂N, is given by the inward pointing vectorfield and by the identity on $\underset{\sim}{\mathbb{R}}^q$. Hence u extends to a nondegenerate q-morphism over the orientable closed surface $\tilde{N} = N \cup_{\partial N} D^2$ with just one singular point at the center of the disk D^2. Thus clearly any nondegenerate vectorfield on $\tilde{N}$ must also have an odd number of singular points (compare with the proof of lemma 2.8). But this contradicts the fact that $\tilde{N}$ must have an even Euler number.

(The stronger claim that $d_1(u_o,u_1)$ generates $\Omega_1(\text{point};\text{trivial}) \cong \mathbb{Z}_2$ follows trivially from the machinery to be devellopped in § 9).

§5. Connections with classical obstructions, and with some classical results concerning the homotopy of Stiefel manifolds.

In this section we apply the homotopy part of our classification theory to the special case when M is a ball.

Assume that $0 \leq k < \min(a,b)$, and let $W^{k+1} = W^{k+1}(\mathbb{R}^a, \mathbb{R}^b)$ denote the space of $b\times a$-matrices of rank $\geq k+1$. A map from a sphere into W^{k+1} can be viewed as a $(k+1)$-morphism and hence leads to an extension problem with an interesting obstruction. Thus for all integers n we obtain a well defined map

$$(5.1) \qquad \sigma : \pi_{n-1}(W^{k+1}) \longrightarrow \Omega_{n-(a-k)(b-k)}(G_{a-k}(\mathbb{R}^a);(b-a)\gamma\oplus\gamma\otimes\gamma)$$

given by $\sigma([u^\partial]) = \omega_{k+1}(D^n \times \mathbb{R}^a, D^n \times \mathbb{R}^b, u^\partial)$. σ is a group homomorphism except for $n = (a-k)=(b-k) = 1$.

<u>Proposition 5.2.</u> <u>If $n < \min\{2(a-k)(b-k)-2,(a-k+1)(b-k)\}$, then σ is an isomorphism, and if n equals this minimum, σ is still onto.</u>

<u>In particular, W^{k+1} is always $((a-k)(b-k)-2)$-connected. Moreover, except when $a=b=k+1$, σ always provides an isomorphism</u>

$$\pi_{(a-k)(b-k)-1}(W^{k+1}) \cong \begin{cases} \mathbb{Z} & \text{if } k=0 \text{ or } a \equiv b(2), \\ \mathbb{Z}_2 & \text{otherwise} \end{cases} ;$$

<u>and any orientation reversing automorphism of $\mathbb{R}^a$, resp. $\mathbb{R}^b$, induces the involution $(-1)^{b-k}\,\mathrm{Id}$, resp. $(-1)^{a-k}\mathrm{Id}$, on this group (which can be identified with the corresponding basepoint-free homotopy group).</u>

<u>Proof.</u> Apply theorem 4.8 to $M = D^{n-1}$, $\alpha = \underset{\sim}{\mathbb{R}}^a$, $\beta = \underset{\sim}{\mathbb{R}}^b$, and to the constant $(k+1)$-morphisms u^∂ and u_o given by the base point of W^{k+1}. To compute the first non-vanishing homotopy group of W^{k+1}, use the fact that $\Omega_o(G_{a-k}(\mathbb{R}^a); (b-a)\gamma\oplus\gamma\otimes\gamma)$ is $\mathbb{Z}$ or $\mathbb{Z}_2$

according to whether $(a+b)\gamma$ is orientable or not; note also the homotopy equivalence $W^{k+1} \sim SO(k+2)$ in the special case when $(a-k)(b-k) = 2$.

Next we digress and (use 5.2 to) indicate the connection between our singularity invariants, and classical cohomology obstructions and characteristic classes. Thus we place ourselves in the situation of § 2, and for simplicity we assume that M is compact.

Write r for $(a-k)(b-k)$ and define the forgetful homomorphism f by the commutative diagram

$$\begin{array}{ccccc} \Omega_{n-r}(G_{a-k}(\alpha);\phi) & \xrightarrow{\mu} & H_{n-r}(G_{a-k}(\alpha);\tilde{\mathbb{Z}}_\phi) & \xrightarrow{p^*} & H_{n-r}(M;\tilde{\pi}_{r-1}\otimes\tilde{\mathbb{Z}}_{TM}) \\ \rotatebox{90}{$\in$} & & & & \text{P.D.}\ \Big\downarrow \cong \\ \omega_{k+1}(\alpha,\beta,u^{\partial}) & & \overset{f}{\searrow} & & H^r(M,\partial M;\tilde{\pi}_{r-1}) \\ & & & & \rotatebox{90}{$\in$} \\ & & & & \bar{c}(u^{\partial}) \end{array}$$

Here we are dealing with singular (co)homology with twisted coefficients. E.g. $\tilde{\mathbb{Z}}_\phi$ is associated with the orientation line bundle of ϕ (or, equivalently, with $(a-k)(w_1(\alpha)+w_1(\beta))+(a+b)w_1(\gamma)+w_1(M)$). $\tilde{\pi}_{r-1}$ stands for the coefficient bundle consisting of the $(r-1)$-dimensional homotopy groups of the fibers of $W^{k+1}(\alpha,\beta)$. According to 5.2 $\tilde{\pi}_{r-1} \cong \pi_{r-1}(W^{k+1}) \otimes \tilde{\mathbb{Z}}_{\bar{\phi}}$, where $\bar{\phi}$ is associated with $(b-k)w_1(\alpha) + (a-k)w_1(\beta)$, and hence $p^*(\tilde{\pi}_{r-1} \otimes \tilde{\mathbb{Z}}_{TM})$ is isomorphic to $\tilde{\mathbb{Z}}_\phi$ or $\tilde{\mathbb{Z}}_2$. The homomorphism μ assigns to $[S,g,\bar{g}]$ the image of the fundamental class of S under g_*, with $\bar{g}$ being used to identify the coefficients. The projection map p, combined, if necessary, with reduction mod 2, induces p_*.

The classical primary obstruction $\bar{c}(u^{\partial})$ (cf. [59], 35.3) to the extension of the partial section u^{∂} of $W^{k+1}(\alpha,\beta)$ can be viewed as a singular cohomology class although originally a triangulation of M is involved in its definition.

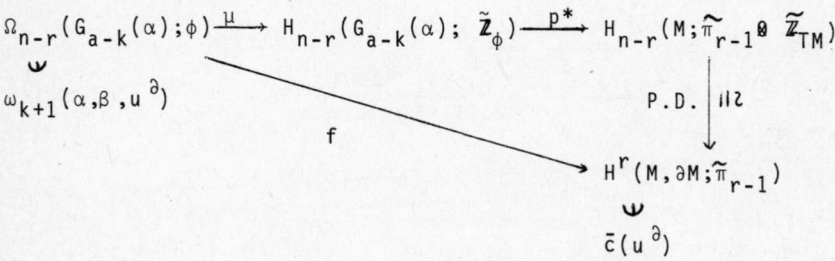

Proposition 5.3.

We have

$$f(\omega_{k+1}(\alpha,\beta,u^{\partial})) = \bar{c}(u^{\partial}).$$

Furthermore, if $\partial M = \emptyset$, *we obtain (when reducing mod 2)*

$$(f(\omega_{k+1}(\alpha,\beta)))_2 = \det\,(w_{b-k-i+j}(\beta-\alpha))_{1\le i,j\le a-k};$$

if in addition we can write $a-k = 2a'$ *and* $b-k = 2b'$, *then*

$$f(\omega_{k+1}(\alpha,\beta)) \equiv \det\,(p_{b'-i+j}(\beta-\alpha))_{1\le i,j\le a'} \quad ,$$

modulo the 2-torsion of $H^{4a'b'}(M;\mathbb{Z})$. *(Here the total Stiefel-Whitney or Pontrjagin class of the virtual bundle* $\beta - \alpha$ *is defined as the obvious quotient).*

Moreover, if M *is closed and* $u : \alpha \longrightarrow \beta$ *is a nondegenerate k-morphism, let* $\mathcal{D}(S)$ *denote the cohomology class on* M *derived from the Thom class of a tubular neighborhood of the singularity* S *of* u *(formed with respect to suitable canonical orientations). Then we have in* $H^{(a-k)(b-k)}(M;\mathbb{Z}_2)$

$$\begin{aligned} \mathcal{D}(S) &= \det(w_{b-k-i+j}(\beta-\alpha))_{1\le i,j\le a-k} \\ &= \det(w_{a-k-i+j}(\alpha-\beta))_{1\le i,j\le b-k} \quad . \end{aligned}$$

If, in addition, again $a-k = 2a'$, $b-k = 2b'$, *then we have in* $H^{4a'b'}(M;\mathbb{Z})$ *mod 2-torsion*

$$\begin{aligned} \mathcal{D}(S) &\equiv \det(p_{b'-i+j}(\beta-\alpha))_{1\le i,j\le a'} \\ &\equiv \pm\det(p_{a'-i+j}(\alpha-\beta))_{1\le i,j\le b'} \quad . \end{aligned}$$

Proof. Let M_q denote the q-skeleton of the smoothly triangulated manifold M, and let $M_q' = M_q \cup \partial M$. It is easy to construct a morphism $u : \alpha \longrightarrow \beta$, extending u^{∂}, such that

(i) $u|M'_r$ is a k-morphism with only non-degenerate singular points in the interior of the r-cells, and

(ii) $s_u : M \longrightarrow \underset{\sim}{\mathrm{Hom}}(\alpha,\beta)$ is transversal to all the manifolds $A^{k-i}(\alpha,\beta)$, $i \geq 0$.

Then the "manifold" $S = s_u^{-1}(\overline{A^k(\alpha,\beta)}) \subset M$ has singularities, but its resolution in $G_{a-k}(\alpha)$ is smooth. The Poincaré dual of the resulting fundamental class can also be derived from the Thom class of the normal bundle of (the non-singular part of) S. When restricted to (M'_r, M'_{r-1}), this Thom class coincides with the obstruction cocycle representing $\bar{c}(u^\partial)$.

Note that this argument still works in the exceptional case $r = 1$, provided we adopt the definition $\pi_0(W^{k+1}) = \tilde{H}_0(W^{k+1}; \mathbb{Z})$ (compare [59], 29.4).

The relations with Stiefel-Whitney and Pontrjagin classes follow from [31], pp. 121, 124. If one applies Poincaré duality directly to $\mu(\omega_{k+1}(\alpha,\beta))$, one can recover further classical characteristic classes.

Now let us specialize proposition 5.2 and similar consequences of §4 to the case when $k+1 = a$, and hence W^{k+1} is homotopy equivalent to the Stiefel manifold $V_{b,a}$ of orthonormal a-frames in b-dimensional euclidean space.

<u>Proposition 5.4.</u> <u>The singularity homomorphism</u>

$$\sigma : \pi_{n-1}(V_{b,a}) \longrightarrow \Omega_{n-1+a-b}(P^{a-1}; \gamma \otimes \underset{\sim}{\mathbb{R}}^{b-a})$$

<u>is an isomorphism for</u> $n < 2(b-a)$ <u>and onto for</u> $n \leq 2(b-a)$.

<u>Example 5.5.</u> When $a = 1$, the homomorphism

$$\sigma : \pi_{n-1}(S^{b-1}) \longrightarrow \Omega_{n-b}(\text{point; trivial coefficients})$$

has the following simple description. Given a smooth map

$u^\partial : S^{n-1} \longrightarrow S^{b-1}$, extend it to a smooth map $u : D^n \longrightarrow D^b$ and pick a regular value y of u in the interior of D^b. Then $\sigma[u^\partial]$ is the bordism class of the framed manifold $u^{-1}(y)$.

Clearly σ commutes with the suspension map

$$E : \pi_{n-1}(S^{b-1}) \longrightarrow \pi_n(S^b), \text{ at least up to sign.}$$

Hence proposition 5.4 gives both the Freudenthal suspension theorem and the usual description of the stable homotopy of spheres in terms of framed bordism.

Next use the splitting $\mathbb{R}^b = \mathbb{R}^{b-a} \oplus \mathbb{R}^a$ to form the stunted projective space P^{b-1}/P^{b-a-1} and to identify it with the Thom space of the bundle $(b-a)\gamma$ over P^{a-1}. For arbitrary n, consider the diagram

$$\begin{array}{ccc} \pi_{n-1}(P^{b-1}/P^{b-a-1}) & \xrightarrow{\ i_*\ } & \pi_{n-1}(V_{b,a}) \\ & {\scriptstyle t_p}\searrow \quad \swarrow{\scriptstyle \sigma} & \\ & \Omega_{n-1+a-b}(P^{a-1};(b-a)\gamma). & \end{array} \tag{5.6}$$

Here i_* is induced by the map i which assigns to $[\ell] \in P^{b-1}/P^{b-a-1}$ the frame $(r_\ell(e_{b-a+1}),\ldots,r_\ell(e_b)) \in V_{b,a}$, where e_j denotes the jth unit vector in $\mathbb{R}^b$ and r_ℓ is the reflection by the hyperplane perpendicular to the line ℓ. Taking transverse inverse images of P^{a-1} leads to the homomorphism t_p which is bijective for $n < 2(b-a)$ and onto for $n \leq 2(b-a)$.

<u>Proposition 5.7. For all integers n we have $\sigma \circ i_* = t_p$ (up to composition with some involution on the target group). Hence i_* is an isomorphism for</u> $n < 2(b-a)$.

The second statement recovers in part a result of I. James ([29], 8.1).

Proof. Given an element in $\pi_{n-1}(P^{b-1}/P^{b-a-1})$, we may represent it by a map $f : S^{n-1} \longrightarrow P^{b-1}/P^{b-a-1}$ which is smooth in a neighborhood of $f^{-1}(P^{a-1})$ and transverse to P^{a-1}. To compute $\sigma \circ i_*[f]$, consider the morphism $u : D^n \times \mathbb{R}^a \longrightarrow D^n \times \mathbb{R}^b$, which corresponds to a linear homotopy between $i \circ f$, given at $\partial D^n = S^{n-1}$, and the inclusion $\mathbb{R}^a \subset \mathbb{R}^{b-a} \times \mathbb{R}^a$, given at the center of D^n. The non-degenerate singularity of u lies at the set $S = \frac{1}{2}(f^{-1}(P^{a-1}))$ in the sphere $S(\frac{1}{2})$ of radius $\frac{1}{2}$, and $\underset{\sim}{\text{Ker}} = ((f \circ 2)|S)^*(\gamma)$. Furthermore the normal structure of the singularity is given by the isomorphism

$$\tilde{j} \circ s_{u_*} = (-f_*) \oplus (-Id) : \nu(S, D^n) = \nu(S, S(\tfrac{1}{2})) \oplus \underset{\sim}{\mathbb{R}}$$

$$\longrightarrow \underset{\sim}{\text{Hom}}(\underset{\sim}{\text{Ker}}, \underset{\sim}{\text{Coker}}) = \underset{\sim}{\text{Hom}}(\underset{\sim}{\text{Ker}}, \underset{\sim}{\mathbb{R}}^{b-a}) \oplus \underset{\sim}{\mathbb{R}},$$

where f_* is induced in the obvious way by the tangent map of f. Thus the diagram (5.6) commutes up to sign and up to the involution on $\Omega_{n-1+a-b}(P^{a-1};(b-a)\gamma)$ which is induced by the negative identity map on the coefficient bundle. ∎

Finally, recall [24] that a continuous map $h : \mathbb{R}^a \longrightarrow \mathbb{R}^b$ is called a skew map if it commutes with $-Id$ and if $h^{-1}\{0\} = \{0\}$. The space $X_{b,a}$ of such skew maps, when endowed with the compact-open topology, contains as a deformation retract the space $Y_{b,a}$ of maps which are linear isometries on each straight line through the origin. Clearly $Y_{b,a}$ is homeomorphic to the space $C^0(S(\underset{\sim}{\text{Hom}}(\gamma, \underset{\sim}{\mathbb{R}}^b)))$ of continuous sections of the sphere bundle of $\underset{\sim}{\text{Hom}}(\gamma, \underset{\sim}{\mathbb{R}}^b)$ over P^{a-1}.

Proposition 5.8. The homomorphism $j_* : \pi_{n-1}(V_{b,a}) \longrightarrow \pi_{n-1}(X_{b,a})$, induced by the inclusion map, is bijective for $n < 2(b-a)$ and onto for $n \leq 2(b-a)$.

This was originally obtained by Haefliger and Hirsch ([24], (1.1)).

<u>Proof.</u> Apply 4.11 and 4.8 to the trivial bundles over $M = D^{n-1}$, to $k = a-1$ and to constant u^{∂} and u_o. When $n = 2(b-a)$, all elements of $\Omega_{n-1+a-b}(D^{n-1} \times P^{a-1};\phi)$ can still occur as difference classes of the form $d_a(u_0,u_1)$; on the other hand, for $a > 1$ these elements classify the non-vanishing sections of $\underset{\sim}{\mathrm{Hom}}(\gamma,\underset{\sim}{\mathbb{R}}^b)$ over $D^{n-1} \times P^{a-1}$.

CHAPTER II. BORDISM CLASSIFICATION.

§6. Fine bordism groups of vector bundle morphisms, and their stability properties.

Let $\phi = (\phi^+,\phi^-)$ be a pair of real vector bundles of dimensions b^+, b^- over a topological space X, and let ξ be a line bundle over X. Also fix an integer k. The subject of this chapter are the bordism groups which we will now assign to these data.

Consider quadruples of the form $(M,g,\overline{g},or)$ such that

(i) M is a smooth closed n-dimensional manifold;

(ii) $g : M \longrightarrow X$ is a continuous map;

(iii) $\overline{g} : TM \oplus g^*(\phi^+) \longrightarrow g^*(\phi^-)$ is a k-morphism, i.e. a vector bundle map of rank $\geq k$ everywhere; and

(iv) $or : \xi_M \cong g^*(\xi)$ is an isomorphism, where ξ_M denotes the orientation bundle $\Lambda\ TM$ of M.

Two such quadruples $(M_\ell,g_\ell,\overline{g}_\ell,or_\ell)$, $\ell = 0,1$, are called bordant if there is

(i) a compact (n+1)-manifold N with $\partial N = M_0 \cup M_1$;

(ii) a continuous map $G : N \longrightarrow X$ extending $g_0 \cup g_1$;

(iii) a (k+1)-morphism

$$\overline{G} : TN \oplus G^*(\phi)^+) \longrightarrow \underset{\sim}{\mathbb{R}} \oplus G^*(\phi^-)$$

which, over ∂N, restricts to

$$\underset{\sim}{\mathbb{R}} \oplus TM_\ell \oplus g_\ell^*(\phi^+) \xrightarrow{Id\oplus\overline{g}_\ell} \underset{\sim}{\mathbb{R}} \oplus g_\ell^*(\phi^-)$$

for $\ell = 0,1$, where $\underset{\sim}{\mathbb{R}}$ embeds into TN via a vector field normal to ∂N and pointing inward along M_0 and outward along M_1; and

(iv) an isomorphism $\xi_N \cong G^*(\xi)$ extending $or_0 \cup or_1$, where $\xi_N | \partial N$ is identified with $\xi_{M_0} \cup \xi_{M_1}$ by the convention in (iii).

As usual, the resulting set of bordism classes has an addition given by disjoint union. When the source or the target bundle of $\bar{g}$ contains a trivial line bundle λ (e.g. when $b^+>0$ or $b^->n$), then the class $[M,g,\bar{g},or]$ has an inverse represented by $(M;g;$ the composition of $\bar{g}$ with $(-Id_\lambda) \oplus Id_{\lambda^\perp}; -or)$; just use $M \times I$ for a zero bordism. More generally, we will see in remark 6.8. below that additive inverses always exist except possibly when $n=b^+=b^-=k=0$ or $n = n+b^+ = b^- = k = 2$. Thus for all other dimension combinations our bordism monoid forms already an abelian group which we denote by

$$\Omega_n(X, \phi, \xi, k) .$$

In the two exceptional cases above, however, we may have to apply the Grothendieck construction to our monoid in order to define the group $\Omega_n(X,\phi,\xi,k)$. E.g. if X is a closed connected oriented surface with Euler number $\chi(X) < -2$, $\phi = (0,TX)$, $\xi = \underset{\sim}{\mathbb{R}}$ and $k = 2$, then the class $[M=X,id,id,or]$ has no inverse. For suppose there were some inverse involving a surface M', with a zero bordism N of $M \cup M'$. Then N has a vectorfield pointing outwards along the boundary, and hence we have from Hopf's theorem (see e.g. [49])

$$0 = 2 \cdot \chi(N) = \chi(M) + \chi(M') .$$

In particular, M' must have a connected component, say M_0', with strictly positive Euler number. On the other hand, the tangent bundle TM_0' is the pullback of TX under a smooth map $g : M_0' \longrightarrow X$. Thus a vectorfield on X with just one zero (of index $\chi(X)$) at a regular value of g, pulls back to a vectorfield on M_0' with index sum $\chi(M_0') = \pm \text{degree}(g) \cdot \chi(X)$.

Since no connected surface has Euler number larger than 2, this leads to the desired contradiction. ∎

Examples.

6.1. Bordism of framefields, and of immersions and k-mersions into Euclidean space.

For $n, m, k \in \mathbb{Z}$, $m \geq 0$, define the unoriented bordism group of n-manifolds M with a k-morphism $\overline{g}: TM \longrightarrow \underset{\sim}{\mathbb{R}}^m$ by

$$\mathfrak{N}_n(m,k) = \Omega_n(BO(1),\ (0,\underset{\sim}{\mathbb{R}}^m),\ \gamma_1,\ k)\ .$$

Here γ_1 denotes the universal line bundle (hence g and or are redundant).

If $k = m \leq n$, we are dealing with manifolds M together with an epimorphism $TM \longrightarrow \underset{\sim}{\mathbb{R}}^m$, or, by dualizing, with an m-framefield. In particular, $\mathfrak{N}_n(n,n)$ is the bordism group of parallelized manifolds.

If $k<m$, results of S.D. Feit [21] allow to interprete $\mathfrak{N}_n(m,k)$ as the bordism group of manifolds with a k-mersion into $\mathbb{R}^m$, i.e. a smooth map whose tangent map has rank $\geq k$ everywhere. (Here we use bordisms with (k+1)-mersions into $\mathbb{R}^m \times I$ which restrict to the obvious product mersions on collar neighborhoods of the boundaries). In particular, when $k = n < m$ we are involved with immersions.

If we replace BO(1) by BSO(1) everywhere, we obtain the corresponding oriented bordism groups $\Omega_n(m,k)$.

6.2. Normal bordism groups.

As a special case the definition above also recovers the normal bordism group $\Omega_n(X,\phi)$ of §2 for any $n \in \mathbb{Z}$ and any virtual bundle ϕ over a space X. Clearly this group can be formed from closed singular manifolds $g: M \longrightarrow X$ in X, together with isomorphisms $\overline{g}: TM \oplus g^*(\phi^+) \oplus \underset{\sim}{\mathbb{R}}^s \xrightarrow{\cong} g^*(\phi^-) \oplus \underset{\sim}{\mathbb{R}}^t$ (s and t being suitable fixed natural numbers). Hence

$$\Omega_n(X,\phi) = \Omega_n(X \times BO(1), (\phi^+ \oplus \underset{\sim}{\mathbb{R}}^s,\ \phi^- \oplus \underset{\sim}{\mathbb{R}}^t),\ \gamma_1,\ b^- + t)\ . \tag{6.3}$$

Furthermore, knowledge of the isomorphism $\overline{g}$ is essentially equivalent to knowing its composition with the inclusion $g^*(\phi^-)\oplus\mathbb{R}^t \subset g^*(\phi^-)\oplus\mathbb{R}^{t+1}$, together with the way $\bar{g}$ relates orientations. Thus we obtain the following alternate description which is more useful for computations:

(6.4) $$\Omega_n(X,\phi) = \Omega_n(X,(\phi^+ \oplus \mathbb{R}^s,\ \phi^- \oplus \mathbb{R}^{t+1}),\ \xi_\phi,\ b^-+t)\ ,$$

where

(6.5) $$\xi_\phi = \Lambda^{b^+}\ \phi^+ \otimes \Lambda^{b^-}\ \phi^-$$

denotes the orientation bundle of ϕ. ■

We start our general discussion of $\Omega_n(X,\phi,\xi,k)$ by showing that most of the time this group does not really depend on the individual bundles ϕ^+ and ϕ^-, but only on their difference in the K-theory of X.

<u>Stability theorem 6.6.</u> <u>The stabilizing map</u>

$$St\ :\ \Omega_n(X,\phi,\xi,k) \longrightarrow \Omega_n(X,(\phi^+ \oplus \mathbb{R},\ \phi^- \oplus \mathbb{R}),\ \xi,\ k+1),$$

<u>defined by $St[M,g,\bar{g},or] = [M,g,\bar{g}\oplus Id,or]$, is an isomorphism for all n and (X,ϕ,ξ,k), except possibly in the following two cases:</u>

<u>(i) $n = 1$, $b^+ = 0$, and $b^- = k = 0$ or 1; (here St is always onto, but may fail to be injective); and</u>

(ii) <u>$n = 2$, $b^+ = 0$, $b^- = k = 1$ or 2; (here St is always injective, but may fail to be onto).</u>

<u>Proof.</u> First we show that St is onto when $b^+>0$ or $b^->n$ or $n>2$; in these cases injectivity follows easily by similar arguments (even for $n=2$).

Let $(M,g,\bar{g},or)$ represent an element in the target group of St. Our goal is to deform the morphism

$$\bar{g}\ :\ TM \oplus g^*(\phi^+) \oplus \mathbb{R} \longrightarrow g^*(\phi^-) \oplus \mathbb{R}$$

(or, for convenience, rather its adjoint with respect to suitable metrics on the vector bundles) until it maps $\mathbb{R}$ identically into $\mathbb{R}$. If this is not feasible, we will have to attach 1- or 2-handles to $M \times I$ in order to obtain the required bordism.

So start with $M \times I$ and $G = g \circ pr_1 : M \times I \longrightarrow X$, where pr_1 is the first projection. Consider the (k+2)-morphisms

$$TM \oplus g^*(\phi^+) \oplus \mathbb{R} \oplus \mathbb{R}_1 \underset{\overline{G}_0'}{\overset{\overline{G}_0 = \overline{g} \oplus Id}{\rightleftarrows}} g^*(\phi^-) \oplus \mathbb{R} \oplus \mathbb{R}_1$$

given over $M \times \{0\}$ and adjoint to one another. Here $\mathbb{R}_1$ stands for $T(M \times I)/TM = \mathbb{R}$.

We want to extend $\overline{G}_0'$ to a (k+2)-morphism $\overline{G}'$. Deform $\overline{G}_0'$ by rotation until it maps $\mathbb{R}_1$ into $\mathbb{R}$ by $-Id$, and also $\mathbb{R}$ into $TM \oplus g^*(\phi^+) \oplus \mathbb{R}_1 = T(M \times I) \oplus g^*(\phi^+)$. After a further homotopy which involves an approximation and which leaves $\mathbb{R}_1$ untouched, we may actually assume that $\mathbb{R}$ gets mapped nontrivially into $T(M \times I)$. Use all these deformations to define $\overline{G}'$ over $M \times [0, \frac{1}{2}]$. Now we can find a vector field v on $M \times [\frac{1}{2}, 1]$ which is given by the image of $1 \in \mathbb{R}$ at $M \times \frac{1}{2}$, points outward at $M \times 1$ and has only non-degenerate zeroes, at points z_ℓ, $\ell = 1, \ldots r$, of the form $z_\ell = (x_\ell, \frac{3}{4})$. Thus by homotopy lifting techniques we obtain $\overline{G}'$ over all of $M \times I$, except at $x_\ell \times [\frac{3}{4}, 1]$, $\ell = 1, \ldots r$.

In order to get rid of these "holes" in the domain of $\overline{G}'$, pick a small ball in $M \times 1$ close to the point $(x_\ell, 1)$ but not containing it. Within this ball attach a handle of the form $D^n \times D^1$ or $D^{n-1} \times D^2$ according to whether the index of v at z_ℓ is $+1$ or -1, and smooth corners. Clearly, both G and $\overline{G}'$ can be extended over the core disk D^1 or D^2. But if we now try to define $\overline{G}'$ over the whole handle in such a way that the extended vector field $v = \overline{G}'(0,1,0)$ points outward also along the new boundary, then v has a new zero z'_ℓ of index -1 or $+1$. (According to lemmas 2.8 and 2.9 this

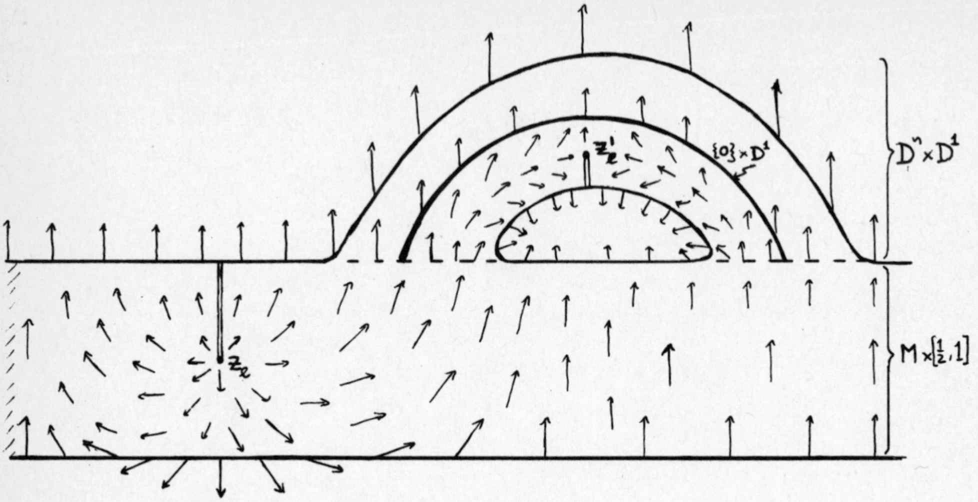

Figure 6.7. A pair of zeroes of v with opposite indices, and the resulting arcs on which $\overline{G}'$ is not defined.

index depends only on the (homotopy) behavior of v on the boundary of the handle; hence it can be computed from the vectorfield Id x -Id on $D^{n+1-i} \times D^i$).

This leads to a new hole in the domain of $\overline{G}'$, again in the form of an arc joining z'_ℓ to the boundary. If $n > 2$, the boundary end points of our arcs starting at z_ℓ and z'_ℓ lie in the same boundary component, and we can isotop the arcs closely together along the boundary. Since the indices of v at z_ℓ and z'_ℓ add up to 0, we can cancel the zeroes of v and thus extend $\overline{G}'$ over the whole bordism $N = M\times I \cup$ handles. After a final rotation, carried out close to the new top boundary $\partial N - M\times\{0\}$, $\overline{G}'$ maps $\underset{\sim}{\mathbb{R}}$ identically into $\underset{\sim}{\mathbb{R}}$ there, and it maps $1 \in \underset{\sim}{\mathbb{R}}_1$ into outward pointing vectors, as desired.

If $b^+>0$ or $b^->n$, rotations and handles (and for $b^->n$, even adjoints) can be avoided in our arguments.

It remains only to consider the case $b^+=0$, $b^-\le n\le 2$. Clearly,

St is bijective for $k < b^-$, since then $\overline{g}$ does not contain any information, the relevant non-degenerate singularities all having negative dimensions. If $n = 2$ and $b^- = k = 0$, the proof above can be modified to show that St is still onto: instead of attaching a handle, we remove an open solid pretzel or ball embedded into MxI around z_ℓ. Similarly, using the surjectivity of the stabilizing homomorphism $\pi_1(SO(2)) \longrightarrow \pi_1(SO(3)) = \mathbb{Z}_2$, we see that St is onto for $n=1$, $b^-=k=0$ or 1. Finally by a direct argument, St is an isomorphism for $n = 0$.

Having proved all the positive statements of the theorem, we now give counterexamples in each of the exceptional dimension combinations. First, if $n = 2$, let X be a closed connected oriented surface with negative Euler number, $\phi = (0, \underset{\sim}{\mathbb{R}}^{b^-})$, $\xi = \underset{\sim}{\mathbb{R}}$ and $b^- = k = 1$ or 2. Pick an epimorphism $\overline{g} : TX \oplus \underset{\sim}{\mathbb{R}} \longrightarrow \underset{\sim}{\mathbb{R}}^{b^-+1}$. Then $[X, id, \overline{g}, or]$ is not in $St(\Omega_n(X,\phi,\xi,k))$. Otherwise, there would be a map from a union of tori to X with degree one (since the mapping degree is a bordism invariant [49]). But it follows easily from Poincaré duality that any map from a closed connected surface into another one of strictly smaller Euler number has degree zero.

Finally, if $n = 1$, let X be a torus with two disjoint open disks removed, $\phi = (0, \mathbb{R}^{b^-})$, $\xi = \underset{\sim}{\mathbb{R}}$ and $b^- = k = 0$ or 1. Then $[\partial X, \text{inclusion}, \overline{g}, or]$ (where $\overline{g}$, or are induced by the orientation of X) is a nontrivial element in the kernel of St. Indeed, $\overline{g} \oplus Id$ extends over the bordism X since a vectorfield on X, pointing outwards along ∂X, has index -2 and since the stabilizing homomorphism $\pi_1(SO(2)) \longrightarrow \pi_1(SO(3))$ annihilates all multiples of 2. On the other hand, suppose there is a zero bordism $(N, G, \overline{G}, or)$ of $(\partial X, \text{inclusion}, \overline{g}, or)$ itself. Then $\overline{G}$ gives a vectorfield, $\chi(N) = 0$, and we may assume N is connected. Glueing two disks both into N and X, and extending G by identical maps, we obtain a map of degree one from the sphere to the torus, which again is not possible.

Remark 6.8. It is easy to see now that, except possibly when $b^+ = 0$ and $n = b^- = k = 0$ or 2, the monoid $\Omega_n(X,\phi,\xi,k)$, formed without a completion in the sense of Grothendieck, is already a group. This follows directly from the theorem except when $n = 2$, $b^+ = 0$, $b^- = k = 1$ or when $n = 1$, but in these cases $[M,g,\bar{g},or]$ has an inverse anyway, since M has vanishing Euler number (cf. [32]) and hence TM contains even a trivial line bundle.

Exercise 6.9. Represent $\Omega_n(X,\phi,\xi,k)$, where $k = \min(n+b^+,b^-)$ and $b^+ > 0$, as a normal bordism group.

§ 7. The exact singularity sequence.

In this section we assume that $b^+ = \dim \phi^+ > 0$. This is almost never a restriction since it can be obtained by stabilizing, which, according to theorem 6.6., rarely ever leads to a loss of information.

We propose now to measure the difference between $\Omega_n(X,\phi,\xi,k+1)$ and $\Omega_n(X,\phi,\xi,k)$. For this purpose it is convenient to introduce relative bordism groups.

So consider quadruples $(M,g,\bar{g},or)$ as in the definition of $\Omega_n(X,\phi,\xi,k)$, except that the compact manifold M is now allowed to have a boundary ∂M, and we require $\bar{g}$ to restrict to a $(k+1)$-morphism over ∂M. A bordism $(N,G,\bar{G},Or)$ between two such quadruples $(M_\ell,g_\ell,\bar{g}_\ell,or_\ell)$, $\ell = 0,1$, is defined as before, with the following modifications. First of all, N is allowed to have corners; more precisely, there is a manifold $\tilde{N}$, possibly with boundary, and a smooth proper map $r : \tilde{N} \longrightarrow \mathbb{R}$ with 0 and 1 being regular values for both r and $r|\partial\tilde{N}$, such that $N = r^{-1}[0,1]$ and $M_\ell = r^{-1}\{\ell\}$, $\ell = 0,1$. Furthermore, $\bar{G}$ has to restrict to a $(k+2)$-morphism over $N \cap \partial\tilde{N}$.

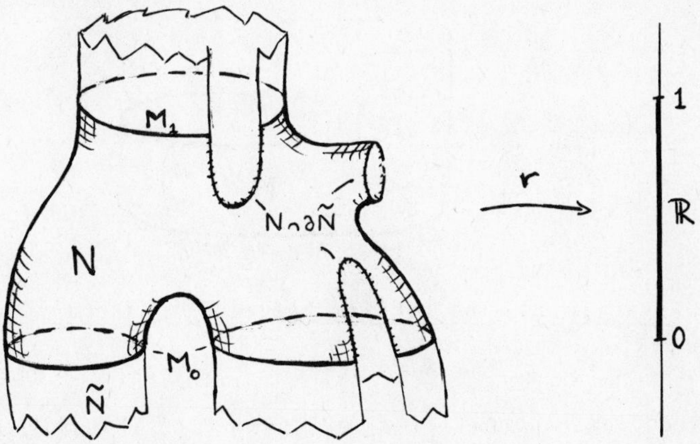

Figure 7.1. A relative bordism

We denote the resulting bordism group by $\Omega_n^{rel}(X,\phi,\xi,k)$. Its significance stems from the sequence

$$\Omega_n(X,\phi,\xi,k+1) \xrightarrow{f} \Omega_n(X,\phi,\xi,k) \xrightarrow{j} \Omega_n^{rel}(X,\phi,\xi,k) \tag{7.2}$$
$$\xrightarrow{\partial} \Omega_{n-1}(X,(\mathbb{R}\oplus\phi^+,\phi^-),\xi,k+1) \xrightarrow{f} \ldots$$

Here f and j are the obvious forgetful maps; we define ∂ by restricting all data to the boundary and using an outward pointing vector field to identify $TM|\partial M$ with $T(\partial M) \oplus \mathbb{R}$.

It is not hard to see that our sequence is exact. To check exactness e.g. at the relative bordism group, note that obviously $\partial \circ j = 0$. On the other hand, given $[M,g,\overline{g},or] \in \ker \partial$, choose a zero bordism $(M',g',\overline{g}',or')$ of $\partial(M,g,\overline{g},or)$. Glue M and M' together along their common boundary ∂M, and fit also the other data together to obtain a quadruple representing a class $\omega \in \Omega_n(X,\phi,\xi,k)$; here we have to use the (easy part of the) arguments in the proof of theorem 6.6 to deform and destabilize $\overline{g}'$ until we obtain a $(k+1)$-morphism extending $\overline{g}|\partial M$ over M'. An obvious bordism, involving $M\times I \cup_{\partial M\times[0,\frac{1}{2}]} M'\times[0,\frac{1}{2}]$ with suitably smoothed corners, now shows that $j(\omega) = [M,g,\overline{g},or]$. Hence the kernel of ∂ equals the image of j.

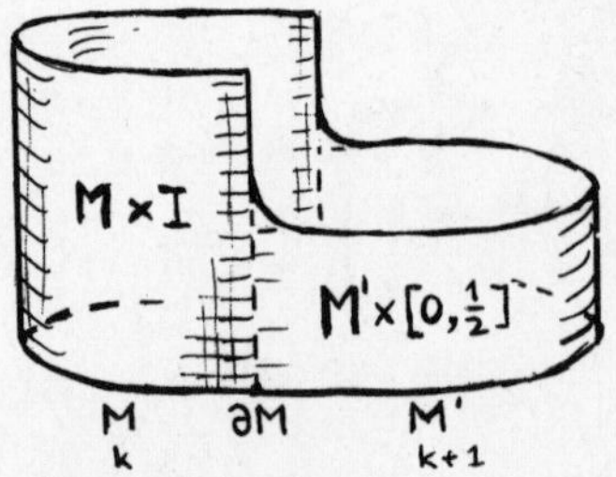

Figure 7.3. The bordism between M and a closed manifold.

Next we exploit the results of § 1 in order to represent our relative group as a normal bordism group. For $[M,g,\overline{g},or] \in \Omega_n^{rel}(X,\phi,\xi,k)$ we may assume that the k-morphism

$$\overline{g} : TM \oplus g^*(\phi^+) \longrightarrow g^*(\phi^-)$$

is non-degenerate (with respect to some differentiable structures on the pullback bundles involved here). Thus the set of points where $\overline{g}$ has precisely rank k forms a closed submanifold S of the interior of M, of codimension $(n+b^+-k)(b^--k)$. Over S we have the vector bundles $\underset{\sim}{Ker}$, $\underset{\sim}{Coker}$ and $\underset{\sim}{Im}$, of dimensions p,q, and k respectively, where we define

(7.4) $$p = n+b^+ - k \quad \text{and} \quad q = b^- - k.$$

Moreover, we obtain isomorphisms

(7.5) $$\overline{h}^+ : TS \oplus (\underset{\sim}{Ker} \otimes \underset{\sim}{Coker}) \oplus g^*(\phi^+) \cong TM \oplus g^*(\phi^+) \cong \underset{\sim}{Ker} \oplus \underset{\sim}{Im}$$

and

(7.6) $$\overline{h}^- : \underset{\sim}{Coker} \oplus \underset{\sim}{Im} \cong g^*(\phi^-)$$

which stabilize and compose to give

(7.7) $$\overline{h} : TS \oplus (\underset{\sim}{Ker} \otimes \underset{\sim}{Coker}) \oplus g^*(\phi^+) \oplus \underset{\sim}{Coker} \cong g^*(\phi^-) \oplus \underset{\sim}{Ker} .$$

In particular, the isomorphism $or:\xi_M \cong g^*(\xi)$ leads to a trivialization

(7.8) $$g^*(\xi) \otimes g^*(\xi_\phi) \otimes \xi_{\underset{\sim}{Ker}} \otimes \xi_{\underset{\sim}{Coker}} \cong \underset{\sim}{\mathbb{R}}$$

over S, where as before (cf. 6.5) ξ_η stands for the orientation bundle of any (virtual) vector bundle η.

Now define

(7.9) $$Y = Y(X, \phi, \xi; p, q)$$

to be the double cover of $X \times BO(p) \times BO(q)$ corresponding to the line bundle $\xi \otimes \xi_\phi \otimes \xi_{\gamma_p} \otimes \xi_{\gamma_q}$, and let ψ denote the pullback

(7.10) $$\psi = \psi(Y,\phi) = \phi^+ \oplus (\gamma_p \otimes \gamma_q) \oplus \gamma_q - (\phi^- \oplus \gamma_p)$$

over Y. (As usually, γ_r stands for the universal bundle over $(BO(r))$.

Clearly $g|S$, together with classifying maps of $\underset{\sim}{Ker}$ and $\underset{\sim}{Coker}$ and with the trivialization (7.8), determines a map $h:S \longrightarrow Y$. Thus our discussion of the singularity of $\overline{g}$ leads to the tripel $(S,h,\overline{h})$ whose normal bordism class depends only on the class $[M,g,\overline{g},or]$ (compare with the proof of lemma 2.8). We obtain a monomorphism

$$\sigma : \Omega_n^{rel}(X,\phi,\xi,k) \longrightarrow \Omega_{n-pq}(Y(X,\phi,\xi,p,q);\psi(Y,\phi)) .$$

(This notation should not lead to confusion with the map defined in (5.1) by a similar approach).

Theorem 7.11. σ is always an isomorphism (provided $\dim \phi^+ > 0$)

Proof. The statement is trivial unless $0 \le k \le \min(n+b^+, b^-)$. It is also obvious when k equals this minimum, since then $\Omega_n^{rel}(X,\phi,\xi,k) = \Omega_n(X,\phi,\xi,k)$ and σ essentially converts mono- or epimorphisms into isomorphisms by adding a suitable bundle to make up for the (co-)kernel.

Thus we may assume that $p,q > 0$. Then

$$\dim S \le \dim(TS \oplus (\underset{\sim}{Ker} \otimes \underset{\sim}{Coker}) \oplus g^*(\phi^+)) - \dim \underset{\sim}{Ker} ,$$

and the information given by h and $\overline{h}$ allows us to reconstruct singularity data as the ones discussed above; to get $\underset{\sim}{Im}$ and $\overline{h}^+$ back, just define $\underset{\sim}{Im}$ as the cokernel of a monomorphism of $\underset{\sim}{Ker}$ into the left hand term in (7.5); moreover, deform and destabilize $\overline{h} \circ ((\overline{h}^+)^{-1} \oplus Id_{\underset{\sim}{Coker}})$ to obtain $\overline{h}^-$. Now apply the model construction of § 1 to obtain a k-morphism

$$\overline{\hat{g}} : T\hat{M} \oplus \hat{g}^*(\phi^+) \cong \underset{\sim}{Ker} \oplus \underset{\sim}{Im} \longrightarrow \underset{\sim}{Coker} \oplus \underset{\sim}{Im} \cong g^*(\phi^-)$$

over the disk bundle $D(\hat{M})$ of $\hat{M} = \underset{\sim}{Hom}(\underset{\sim}{Ker}, \underset{\sim}{Coker})$. Here $g : D(\hat{M}) \longrightarrow X$ is the obvious extension of the map from S to X given by h. Note that $\overline{\hat{g}}$ is a (k+1)-morphism over the boundary sphere bundle of $D(\hat{M})$, and also that the orientation information contained in h leads to an isomorphism $or : \xi_{\hat{M}} \cong \hat{g}^*(\xi)$. Since this construction can be carried out for all $[S,h,\overline{h}]$ in the target group of σ, and since

$\sigma[D(\hat{M}),\hat{g},\bar{\hat{g}},or] = [S,h,\bar{h}]$, σ is onto. (Actually, we have described σ^{-1})).

On the other hand, given a class $[M,g,\bar{g},or]$ in the kernel of σ, we can apply the same construction to a suitable zero bordism of the singularity. In view of the structure lemma 1.10, we can fit the resulting $D(\hat{M})$ and $M\times I$ together to get a zero bordism for $[M,g,\bar{g},or]$. Thus σ is also injective.∎

Observe the analogy of the main argument in this proof and in the proof of theorem 3.1. However, restrictions on n are unnecessary in 7.11 since no embedding and unlinking questions enter the picture here.

All the homomorphisms in the sequence (7.2) are compatible with the obvious (absolute and relative) stabilization isomorphisms (defined as in theorem 6.6); and so is σ, provided we choose the sign convention in (7.8) suitably. Hence (7.2) can be extended to a long exact sequence which, in view of the last theorem, takes the form

(7.12)
$$\cdots \xrightarrow{f} \Omega_{n+1}(X,(\phi^+,\phi^-\oplus\underset{\sim}{\mathbb{R}}),\xi,k+1) \xrightarrow{\sigma\circ j}$$
$$\to \Omega_{n+1-pq}(Y(X,\phi,\xi,p,q);\psi(Y,\phi)) \underset{\rho}{\overset{\delta=\partial\circ\sigma^{-1}}{\rightleftarrows}} \Omega_n(X,\phi,\xi,k+1) \xrightarrow{f} \Omega_n(X,\phi,\xi,k) \xrightarrow{\sigma\circ j}$$
$$\to \Omega_{n-pq}(Y(X,\phi,\xi,p,q);\psi(Y,\phi)) \underset{\rho}{\overset{\delta}{\rightleftarrows}} \Omega_{n-1}(X,(\underset{\sim}{\mathbb{R}}\oplus\phi^+,\phi^-),\xi,k+1) \to \cdots$$

where ∂ is understood to possibly contain an obvious destabilization. Since the kernel and cokernel dimensions p and q are the same for all the singularity terms in our exact sequence, we can identify the spaces Y and the coefficient bundle ψ involved in these terms. Note here that Y and ψ do not really depend on the bundles ϕ^+ and ϕ^- individually, but only on their difference in the reduced K-theory of X.

To define the homomorphism ρ in (7.12), consider the diagram

(7.13)

$$\begin{array}{ccc}
 & \Omega^{rel}_{n+1}(X,(\phi^+,\phi^-\oplus \underset{\sim}{\mathbb{R}}),\xi,k+1) & \\
\swarrow^{\cong\,\sigma} & & \searrow^{\partial}\ \nwarrow^{\rho'} \\
\Omega_{n+1-pq}(Y(X,\phi,\xi;p,q);\psi(Y,\phi)) & \underset{\rho}{\overset{\delta}{\rightleftarrows}} & \Omega_n(X,\phi,\xi,k+1) \\
\nwarrow^{\Delta} & & \swarrow_{\sigma\circ j} \\
 & \Omega_{n-pq+p+q-1}(Y(X,\phi,\xi;p-1,q-1);\phi+\gamma_{p-1}\otimes\gamma_{q-1}+\gamma_{q-1}-\gamma_{p-1}) . &
\end{array}$$

Here δ is as in (7.12), and $\sigma\circ j$ comes from the corresponding exact sequence when k is replaced by $k+1$. Moreover, Δ is defined as follows. Given an element $z=[S,h,\overline{h}]$ in the domain of Δ, let $S_1 \subset S$ be the zero set of a non-degenerate section in the pullback of $\gamma_{p-1} \oplus \gamma_{q-1}$. Then S_1, together with $h|S_1$ and the isomorphism

$$TS_1 \oplus \phi^+ \oplus ((\gamma_{p-1}\oplus \underset{\sim}{\mathbb{R}}) \otimes (\gamma_{q-1}\oplus \underset{\sim}{\mathbb{R}})) \oplus (\gamma_{q-1}\oplus \underset{\sim}{\mathbb{R}}) \cong$$

$$TS|S_1 \oplus \phi^+ \oplus \gamma_{p-1} \otimes \gamma_{q-1} \oplus \gamma_{q-1} \oplus \underset{\sim}{\mathbb{R}}^2 \xrightarrow{\ \overline{h}|\ } \phi^- \oplus (\gamma_{p-1}\oplus \underset{\sim}{\mathbb{R}}) \oplus \underset{\sim}{\mathbb{R}}$$

gives rise to $\Delta(z)$. Now put

(7.14) $\qquad \rho = \Delta \circ (\sigma\circ j)$.

There is another simple description of ρ. For $w=[M,g,\overline{g},or] \in \Omega_n(X,\phi,\xi,k+1)$ define

$$\rho'(w) = [M\times I,\ g\circ pr_1,\ pr_1^*(\overline{g})\oplus\overline{g}_2,\ c\cdot pr_1^*(or)] ,$$

where at

plication by $2t-1$. Then for suitable conventions for the sign $c=\pm 1$, and for the first isomorphism in the construction of Δ above, we have

(7.14') $\qquad \rho = \sigma \circ \rho'$.

In other words, we can obtain the singularity data of $\rho'(w)$ from those of w. The correction term Δ makes up for the fact that the external product $pr^*(\overline{g}) \oplus \overline{g}_2$ may be degenerate, even if its factors are not.

Indeed, assume the (k+1)-morphism $\bar{g}$ has its non-degenerate singularity at S. A non-degenerate section of $\mathrm{Ker}(\bar{g}) \oplus \mathrm{Coker}(\bar{g})$ with zero sets, leads to maps $\bar{g}_{12} : TM \oplus g^*(\phi^+) \xrightarrow{\mathrm{proj}_*} \mathrm{Ker}(\bar{g}) \longrightarrow \mathbb{R}$, and $\bar{g}_{21} : \mathbb{R} \longrightarrow \mathrm{Coker}(\bar{g}) \hookrightarrow g^*(\phi^-)$ over $S\times\{\frac{1}{2}\}$. If we extend them suitably over all of $M\times I$ and add them in the obvious way to $pr_1^*(\bar{g}) \oplus \bar{g}_2$, the resulting (k+1)-morphism has its non-degenerate singularity at $S_1\times\{\frac{1}{2}\}$. To get (7.14'), just compare the singularity data.

Now let i denote the involution on the bordism groups which replaces an orientation isomorphism or by -or. The following is an immediate consequence of (7.14').

Proposition 7.15. *We have* $\delta\circ\rho = \mathrm{Id} + i$.

The long exact sequence (7.12), together with the last proposition, provide a powerful tool for calculating fine bordism groups and elements in them. The following lemma will also be useful.

Lemma 7.16. *Assume X is path-connected and $n>0$, $b^+>0$. Then every class $[M,g,\bar{g},or] \in \Omega_n(X,\phi,\xi,k)$ has a representative with connected underlying manifold M, except possibly when $k= n+b^+ = b^-$ and the orientation bundle ξ_ϕ is isomorphic to ξ over every loop in X.*

In particular, every strictly positive dimensional normal bordism class in X (with coefficients in an arbitrary virtual vector bundle) has such a representative.

Proof. Attach handles of the form $D^n\times D^1$ to the upper end of $M\times I$ to get a bordism from $M=M\times\{0\}$ to a connected manifold M'. The attaching maps have to be chosen carefully so that g, $\bar{g}$ and or can be extended over the whole bordism. Most of the time, we need only to make the attaching maps compatible with the orientation isomorphism or; the extension of $\bar{g}$ will present no problem, since the relevant singularities will have higher codimension than 1, which is the dimension of the core $\{0\}\times D^1$ of the handle. Moreover, if $k=n+b=b^-$, then $\bar{g}$ is an isomorphism inducing a map $\xi_M \cong g^*(\xi_\phi)$, which may conflict with or : $\xi_M \cong g^*(\xi)$ at the endpoints of $\{0\}xD^1$. In order to extend our data over all of $\{0\}xD^1$ anyway, we may have to use (e.g. for the extension of g over this core) a loop in X over which ξ_ϕ and ξ are not isomorphic. Finally, an argument as in the easy part of the proof of the stability theorem 6.6 is needed to deform the extension of $\bar{g}$ at M' until it is suitably tangent to this new boundary.■

So far in this section we have assumed that $b^+ > 0$. We will see now that this presents no serious restriction in the case of the bordism groups defined in example 6.1.

Proposition 7.17. For all $n, m, k \in \mathbb{Z}$, $m \geq 0$, the stabilizing maps (cf. 6.6)

$$\mathfrak{N}_n(m,k) = \Omega_n(BO(1),(0,\mathbb{R}^m),\gamma_1,k) \xrightarrow{St} \Omega_n(BO(1),(\mathbb{R},\mathbb{R}^{m+1}),\gamma_1,k+1)$$

and

$$\Omega_n(m,k) = \Omega_n(\text{point},(0,\mathbb{R}^m),\mathbb{R},k) \xrightarrow{St} \Omega_n(\text{point},(\mathbb{R},\mathbb{R}^{m+1}),\mathbb{R},k+1)$$

are isomorphisms.

Proof. In view of the stability theorem 6.6 it remains only to prove injectivity when $n = 1$, $m = k = n-\ell$, where in both cases $\ell = 0$ or 1. Moreover, note that for arbitrary $n > 0$ there are isomorphisms $\mathfrak{N}_n(n,n) \cong \Omega_n(n-1,n-1)$ (compare with (6.4)) and $\Omega_n(n,n) \cong \Omega_n(n-1,n-1) \oplus \Omega_n(n-1,n-1)$ (by splitting up into orientation preserving and reserving components of a framing), as well as corresponding isomorphisms on the stable level, all of them compatible with stabilization. Hence we need only consider the case $\ell = 1$.

The Möbius strip, together with a suitable vectorfield (cf. figure 7.18), provides a bordism which shows that $\mathfrak{N}_1(0,0) = 0$; hence there is nothing to prove here. Furthermore, using an appropriate vectorfield on S^1xI, we see that there are at most two elements in $\Omega_1(0,0)$.

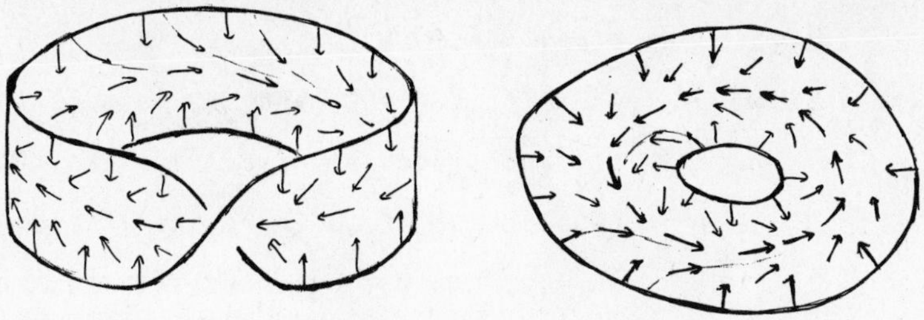

Figure 7.18. Bordisms showing that $\mathfrak{N}_1(0,0) = 0$ and $2 \cdot \Omega_1(0,0) = 0$.

Now, for $n=1,2$, we have the exact sequence (cf. 7.12)

$$\begin{array}{ccccc}
\cdots \to \Omega_{n+1}(B(S)O(1),(\underset{\sim}{\mathbb{R}},\underset{\sim}{\mathbb{R}}^{n+1}),\gamma_1,n) & \xrightarrow{j} & \Omega^{rel}_{n+1}(B(S)O(1)(\underset{\sim}{\mathbb{R}},\underset{\sim}{\mathbb{R}}^{n+1}),\gamma_1,n) & \xrightarrow{\partial} & \Omega_n(B(S)O(1)(\mathbb{R},\mathbb{R}^n),\gamma_1,n) \xrightarrow{f} \cdots \\
\text{forgetful}\Big\downarrow\cong & & \sigma\Big\downarrow\cong & & \\
\mathfrak{N}_{n+1} \quad \text{resp.} \quad \Omega_{n+1} & & \Omega_{n-1}(Y;\gamma_2\otimes\gamma_1+\gamma_1-\gamma_2) & &
\end{array}$$

where we identify Y via the obvious projection with $BO(2) \times BO(1)$ in the unoriented case, resp. with $BO(2)$ in the oriented case (then γ_1 stands for ξ_{γ_2}).

If $n=1$, we have $\partial \neq 0$ in the oriented setting; hence $\Omega_1(\text{point},(\underset{\sim}{\mathbb{R}},\underset{\sim}{\mathbb{R}}^1),\underset{\sim}{\mathbb{R}},1) = St(\Omega_1(0,0))$ is non-trivial, and St must be an isomorphism here. Note also that $\Omega_1(0,0) = \mathbb{Z}_2$.

If $n=2$, ∂ is an isomorphism. To see this in the unoriented setting, observe that for any class z in $\Omega_2(BO(1),(\underset{\sim}{\mathbb{R}},\underset{\sim}{\mathbb{R}}^2),\gamma_1,2)$ the tangent bundle of a representing manifold M is stably equivalent to a line bundle; hence $w_2(M)$ vanishes, and so does $f(z)$. Next, interprete the solid torus $\hat{M}_1$, and the solid Klein bottle $\hat{M}_2$ as disk bundles over S^1 and consider a "horizontal" vectorfield which is always parallel to the core circle, as well as a "vertical" vectorfield which points radially away from it. By taking adjoints and stabilizing suitably we obtain classes $z_1, z_2 \in \Omega^{rel}_3(BO(1),(\underset{\sim}{\mathbb{R}},\underset{\sim}{\mathbb{R}}^3),\gamma_1,2)$ such that clearly $\partial(z_1),\partial(z_2) \in St(\mathfrak{N}_2(1,1))$. In order to see that St is onto here, we only need to show that $\sigma(z_1),\sigma(z_2)$ generate $\Omega_1(Y;\gamma_2\otimes\gamma_1+\gamma_1-\gamma_2)$. Thus, consider a class $[S,h,\bar{h}]$ in this group. After possibly subtracting $\sigma(z_2)$ to make $w_1(h^*(\gamma_2))[S] = 0$ and applying lemma 7.16, we may assume that $S=S^1$ and h is constant (since $h^*(\gamma_2)$ is now trivial, and since $\bar{h}$ always implies that $h^*(\gamma_1)$ is trivial). But $\pi_1(SO)$ being $\mathbb{Z}_2$, there are at most two such classes, namely $\sigma(z_1)$ and 0. In the oriented setting, $h^*(\gamma_2)$ is also automatically trivial, and it is not necessary (nor possible) to use $\sigma(z_2)$. As a consequence, the only possibly non-trivial class in $\mathfrak{N}_2(2,2)(\cong\Omega_2(1,1))$ can be represented by a torus S^1xS^1 with the obvious product framing.■

Exercise 7.19. Prove that $\mathfrak{N}_2(2,2)$ is indeed $\mathbb{Z}_2$.

a.) Recall from the last proof that $\mathfrak{N}_2(2,2)\cong\Omega_1(BO(2),\gamma_2\otimes\xi_{\gamma_2}+\xi_{\gamma_2}-\gamma_2)$. Use the description (6.4) of this group and write down the relevant piece of the exact sequence (7.12).

b.) Using proposition 5.3, express $\sigma\circ j$ by a Stiefel-Whitney number and show that it vanishes.

c.) Conclude that δ is an isomorphism, giving the desired result.■

Actually, this kind of program can be carried through successfully in much greater generality (see e.g. § 9 - 12).

Next we spell out a few consequences of the last proposition.

Corollary 7.20. For all $n>0$ the stable homotopy group $\pi_n^s=\lim_{r\to\infty}\pi_{n+r}(S^r)$ can be identified via the Thom-Pontrjagin construction (cf. example 5.5) and destabilization, with (not stably!) parallelized bordism of parallelized n-manifolds.

Example 7.21. For $n=1$ or 2 $\pi_n^s=\mathbb{Z}_2$ and the generator can be represented by the n-dimensional torus with an invariant framing; cf. the last proof and exercise.

Corollary 7.22. Let M be a closed smooth manifold whose stable span is $\geq m$ (i.e. for some number N there are m+N linearly independent sections of $TM \oplus \mathbb{R}^N$). Then M is bordant (in the classical sense of Thom) to a manifold M' which allows at least m linearly independent vectorfields. If M is oriented, we may choose M' and the bordism to be oriented, too. ∎

Next, we apply the general theory of this section to the right hand groups in proposition 7.17. (thus the requirement $b^+>0$ is satisfied), and we use the isomorphisms St to translate back. We obtain the long exact sequence

$$\begin{aligned} &.. \xrightarrow{f} \mathfrak{N}_{n+1}(m+1,k) \xrightarrow{\sigma\circ j} \Omega_{n+1-pq}(Y;\gamma_p \otimes \gamma_q + \gamma_q - \gamma_p) \underset{\rho}{\overset{\delta}{\rightleftarrows}} \mathfrak{N}_n(m,k) \\ (7.23) \qquad &\xrightarrow{f} \mathfrak{N}_n(m,k-1) \xrightarrow{\sigma\circ j} \dots \\ &.. \xrightarrow{f} \mathfrak{N}_{n-m}(0,k-1-m) \longrightarrow \Omega_{n-m-pq}(Y;\gamma_p \otimes \gamma_q + \gamma_q - \gamma_p) \longrightarrow 0 \quad , \end{aligned}$$

and a precise analogue for the oriented groups $\Omega_n(m,k)$. Here

$$(7.24) \qquad p = n-k+1 \qquad \text{and} \qquad q = m-k+1 \ ;$$

Y stands for $BO(p) \times BO(q)$ in the unoriented case (the original factor $BO(1)$ is made redundant by the formation of double covers), and for the double cover belonging to $\xi_{\gamma_p} \otimes \xi_{\gamma_q}$ in the oriented case. The homomorphisms f, $\sigma\circ j$ and δ (as well as ρ, ρ', Δ (cf. 7.13)) and i) can all be defined also in the obvious direct way, except that at the very end of the construction of δ we may need to refer to proposition 7.17 for a non-trivial destabilization. The relations (7.14), (7.14') and (7.15) still hold. When $\delta \neq 0$, even σ^{-1} can be described as in the proof of theorem 7.11. (observe that the condition $b^+>0$ is not needed there unless $k=\min(n+b^+,b^-)$).

Finally, the singularitiy sequence allows us to extend a well-known fact about normal bordism to general fine bordism groups.

Theorem 7.25. The abelian group $\Omega_n(X,\phi,\xi,k)$ is finitely generated, provided X is a CW-complex with compact skeletons in all dimensions and provided $\dim \phi^+>0$.

Also, for all $n,m,k \in \mathbb{Z}$, $\mathfrak{N}_n(m,k)$ and $\Omega_n(m,k)$ are finitely generated.

Proof. Given any CW-complex Y, any virtual bundle ψ over Y and any $r \in \mathbb{Z}$, the inclusion of the $(r+1)$-skeleton Y_{r+1} into Y induces an isomorphism $\Omega_r(Y_{r+1},\psi|Y_{r+1}) \cong \Omega_r(Y,\psi)$; this follows from a successive transversality argument with respect to the midpoints of the cells of dimensions $>r+1$. If in addition Y_{r+1} is compact, then the isomorphism (2.2) identifies $\Omega_r(Y_{r+1},\psi|Y_{r+1})$ with a stable homotopy group of a compact Thom complex, which is known to be finitely generated.

Now recall that the Grassmannians $BO(i)$ are CW-complexes with compact skeletons [46], thus if X is also such a complex, the methods above apply to $Y=Y(X,\phi,\xi,p,q)$. Hence we can use the sequence (7.12) and descending induction over k to prove the first statement of the theorem.

The second statement follows now from proposition 7.17. ∎

§8. Rational bordism of framefields and of k-mersions.

As a first illustration of the strength of the methods developped in the last section, we compute the groups $\mathfrak{N}_n(m,k) \otimes \mathbb{Q}$ and $\Omega_n(m,k) \otimes \mathbb{Q}$ (compare example 6.1).

We start with the unoriented case. The following result gives the complete solution of a question which was raised, and answered in a number of cases, by J. Maffei [43]. It also overlaps with some of the work of R. Wells [68] who computed the immersion groups $\mathfrak{N}_n(m,n) \otimes \mathbb{Q}$.

Theorem 8.1. For all integers n, m, k the group $\mathfrak{N}_n(m,k) \otimes \mathbb{Q}$ vanishes except possibly when $n \equiv m \equiv k(2)$ and $n, m \geq 0$. In this case, however, we have an isomorphism

$$\rho_{\mathbb{Q}} : \mathfrak{N}_n(m,k) \otimes \mathbb{Q} \cong H_{n-(n-k)(m-k)-(n-k)-(m-k)}(BO(n-k+1)\times BO(m-k+1);\mathbb{Q})$$

defined as follows. Given $z = [M,\bar{g}] \in \mathfrak{N}_n(m,k)$, let $S \subset M$ be the non-degenerate singularity of $\bar{g}$, and let $S_1 \subset S$ be the zero set of a non-degenerate section of $\mathrm{Ker}(\bar{g}) \oplus \mathrm{Coker}(\bar{g})$. Then S_1 can be oriented canonically, and the image of its fundamental cycle, under the classifying maps of $\mathrm{Ker}(\bar{g})$ and $\mathrm{Coker}(\bar{g})$, gives rise to $\rho_{\mathbb{Q}}(z \otimes 1_{\mathbb{Q}})$.

Proof. Note that $\delta \circ \rho = 2 \cdot \mathrm{Id}$ in (7.23) (see proposition 7.15); hence the sequence splits when tensored with the rationals $\mathbb{Q}$. Together with the (rational) Hurewicz isomorphism theorem of classical algebraic topology (see [15], proposition I.5.2. and our last proof), this leads to an isomorphism for $m \geq 0$,

$$(8.2) \quad \mathfrak{N}_n(m,k)\otimes\mathbb{Q} \oplus \mathfrak{N}_{n+1}(m+1,k)\otimes\mathbb{Q} \cong H_{n+1-pq}(BO(p)\times BO(q);\tilde{\mathbb{Q}}_{p,q}) ,$$

where $p = n-k+1$, $q = m-k+1$, and $\tilde{\mathbb{Q}}_{p,q}$ denotes the rational coefficient bundle associated with the orientation bundle (cf. 6.5) of

$\gamma_p \otimes \gamma_q + \gamma_q - \gamma_p$, or, equivalently, with the cohomology class

(8.3) $\quad (q+1)w_1(\gamma_p) + (p+1)w_1(\gamma_q) \quad \in \quad H^1(BO(p) \times BO(q); \mathbb{Z}_2)$.

In order to compute twisted rational (co)homology groups such as the right hand term in (8.2), we will use the following general fact. Given a double covering $\tilde{X}$ of space X, let $\tilde{\mathbb{Q}}^\ell$ be the ℓth tensor power of the corresponding rational coefficient bundle $\tilde{\mathbb{Q}}$ over X, $\ell=0,1$, and write c for the involution on $\tilde{X}$ which interchanges the two sheets. Then, if $\tilde{\mathbb{Q}}_1$ denotes a rational coefficient bundle (associated with some $\mathbb{Z}_2$-bundle) over X (as well as its pullback to $\tilde{X}$), we have a canonical isomorphism

(8.4) $\quad H_*(X; \tilde{\mathbb{Q}}^\ell \otimes \tilde{\mathbb{Q}}_1) \cong \{z \in H_*(\tilde{X}; \tilde{\mathbb{Q}}_1) \mid (-1)^\ell c_*(z) = z\}$,

and similarly in cohomology. Indeed, such an isomorphism exists already at the chain level; it relates a singular simplex in X to its two liftings to $\tilde{X}$.

Now recall that $BSO(i)$ can be realized as the double covering space over $BO(i)$ associated with the orientation bundle of the universal i-plane bundle; again denote the involution interchanging the sheets by c. Applying formula (8.4) twice, we obtain for $r \in \mathbb{Z}$,

(8.5)

$$H_r(BO(p) \times BO(q); \tilde{\mathbb{Q}}_{p,q})$$
$$\cong \{z \in H_r(BSO(p) \times BSO(q); \mathbb{Q}) \mid z = (-1)^{q+1}(c \times Id)_*(z) = (-1)^{p+1}(Id \times c)_*(z)\}$$
$$= \sum_{s+t=r} \{z' \in H_s(BSO(p);\mathbb{Q}) \mid c_*(z') = (-1)^{q+1}z'\} \otimes \{z'' \in H_t(BSO(q);\mathbb{Q}) \mid c_*(z'') = (-1)^{p+1}z''\}$$

Furthermore, recall ([46], theorem 32) that $H^*(BSO(2j);\mathbb{Q})$ (resp. $H^*(BSO(2j+1);\mathbb{Q})$) is a polynomial ring generated by the Pontrjagin classes $p_1,\ldots,p_{j-1}$ and the Euler class χ of the universal oriented bundle (resp. by $p_1,\ldots,p_j$). Since the involution c can be covered by an orientation reversing automorphism of the universal bundle, we have $c^*(p_\ell) = p_\ell$ for all ℓ, but $c^*(\chi) = -\chi$. Clearly, the subspace fixed by $-c^*$ in the cohomology of $BSO(2j)$ is $\mathbb{Q}[p_1,\ldots,p_{j-1},\chi^2]\cdot\chi$. Thus we have

$$(8.6) \qquad \{z\in H^*(BSO(i);\mathbb{Q}) \mid c^*(z)=-z\} \cong \begin{cases} H^{*-i}(BSO(i+1);\mathbb{Q}) & i \text{ even} \\ 0 & i \text{ odd} \end{cases},$$

and similarly in homology.

We apply this to (8.5) and, ultimately, to (8.2). If $n \not\equiv m(2)$, then $p \not\equiv q(2)$, and either p and $q+1$, or q and $p+1$, are both odd; thus $H_{n+1-pq}(BO(p)\times BO(q);\tilde{\mathbb{Q}}_{p,q})$ vanishes in this case, and so does $\mathfrak{N}_n(m,k)\otimes\mathbb{Q}$.

If $n \equiv m \not\equiv k(2)$, then p and q are even; and

$$H_{n+1-pq}(BO(p)\times BO(q);\tilde{\mathbb{Q}}_{p,q}) \cong H_{n+1-pq-p-q}(BSO(p+1)\times BSO(q+1);\mathbb{Q})$$

or equivalently,

$$\mathfrak{N}_n(m,k)\otimes\mathbb{Q} \oplus \mathfrak{N}_{n+1}(m+1,k)\otimes\mathbb{Q} \cong \mathfrak{N}_{n+1}(m+1,k)\otimes\mathbb{Q} \oplus \mathfrak{N}_{n+2}(m+2,k)\otimes\mathbb{Q})$$

Therefore, for large N

$$\mathfrak{N}_n(m,k)\otimes\mathbb{Q} \cong \mathfrak{N}_{n+2}(m+2,k)\otimes\mathbb{Q} \cong \ldots \cong \mathfrak{N}_{n+2N}\otimes\mathbb{Q} = 0 .$$

It follows that $\mathfrak{N}_n(m,k)\otimes\mathbb{Q}$ can be nontrivial only when $n \equiv m \equiv k(2)$ and $n,m \geq 0$, in which case (8.2) essentially gives the required isomorphism composed of $g\otimes Id_{\mathbb{Q}}$ and the Hurewicz isomorphism. ∎

In order to describe $\Omega_n(m,k)\otimes\mathbb{Q}$, it will be convenient first to discuss a few facts about the universal model for singularities which was studied in [31]. Given an infinite dimensional real Hilbert space H and integers $p,q\geq 0$, denote by F_{p-q} the space of Fredholm operators with index $p-q$, i.e. of linear continuous maps $h : H \longrightarrow H$ with finite kernel and cokernel dimensions whose difference is $p - q$. Define the open subspace

$$(8.7) \qquad W_{p,q} = \{h \in F_{p-q} \mid \dim\ker(h) \leq p, \ \dim\operatorname{coker}(h) \leq q\}$$

(here we depart from the notation $V_{p,q}$ used in [31], in order to avoid confusion with Stiefel manifolds), and its "singularity submanifold"

$$A_{p,q} = \{h \in F_{p-q} \mid \dim \ker(h) = p,\ \dim \operatorname{coker}(h) = q\} .$$

Recall Kuiper's result [40] that the general linear group $GL(H)$ of H is contractible. This leads to canonical homotopy equivalences

(8.8) $$F_{p-q} \sim BO \quad \text{and} \quad (\mathrm{Ker}, \mathrm{Coker}) : A_{p,q} \xrightarrow{\sim} BO(p) \times BO(q) .$$

It implies also that every $GL(H)$-bundle over a paracompact space X is trivial. In particular, given a vector bundle map $\overline{g} : E \longrightarrow F$ over X such that $\dim E - \dim F = p-q$, "infinite stabilization" leads to a well-defined homotopy class

$$St_{\infty}(\overline{g}) \in [X, F_{p-q}]$$

determined by the infinite-dimensional bundle map

(8.9) $$X \times H \cong E \oplus (X \times H) \xrightarrow{\overline{g} \oplus \mathrm{Id}} F \oplus (X \times H) \cong X \times H .$$

$St_{\infty}(\overline{g})$ depends actually on the difference $E-F$ in the vector bundle K-theory of X.

Recall also that $H^{*}(F_{p-q};\mathbb{Q})$ is a polynomial algebra generated by the universal Pontrjagin classes

(8.10) $$p_k \in H^{4k}(F_{p-q};\mathbb{Q}), \quad k = 1,2,\ldots .$$

$St_{\infty}(\overline{g})^{*}$ is determined by the formula

(8.11) $$St_{\infty}(\overline{g})^{**}(1+p_1+p_2+\ldots+p_k\ldots) = \frac{(1+p_1(E)+p_2(E)+\ldots)}{(1+p_1(F)+p_2(F)+\ldots)} ,$$

which holds in $H^{**}(X;\mathbb{Q}) = \prod_{i \geq o} H^{i}(X;\mathbb{Q})$.

Example 8.12. If $X = A_{p,q}$ and $\overline{g}$ is the zero morphism from $E = \mathrm{Ker}$ to $F = \mathrm{Coker}$, then the homotopy class $St_{\infty}(\overline{g})$ can be represented by the map

$$i_{p,q} : A_{p,q} \subset F_{p-q} .$$

If we denote the Pontrjagin classes of Ker and Coker by p'_i and p''_i respectively, the induced homomorphism

$$i^*_{p,q}: H^*(F_{p-q};\mathbb{Q}) \cong \mathbb{Q}[p_1,\ldots,p_k,\ldots] \longrightarrow H^*(A_{p,q},\mathbb{Q}) \cong \mathbb{Q}[p'_1,\ldots p'_{[\frac{p}{2}]};p''_1,\ldots,p''_{[\frac{q}{2}]}]$$

is determined by

$$i^{**}_{p,q}(1+p_1+p_2+\ldots+p_k+\ldots) = \frac{1+p'_1+\ldots+p'_{[\frac{p}{2}]}}{1+p''_1+\ldots+p''_{[\frac{q}{2}]}} .$$

More details and references concerning this example and the preceding discussion can be found in [31].

Proposition 8.13. Given any integers $p,q \geq 0$, let the homomorphisms

$$i^*_{p,q} : H^*(F_{p-q};\mathbb{Q}) \xrightarrow{\bar{i}^*_{p,q}} H^*(W_{p,q};\mathbb{Q}) \xrightarrow{\bar{\bar{i}}^*_{p,q}} H^*(A_{p,q};\mathbb{Q})$$

be induced by inclusions. Then $\bar{i}^*_{p,q}$ is onto, and $\bar{\bar{i}}^*_{p,q}$ is injective.

Proof. Recall from [31] that $A_{p,q}$ is a closed analytic submanifold of $W_{p,q}$ with normal bundle isomorphic to Hom (Ker, Coker). Because of the linear structure of the Banachspace $L(H)$ in which $W_{p,q}$ lies as an open set, $A_{p,q}$ actually has a tubular neighborhood in $W_{p,q}$. Thus the Thom isomorphism together with excision, gives the vertical isomorphism in the following commutative diagram

$$(8.14)\quad \begin{array}{ccccccc} \xrightarrow{\partial} & H^*(W_{p,q},W_{p-1,q-1};\mathbb{Q}) & \longrightarrow & H^*(W_{p,q};\mathbb{Q}) & \xrightarrow{j^*} & H^*(W_{p-1,q-1};\mathbb{Q}) & \xrightarrow{\partial} \\ & \updownarrow & & \downarrow \bar{\bar{i}}^*_{p,q} & & & \\ & H^{*-pq}(A_{p,q};\tilde{\mathbb{Q}}) & & H^*(A_{p,q};\mathbb{Q}) & & & \\ & \downarrow & & \downarrow & & & \\ & H^{*-pq}(BSO(p)\times BSO(q);\mathbb{Q}) & \xrightarrow{\cdot\bar{\chi}} & H^*(BSO(p)\times BSO(q);\mathbb{Q}) & & & \end{array}$$

Here $\tilde{\mathbb{Q}}$ denotes the rational coefficient system also associated with the orientation bundle of Hom (Ker, Coker), i.e. with the tensor product of q times the orientation bundle of Ker with p times that of Coker. The bottom vertical injections are obtained from the homotopy equivalence (8.8) and two-fold application of formula (8.4); thus e.g. $H^{*-pq}(A_{p,q};\tilde{\mathbb{Q}})$ is injected onto the joint fixed space of $(-1)^q(c\times Id)^*$ and $(-1)^p(Id\times c)^*$, where c interchanges the leaves in the double cover $BSO(r)$ over $BO(r)$. The top line is the obvious exact sequence. The bottom horizontal arrow denotes multiplication by the Euler class $\overline{\chi}$ of the (suitably oriented) bundle $\mathrm{Hom}(\tilde{\gamma}_p,\tilde{\gamma}_q)$. If p or q is even, $\pm\overline{\chi}$ restricts to a non-vanishing class on $BSO(p)\times *$ or on $*\times BSO(q)$ (namely to a power of the Euler class of the canonical bundle); hence multiplication with $\overline{\chi}$ is injective in this case, since it takes place in a polynomial algebra. On the other hand, if p and q are both odd, then $H^{*-pq}(A_{p,q};\mathbb{Q}) = 0$, as can be seen from the eigenspace representation. Therefore, in any case $\partial = 0$, j^* is onto, and $\bar{\bar{i}}^*_{p,q}$ is injective on the kernel of j^*. Applying the second statement to the inclusions

$$W_{p,q} \subset W_{p+1,q+1} \subset \dots \subset \bigcup_{i\geq 0} W_{p+i,q+i} = F_{p-q}$$

we see that $\overline{i}^*_{p,q}$ is onto.

Next consider the diagram (compare with the upper right hand corner of 8.14)

$$\begin{array}{ccc} & H^*(F_{p-q};\mathbb{Q}) & \\ {\scriptstyle \overline{i}^*_{p,q}}\swarrow & & \searrow{\scriptstyle \overline{i}^*_{p-1,q-1}} \\ H^*(W_{p,q};\mathbb{Q}) & \xrightarrow[j^*]{} & H^*(W_{p-1,q-1};\mathbb{Q}) \\ \downarrow{\scriptstyle \bar{\bar{i}}^*_{p,q}} & & \downarrow{\scriptstyle \bar{\bar{i}}^*_{p-1,q-1}} \\ H^*(A_{p,q};\mathbb{Q}) & \xrightarrow[\overline{j}^*]{} & H^*(A_{p-1,q-1};\mathbb{Q}) \end{array} \tag{8.15}$$

induced by natural inclusions and by the map $\overline{j} : A_{p-1,q-1} \longrightarrow A_{p,q}$

which sends $g \in A_{p-1,q-1}$ to the composed operator
$\bar{j}(g) : H \cong H \times \mathbb{R} \xrightarrow{g \times 0} H \times \mathbb{R} \cong H$ (involving a fixed isomorphism $H \cong H \times \mathbb{R}$). In view of example 8.12, $\bar{j}$ commutes up to homotopy with the inclusions into F_{p-q}, and so does j obviously. Since $\bar{i}^*_{p,q}$ is onto, all parts of our diagram are commutative.

Now we can establish the injectivity of $\bar{\bar{i}}^*_{p,q}$ inductively. If p or q is zero, then $W_{p,q} = A_{p,q}$ and there is nothing to prove. So assume that $\bar{\bar{i}}^*_{p-1,q-1}$ is injective. Then $\bar{\bar{i}}^*_{p,q}$ maps a complement of $\ker j^*$ in $H^*(W_{p,q};\mathbb{Q})$ injectively into a complement of $\ker \bar{j}^*$ in $H^*(A_{p,q};\mathbb{Q})$; $\ker j^*$ itself gets mapped into $\ker \bar{j}^*$, and injectively so, as we have noted above. Thus $\bar{\bar{i}}^*_{p,q}$, as a whole, is a monomorphism.■

<u>Definition 8.16.</u> Given integers $a,b \geq 0$, let $S(a,b)$ denote the set of all $(b+1)$-tupels $s=(s(1), s(2),..,s(b+1))$ of integers such that $s(1) \geq s(2) \geq ... \geq s(b+1) \geq a+1$. For $s \in S(a,b)$ define a polynomial

$$\mathcal{H}_s = \mathcal{H}_s(p_1,\ldots p_k,\ldots) \in H^{4(s(1)+s(2)+\ldots+s(b+1))}(F_{p-q};\mathbb{Q})$$

in the universal Pontrjagin classes (cf. 8.10) by the $(b+1)\times(b+1)$-determinant

(8.17)

$$\mathcal{H}_s(p_1,..)=\det(p_{s(i)-i+j}) = \begin{vmatrix} p_{s(1)} & p_{s(1)+1} & \cdots\cdots & & p_{s(1)+b} \\ p_{s(2)-1} & p_{s(2)} & \cdots\cdots & & p_{s(2)+b-1} \\ \vdots & & \ddots & & \vdots \\ \cdot & & p_{s(i)} & & \cdot \\ \vdots & & & \ddots & \vdots \\ \cdot & & & p_{s(b)} & p_{s(b)+1} \\ p_{s(b+1)-b} & & \cdots\cdots & & p_{s(b+1)} \end{vmatrix}$$

(where we put $p_o = 1$, $p_{-1} = p_{-2} = ... = 0$).

<u>Remark.</u> Similar determinants occur also in the work of Ruget ([56] p. 302).

<u>Proposition 8.18.</u> <u>For all integers $p,q \geq 0$,</u> $\ker \bar{i}^*_{p,q} = \ker i^*_{p,q}$ <u>equals the ideal in</u> $H^*(F_{p-q};\mathbb{Q}) = \mathbb{Q}[p_1,\ldots,p_k,..]$ <u>generated by the set</u> $\{\mathcal{H}_s(p_1\ldots,p_\ell,..) \mid s \in S([\frac{p}{2}],[\frac{q}{2}])\}$

<u>Proof.</u> Since $\bar{\bar{i}}^*_{p,q}$ is injective by the last proposition, $\bar{i}^*_{p,q}$ and $i^*_{p,q}$ have the same kernel. Its analysis is reduced to a purely algebraic question by the description of $i^*_{p,q}$ in example 8.12.

Write a for $[\frac{p}{2}]$, b for $[\frac{q}{2}]$, $\hat{p}_k$ for $i^*_{p,q}(p_k)$, $k \in \mathbb{Z}$; and recall that the graded algebra homomorphism $i^*_{p,q}:\mathbb{Q}[p_1,\ldots,p_k,\ldots] \longrightarrow \mathbb{Q}[p'_1,\ldots,p'_a;p''_1,\ldots,p''_b]$ is characterized by the set of equations

$$p'_k = p''_b \cdot \hat{p}_{k-b} + \ldots + p''_{k-j} \cdot \hat{p}_j + \ldots + p''_1 \cdot \hat{p}_{k-1} + \hat{p}_k, \quad k = 0,1,2,\ldots . \tag{8.19}$$

In particular, if $k > a$ this expression vanishes. Thus, when we apply $i^*_{p,q}$ to the matrix in (8.17), the last column becomes a linear combination of the previous ones, and the determinant has to vanish. This proves that the ideal $I(a,b)$ generated in $\mathbb{Q}[p_1,\ldots,p_k,\ldots]$ by the $\mathcal{H}_s$, $s \in S(a,b)$, lies in the kernel of $i^*_{p,q}$.

To prove the converse, we will show that

(i) there is an (additive) direct sum decomposition

$$\mathbb{Q}[p_1,\ldots,p_k\ldots] = I(a,b) \oplus \bigoplus_r \mathbb{Q}[p_1,..,p_a] \cdot p^r , \tag{8.20}$$

where $\{p^r\}$ is the set of all products of the variables $p_{a+1}, p_{a+2},\ldots$ with at most b (not necessarily distinct) factors; and

(ii) $i^*_{p,q}$ is injective on the complement of $I(a,b)$ in this decomposition.

Clearly, this determines both the kernel and the image of $i^*_{p,q}$.

Given a product of the variables $p_{a+1},\ldots$ with more than b factors, it is divisible by the product $p_{s(1)} \cdot p_{s(2)} \cdots p_{s(b+1)}$ for a suitable $s \in S(a,b)$. This expression equals $\mathcal{H}_s$ plus a sum of terms of the form $\pm p_{t(1)} \cdot p_{t(2)} \cdots p_{t(b+1)}$ with $t(1) \geq t(2) \geq \ldots \geq t(b+1)$ and

$\sum i\cdot t(i) < \sum i\cdot s(i)$. (Prove by induction over $n \geq 1$ that $\sum i(\sigma(i)-i) < 0$ for each non-identical permutation σ of the set $\{1,2,\ldots,n\}$). After a finite sequence of such substitutions, $t(b+1)$ in each product will be smaller or equal to a. Iterating this procedure, we see that $\mathbb{Q}[p_1,\ldots p_k,\ldots]$ is generated additively by the expression on the right hand side of equation (8.20).

To establish the remainder of the claims (i) and (ii) above, it suffices to show that the corresponding products $\hat{p}^r$ in $\mathbb{Q}[p_1',\ldots,p_a';p_1'',\ldots,p_b'']$ are linearly independent with respect to coefficients in the subring $\mathbb{Q}[\hat{p}_1,\ldots,\hat{p}_a]$. To this end observe first that $\mathbb{Q}[p_1',\ldots,p_a';p_1'',\ldots,p_b'']$ can also be viewed as a polynomial ring generated by the elements $\hat{p}_1,\ldots,\hat{p}_a;p_1'',\ldots,p_b''$. This follows from formula (8.19) applied to $k=1,\ldots,a$. The same formula also implies, via induction over $i \geq 0$, that

(8.21) $$\hat{p}_{a+1} = \bar{p}_i'' \cdot \hat{p}_a + \alpha(\hat{p}_1,\ldots,\hat{p}_{a-1};p_1'',\ldots,p_b'') ,$$

where the polynomial α has graded degree $>4i$ with respect to the variables $p_1'',\ldots,p_b''$; $\bar{p}_i''$ denotes the $4i$-th homogeneous part of the multiplicative inverse of $1+p_1''+\ldots p_b''$, i.e. it is characterized by the formulas

(8.22) $$0 = p_b'' \cdot \bar{p}_{i-b}'' + \ldots p_{i-j}'' \cdot \bar{p}_j'' + \ldots + p_1'' \cdot \bar{p}_{i-1}'' + \bar{p}_i'', \quad i = 1,2\ldots .$$

Assume now that there is a non-trivial relation

$$0 = \beta_1 \cdot \hat{p}^{r_1} + \ldots + \beta_k \cdot \hat{p}^{r_k} + \beta_{k+1} \cdot \hat{p}^{r_{k+1}} + \ldots + \beta_\ell \cdot \hat{p}^{r_\ell}$$

between products of the type $\hat{p}^r = \hat{p}_{a+r(1)} \cdot \hat{p}_{a+r(2)} \cdots \hat{p}_{a+r(c)}$ $(r(1) \geq r(2) \geq \ldots \geq r(c) > 0,\ c \leq b)$, with coefficients $\beta_1,\ldots,\beta_\ell \in \mathbb{Q}[\hat{p}_1,\ldots,\hat{p}_a]$. We may assume that the first few coefficients $\beta_1,\ldots,\beta_k$ are non-trivial and that the corresponding r's are just the ones with minimal sum $r(1)+r(2)+\ldots+r(c)$ among the occurring r's.

Then (8.21) implies

$$0 = \beta_1 \cdot \hat{p}_a^{c_1} \cdot \bar{p}''^{r_1} + \ldots + \beta_k \cdot \hat{p}_a^{c_k} \cdot \bar{p}''^{r_k} \quad + \text{ terms,}$$

which are of higher degree with respect to $p''_1,\ldots,p''_b$; here $\bar{p}''^{r}$ stands for $\bar{p}''_{r(1)} \cdot \bar{p}''_{r(2)} \cdots \bar{p}''_{r(c)}$. This leads to a contradiction, and the proposition is proved, as soon as we have established the following result.

Lemma 8.23. Given any integer $b \geq 0$, form the polynomial algebra $\mathbb{Q}[p''_1,\ldots,p''_b]$ in any variables $p''_1,\ldots,p''_b$ and consider the elements $\ldots p''_{-2} = \bar{p}''_{-1} = 0$, $\bar{p}''_0 = 1$, $\bar{p}''_1,\ldots \bar{p}''_i,\ldots$ defined inductively by the equations (8.22). Then the set of products $\bar{p}''^{r}$ of b factors from among $\bar{p}''_0, \bar{p}''_1,\ldots,\bar{p}''_i,\ldots$, that is the set $\{\bar{p}''_{r(1)} \cdots \bar{p}''_{r(b)} \mid r(1) \geq r(2) \geq \ldots \geq r(b) \geq 0\}$, forms a basis of the $\mathbb{Q}$-vectorspace $\mathbb{Q}[p''_1,\ldots,p''_b]$.

Proof. This corresponds to the special case of the previous situation when $p=0$. Note that then $i^*_{p,q}$ is onto, since it is composed of the involution on $\mathbb{Q}[p_1,\ldots,p_i,\ldots]$ taking p_i to the 4i-th homogenious part of the multiplicative inverse of $1+p_1+p_2+\ldots$, and of the obvious projection onto $\mathbb{Q}[p''_1,\ldots,p''_b]$. Therefore, the discussion following (8.20) implies that the products $\bar{p}''^{r} = \bar{p}''_{r(1)} \cdots \bar{p}''_{r(b)}$, $r(1) \geq \ldots \geq r(b) \geq 0$, generate the $\mathbb{Q}$-vectorspace $\mathbb{Q}[p''_1,\ldots,p''_b]$. For each such b-tupel r, we can also define the monomial

$$(8.24) \qquad p''^{\,r(1)-r(2)}_1 \cdot p''^{\,(r(2)-r(3))}_2 \cdot \ldots \cdot p''^{\,r(b-1)-r(b)}_{b-1} \cdot p''^{\,r(b)}_b$$

of the same degree as $\bar{p}''^{r}$, namely

$$4(r(1)+r(2)+\ldots+r(b)) = 4((r(1)-r(2)) + 2(r(2)-r(3))+\ldots+b \cdot r(b)).$$

Since these monomials are linearly independent, so are the $\bar{p}''^{r}$'s for dimension reasons.

Acknowledgement. I would like to thank Hans-Peter Kraft for a discussion concerning the algebra involved in the proof of the previous proposition.■

Our interest in the spaces $W_{p,q}$ and F_{p-q} will be better understood once it is seen how closely their rational homology groups are related to the groups $\Omega_n(m,k)\otimes\mathbb{Q}$.

Definition 8.25. Let M be a closed oriented smooth manifold of dimension n, and let m,k be two integers. We will say that M satisfies the condition $H(m,k)$ if all those (linear combinations of) Pontrjagin numbers of M vanish which contain a $[\frac{1}{2}(m-k+2)] \times [\frac{1}{2}(m-k+2)]$-determinant of the form

$$\det(p_{s(i)-i+j}(M)) \text{ , with } s(1) \geq s(2) \geq \ldots \geq s([\tfrac{1}{2}(m-k+2)]) > \tfrac{1}{2}(n-k)$$

(cf. 8.17), as a factor. (If $m<k$, we require only divisibility by 1).

Theorem 8.26. For all integers n,m,k with $m \geq 0$ there is an isomorphism

$$\Omega_n(m,k)\otimes\mathbb{Q} \cong \mathfrak{N}_n(m,k)\otimes\mathbb{Q} \oplus \left\{[M]\otimes c \in \Omega_n\otimes\mathbb{Q} \,\middle|\, \begin{array}{l} M \text{ satisfies condition} \\ H(m,k);\ c \in \mathbb{Q} \end{array}\right\}$$

given by the obvious forgetful maps from $\Omega_n(m,k)$ into $\mathfrak{N}_n(m,k)$ and into Thom's oriented bordism group Ω_n. The two factors correspond to the eigenspaces of the orientation reversing involution.

The first factor has been determined in theorem 8.1. We will soon see that the second factor is isomorphic to $H_n(W_{n-k,m-k};\mathbb{Q})$; hence it can be computed inductively from (8.14).

Our theorem overlaps with work of O. Burlet [12] who considered rational bordism of embeddings and immersions.

Proof. Tensorize the oriented analogue of (7.23) with $\mathbb{Q}$ to obtain the exact sequence

$$
\begin{array}{ccccccc}
 & & & & \Omega_n^{rel}(\text{point},(\underset{\sim}{\mathbb{R}},\underset{\sim}{\mathbb{R}}^{m+1}),\underset{\sim}{\mathbb{R}},k+1)\otimes\mathbb{Q} & & \\
 & & & \nearrow & \cong\big\downarrow \sigma\otimes Id_{\mathbb{Q}} & \searrow & \\
\ldots\to \Omega_n(m,k+1)\otimes\mathbb{Q} & \longrightarrow & \Omega_n(m,k)\otimes\mathbb{Q} & \longrightarrow & \Omega_{n-pq}(Y;\gamma_p\otimes\gamma_q+\gamma_q-\gamma_p)\otimes\mathbb{Q} & \longrightarrow & \ldots \\
 & & & \searrow & \cong\big\downarrow \mu & \nearrow & \\
 & & & & H_{n-pq}(Y;\tilde{\mathbb{Q}}^{q+1}_{\gamma_p}\otimes\tilde{\mathbb{Q}}^{p+1}_{\gamma_q}) & &
\end{array}
$$

where we put

$$p = n-k \quad \text{and} \quad q = m-k$$

(this differs from our previous notation), and μ denotes the Hurewicz-isomorphism (cf. [15], proposition I.5.2); $\tilde{\mathbb{Q}}_{\gamma_p}$ and $\tilde{\mathbb{Q}}_{\gamma_q}$ are the rational coefficient bundles associated with the (pullbacks of the) orientation bundles ξ_{γ_p} and ξ_{γ_q}. Y denotes the double cover over $BO(p)\times BO(q)$ belonging to $\xi_{\gamma_p}\otimes\xi_{\gamma_q}$. Interchanging the sheets of this double cover induces involutions on the two lower groups at the right hand which, together with the orientation reversing involutions on the remaining groups, define a compatible involution of the whole diagram above. Therefore, we can decompose according to the eigenvalues $+1$ and -1. We denote the eigenspaces in $\Omega_n(m,k)\otimes\mathbb{Q}$ by $\Omega_n^+(m,k)$ and $\Omega_n^-(m,k)$; also, by (8.4), we can identify the factors of the homology group at the bottom with homology groups of $BO(p)\times BO(q)$. Thus we obtain the top double line in the following commuting diagram of horizontal exact sequences.

$$\begin{array}{ccccccc}
\cdots \xrightarrow{\delta^+} \Omega_n^+(m,k+1) & \xrightarrow{f^+=0} & \Omega_n^+(m,k) & \longrightarrow & H_{n-pq}(BO(p)\times BO(q);\tilde{\mathbb{Q}}_{\gamma_p}^{q+1}\otimes\tilde{\mathbb{Q}}_{\gamma_q}^{p+1}) \xrightarrow{\delta^+} \\
\oplus & & \oplus & & \oplus \\
\cdots \xrightarrow{\delta^-} \Omega_n^-(m,k+1) & \xrightarrow{f^-} & \Omega_n(m,k) & \longrightarrow & H_{n-pq}(BO(p)\times BO(q);\tilde{\mathbb{Q}}_{\gamma_p}^{q}\otimes\tilde{\mathbb{Q}}_{\gamma_q}^{p}) \xrightarrow{\delta^-} \\
\Big\downarrow st_{n,m,k+1} & & \Big\downarrow st_{n,m,k} & & \Big\downarrow st^{rel}_{n,m,k} \\
\cdots \xrightarrow{\partial=0} H_n(W_{p-1,q-1};\mathbb{Q}) & \longrightarrow & H_n(W_{p,q};\mathbb{Q}) & \longrightarrow & H_n(W_{p,q},W_{p-1,q-1};\mathbb{Q}) \xrightarrow{\partial=0}
\end{array}$$

Let us first discuss the sequence corresponding to the eigenvalue $+1$. It follows from proposition 7.15 that δ^+ is onto, hence $f^+ = 0$ and

$$\Omega_n^+(m,k) \oplus \Omega_{n-1}^+(m-1,k) \cong H_{n-pq}(BO(p)\times BO(q);\ \tilde{\mathbb{Q}}_{\gamma_p}^{q+1} \otimes \tilde{\mathbb{Q}}_{\gamma_q}^{p+1}).$$

In view of (8.2), this homology group is also isomorphic to $\mathfrak{N}_n(m,k)\otimes\mathbb{Q} \oplus \mathfrak{N}_{n-1}(m-1,k)\otimes\mathbb{Q}$. Now consider the homomorphisms

$$\Omega_n^+(m,k) \underset{d}{\overset{r}{\rightleftarrows}} \mathfrak{N}_n(m,k) \otimes \mathbb{Q} \qquad ,$$

where r forgets about orientations and d assigns to a k-morphism over a manifold M its pullback over the orientation double cover of M. Clearly we have $d\cdot r = 2\cdot \mathrm{Id}$, by the very definition of $\Omega_n^+(m,k)$. From a dimension argument involving the isomorphisms above, we see that r is a bijection with inverse $\frac{1}{2}d$.

Next we discuss the lower part of diagram (8.27). Given a k-morphism $\overline{g} : TM \longrightarrow \mathbb{R}^m$ over a compact oriented n-manifold M (with $\overline{g}|\partial M$ being a (k+1)-morphism), infinite stabilization (see 8.9) defines a map

$$St_\infty(\overline{g}) : (M,\partial M) \longrightarrow (W_{p,q},W_{p-1,q-1})$$

up to homotopy; the image of the fundamental class of M under the induced homomorpism $St_\infty(\bar{g})_*$ depends only on the (absolute or relative) bordism class $[M,\bar{g}]$. This construction gives rise to the vertical arrows in (8.27). We denote their restrictions to the (-1)-eigenspaces by the symbol st, with the appropriate sub- and superscripts to indicate their domains.

These homomorphisms also fit into the following commuting diagram

$$
(8.28)\qquad
\begin{array}{ccccccc}
\Omega_n^-(m,k+1) & \xrightarrow{f^-} & \Omega_n^-(m,k) & \longrightarrow & \dots \longrightarrow & \Omega_n \otimes \mathbb{Q} \\
\Big\downarrow st_{n,m,k+1} & & \Big\downarrow st_{n,m,k} & & & \Big\downarrow st_{n,m} \\
H_n(W_{n-k-1,m-k-1};\mathbb{Q}) & \longrightarrow & H_n(W_{n-k,m-k};\mathbb{Q}) & \longrightarrow & \dots \longrightarrow & H_n(F_{n-m};\mathbb{Q}) .
\end{array}
$$

Here $st_{n,m}$ is formed by applying the above construction to the zero morphisms $TM \longrightarrow \mathbb{R}^m$ over closed oriented n-manifolds M. Now Thom [64] has shown that $\Omega_* \otimes \mathbb{Q}$ is a polynomial algebra with generators $[M_4]$, $[M_8]$,...$[M_{4k}]$, of indicated dimensions, and that the Pontrjagin numbers form a complete set of invariants for elements in $\Omega_* \otimes \mathbb{Q}$. Hence (8.10) and (8.11) imply that $st_{n,m}$ is an isomorphism. In the very special case when

$$n + 1 < (n - k)\cdot(m - k) \quad \text{and} \quad n - k,\ m - k \geq 0 ,$$

all the other arrows in (8.28) are also isomorphisms, since then the third terms in sequences like (8.27) vanish for dimension reasons; here we use already the following remark.

For general n, m, k we can deduce from the discussion centering around (8.14) that the boundary operators in the bottom sequence in diagram (8.27) vanish, and that the two homology groups connected by $st^{rel}_{n,m,k}$ have same dimensions. Therefore, if $st_{n,m,k}$ and $st_{n+1,m+1,k+1}$ are isomorphisms, then $st^{rel}_{n,m,k}$ and $st^{rel}_{n+1,m+1,k+1}$ are onto and hence bijective, and so is $st_{n,m,k+1}$. By the same token, if $st_{n,m,k}$ failed to be an isomorphism, then so would $st_{n,m,k-1} \oplus st_{n+1,m+1,k}$, and also, by induction,

$$St_{(N)} := \bigoplus_{j=0}^{N} st_{n+j,\ m+j,\ k-N+j}$$

for all natural numbers N. But if N is sufficiently large, we have

$$(n+j+1 \leq)\ n+N+1 < (n-k+N)(m-k+N)\ \big(= ((n+j)-(k-N-j))(m+j-(k-N+j))\big),$$

and all the factors of $St_{(N)}$ are isomorphisms, as we saw above.

We conclude that all the vertical arrows in the diagrams (8.27) and (8.28) are always isomorphisms, and that all the horizontal arrows in (8.28) are injective. Therefore, the obvious forgetful map identifies $\Omega_n^-(m,k)$ with the subspace of $\Omega_n \otimes \mathbb{Q}$ whose image under $st_{n,m}$ consists precisely of those homology classes which are annihilated by the kernel of

$$\overline{i}^*_{n-k,m-k} : H^n(F_{n-m};\mathbb{Q}) \longrightarrow H^n(W_{n-k,m-k};\mathbb{Q}) .$$

Proposition 8.18 and formula (8.11) now imply the theorem. ∎

Corollary 8.29. Let M be a closed oriented smooth n-dimensional manifold, and let m, k be integers, $m \geq 0$. Then M satisfies condition H(m,k) (see def. 8.25) if and only if some positive multiple of M is (oriented) bordant to a manifold M' which admits a k-morphism from TM' to $\mathbb{R}^m$.

In particular, the following are equivalent for all non-negative $q < n$:

(i) a positive multiple of M is oriented bordant to a manifold which admits a q-framefield (resp. an immersion into $\mathbb{R}^{2n-q}$),

(ii) all Pontrjagin numbers of M involving some $p_i(M)$ (resp. $\overline{p}_i(M)$) with $2i > n-q$ as a factor, vanish. (Here $\overline{p}(M)$ denotes the multiplicative inverse of $p(M)$.)

This follows from theorem 8.26 and the second part of its proof (see diagram 8.28).∎

Finally, note that the obvious product makes

$$\mathfrak{N}_*(*,*) = \sum_{n,m,k} \mathfrak{N}_n(m,k) \quad \text{and} \quad \Omega_*(*,*) = \sum_{n,m,k} {}_n(m,k)$$

and their rational analogues into tri-graded rings.

Exercise 8.30. In [68] Robert Wells proved that the "immersion ring" $\sum_{n<m} \mathfrak{N}_n(m,n) \otimes \mathbb{Q}$ is a bigraded polynomial algebra with generators f_0, f_1, f_2 such that $f_i \in \mathfrak{N}_{4i+2}(4i+4;4i+2)$. Check that this is compatible with our computations of the additive structure.

Problem 8.31. Determine the multiplicative structure of $\mathfrak{N}_*(*,*) \otimes \mathbb{Q}$ and $\Omega_*(*,*) \otimes \mathbb{Q}$.

§9. Calculation of low-dimensional normal bordism groups.

Throughout this section let X denote a path-connected paracompact space which is homotopy equivalent to a CW-complex with compact skeletons in all dimensions. (Note that every compact connected C^∞-manifold or CW-complex M is such a space, and so is the product $M \times X$ with every X as above [48], [70]). Also, let $\phi = (\phi^+, \phi^-)$ be a pair of real vector bundles of strictly positive dimensions b^+, b^- over X, and let n be an integer. For convenience we will assume sometimes that $n + b^+ = b^-$.

Define $\bar{\Omega}_n(X,\phi)$ to be the group of bordism classes of tripels (M, g, or) where M is a smooth closed n-manifold, $g : M \longrightarrow X$ is continuous, and $or : \xi_M \cong g^*(\xi_\phi)$ is an isomorphism (see also the beginning of §6, and formula (6.5)). This group can often be computed. Our goal is to compare it to the full normal bordism group $\Omega_n(X,\phi)$ via the map

$$\bar{f} : \Omega_n(X,\phi) \longrightarrow \bar{\Omega}_n(X,\phi)$$

which forgets about an isomorphism $\bar{g} : TM \oplus g^*(\phi^+) \longrightarrow g^*(\phi^-)$ and retains only the orientation information underlying $\bar{g}$.

Our first result is mainly of theoretical interest.

<u>Proposition 9.1.</u> <u>The kernel of the homomorphism $\bar{f}$ is a finite group.</u>

<u>Proof.</u> There is a Hurewicz isomorphism $\Omega_n(X;\phi) \otimes \mathbb{Q} \cong H_n(X;\tilde{\mathbb{Q}})$, deduced from the Thom-Pontrjagin construction, the classical Hurewicz map and the Thom-Gysin isomorphism (see [15], proposition I,5.2, and also the proof of our theorem 7.25). Moreover, it is not hard to see that we can factorize through the map $\bar{f} \otimes Id_{\mathbb{Q}}$ which therefore is injective. In view of 7.25, this completes the proof. ∎

Next recall that an isomorphism $\bar{g} : TM \oplus g^*(\phi^+) \longrightarrow g^*(\phi^-)$ is essentially known as soon as we know its composition with the projection $TM \oplus g^*(\phi^+) \oplus \underset{\sim}{\mathbb{R}} \longrightarrow TM \oplus g^*(\phi^+)$, as well as the way $\bar{g}$ relates orientations. Hence, we can identify the normal bordism group $\Omega_n(X,\phi)$ with $\Omega_n(X,(\phi^+ \oplus \underset{\sim}{\mathbb{R}},\phi^-),\xi_\phi,b^-)$ and decompose $\bar{f}$ into a string of forgetful maps

$$\Omega_n(X,\phi) \xrightarrow{f} \Omega_n(X,(\phi^+ \oplus \underset{\sim}{\mathbb{R}},\phi^-),\xi_\phi,b^- -1) \xrightarrow{f} \ldots \xrightarrow{f} \Omega_n(X,(\phi^+ \oplus \underset{\sim}{\mathbb{R}},\phi^-),\xi_\phi,b^- -j) \longrightarrow \bar{\Omega}_n(X,\phi) .$$

Note that the kernel and cokernel of f here can be expressed in terms of lower dimensional normal bordism groups, since in the corresponding exact sequences (7.12) all singularities have codimension ≥ 2. (This is the main reason for choosing the above description of $\Omega_n(X,\phi)$). Also observe, that the right hand arrow is bijective, if $(j+2)\cdot(j+1) > n+1$, again because of the high codimensions of the relevant singularities. In particular, for $n \leq 4$

$$f = \bar{f} : \Omega_n(X,\phi) \longrightarrow \Omega_n(X,(\phi^+ \oplus \underset{\sim}{\mathbb{R}},\phi^-),\xi_\phi,b^- -1) = \bar{\Omega}_n(X,\phi)$$

fits into an exact singularity sequence with third terms

$$\Omega_i(Y(X,(\phi^+ \oplus \underset{\sim}{\mathbb{R}},\phi^-),\xi_\phi;2,1);\psi) \quad = \quad \Omega_i(X \times BO(2);\phi+\Gamma) ,$$

where

(9.2) $$\Gamma = \gamma_2 \otimes \xi_{\gamma_2} + \xi_{\gamma_2} - \gamma_2 .$$

(Formula 7.8 identifies the 1-dimensional cokerbundle at a singularity with the orientation bundle of $\underset{\sim}{\text{Ker}}$). This will often allow us to compute low-dimensional normal bordism groups, by induction over n.

Given a linear map $\ell : H^i(Y;\mathbb{Z}_2) \longrightarrow H^j(Z;\mathbb{Z}_2)$ between cohomology groups of finite $\mathbb{Z}_2$-dimensions, we define the <u>adjoint of ℓ</u> to be the unique linear map $\ell_* : H_j(Z;\mathbb{Z}_2) \longrightarrow H_i(Y;\mathbb{Z}_2)$ satisfying

$(\ell(y))(z) = y(\ell_*(z))$ for all $y \in H^i(Y;\mathbb{Z}_2)$, $z \in H_j(Z;\mathbb{Z}_2)$. Also, in the following, the symbol μ stands for Hurewicz homomorphisms from bordism to homology (sometimes composed with obvious forgetful maps).

<u>Theorem 9.3.</u> <u>Let X be pathconnected, paracompact and homotopy equivalent to a CW-complex with compact skeletons in all dimensions, and let $\phi = \phi^+ - \phi^-$ be a virtual vectorbundle over X. Then there is the following commuting diagram of horizontal and vertical exact sequences.</u>

$$\begin{array}{ccccccccc}
 & & & & & & 0 & & \\
 & & & & & & \downarrow & & \\
 & & & & & & \ker f_2' & & \\
 & & & & & & \downarrow & & \\
\Omega_3(X\times BO(2);\phi+\Gamma) & \xrightarrow{\delta_4} & \Omega_4(X;\phi) & \xrightarrow{f_4} & \overline{\Omega}_4(X;\phi) & \xrightarrow{\sigma\circ j_4} & \Omega_2(X\times BO(2);\phi+\Gamma) & \xrightarrow{\delta_3} & \\
 & & & & & & \downarrow f_2' & & \\
 & & & & & & (\ker : H_2(X;\mathbb{Z}_2) \xrightarrow{w_2(\phi)} \mathbb{Z}_2) \oplus \mathbb{Z}_2 \oplus \begin{cases} \mathbb{Z} & \text{if } w_1(\phi)=0 \\ \mathbb{Z}_2 & \text{if } w_1(\phi)\neq 0 \end{cases} & & \\
 & & & & & & \downarrow & & \\
 & & & & & & 0 & &
\end{array}$$

$$\begin{array}{ccccccccc}
 & & & & & & \mathbb{Z}_2 & & \\
 & & & *) & & & \downarrow \delta_1' & & *) \\
 & \xrightarrow{\delta_3} & \Omega_3(X;\phi) & \xrightarrow{f_3} & \overline{\Omega}_3(X;\phi) & \xrightarrow{\sigma\circ j_3} & \Omega_1(X\times BO(2);\phi+\Gamma) & \xrightarrow{\delta_2} & \\
 & & & & \downarrow \mu & & \downarrow \text{proj}_*\circ\mu = f_1' \quad (\cong \mathbb{Z} \text{ iff } w_2(\phi)\neq 0) & & \\
 & & & & H_3(X;\mathbb{Z}_2) & \xrightarrow[\text{of } w_2(\phi)\cdot]{\text{adjoint}} & H_1(X;\mathbb{Z}_2) & & \\
 & & & & & & \downarrow & & \\
 & & & & & & 0 & &
\end{array}$$

$$\xrightarrow{\delta_2} \Omega_2(X;\phi) \xrightarrow{f_2} \overline{\Omega}_2(X;\phi) \xrightarrow{\sigma\circ j_2} \mathbb{Z}_2 \xrightarrow{\delta_1} \Omega_1(X;\phi) \xrightarrow{f_1} \overline{\Omega}_1(X;\phi) \to 0.$$

$$[N,g,or] \longrightarrow g^*(w_2(\phi))[N]$$

<u>Moreover</u>, $$\Omega_o(X;\phi) \underset{\cong}{\xrightarrow{f_o}} \overline{\Omega}_o(X;\phi) \cong \begin{cases} \mathbb{Z} & \text{if } w_1(\phi) = 0 \\ \mathbb{Z}_2 & \text{if } w_1(\phi) \neq 0 . \end{cases}$$

---

*) Using classical obstruction theory, Wu relations and the fact that $\pi_2(SO) = 0$, C. Olk [**50**] has shown that every $[N,g,or] \in \overline{\Omega}_3(X;\phi)$ such that $g^*(w_2(\phi)) = 0$, lies in the image of f_3. In particular, if $w_2(\phi) = 0$, then δ_2 is injective.

Here $\delta_1(1)$ and $\delta_1'(1)$ can be represented by the unit circle with a constant map and the standard parallelization, suitably stabilized. Given $x = [S,h=(h_1,h_2),\bar{h}] \in \Omega_2(X\times BO(2);\phi+\Gamma)$, we have $f_2'(x) = (h_{1*}([S]);[S];W_2(x))$, where $W_2(x)$ denotes the (algebraic) number of zeroes of a non-degenerate section of $h_2^*(\gamma_2)$ over S. (This is a well-defined integer when $w_1(\phi) = 0$, since then $\bar{h}$ induces an isomorphism $\xi_S \cong h_2^*(\xi_{\gamma_2})$ of orientation bundles).

For all $[N,g,or] \in \bar{\Omega}_4(X,\phi)$ the three components of $f_2'\circ\sigma\circ j_4[N,g,or]$ are given by

$$f_2'\circ\sigma\circ j_4([N,g,or])_1 = (Sq^2 + w_2(\phi)\cdot)_*(g_*[N]) ;$$
$$f_2'\circ\sigma\circ j_4([N,g,or])_2 = (w_4(N)+g^*(w_1(\phi)^4+w_2(\phi)^2+w_1(\phi)w_3(\phi)+w_4(\phi)))[N];$$
$$f_2'\circ\sigma\circ j_4([N,g,or])_3 = \begin{cases} \pm(p_1(N) + g^*(p_1(\phi)))[N] & \text{if } w_1(\phi) = 0 \\ (w_2(N)^2 + g^*(w_1(\phi)^4 + w_2(\phi)^2))[N] & \text{if } w_1(\phi) \neq 0 . \end{cases}$$

(Here we have identified $\mathfrak{X}_2$ with $\mathbb{Z}_2$; if $w_1(\phi) = 0$, we use a fixed trivialization of ξ_ϕ to orient N).

Finally, if $w_1(\phi) \neq 0$, then $\ker f_2'$ is canonically isomorphic to the quotient of $H_1(X;\mathbb{Z}_2)$ by its subgoup $\{x\in H_1(X;\mathbb{Z}_2)\,|\,c(x) = 0$ for all $c\in H^1(X;\mathbb{Z}_2)$ such that $w_2(\phi)\cdot c = w_1(\phi)c^2\}$; thus in this case f_2' is an isomorphism iff $w_2(\phi)\cdot c + w_1(\phi)\cdot c^2 \neq 0$ for all $c \neq 0$ in $H_1(X;\mathbb{Z}_2)$, If $w_1(\phi) = 0$ and $w_2(\phi) \neq 0$, then $\ker f_2'$ is isomorphic to $H_1(X\times BO(2);\mathbb{Z}_2)/G$, where

$$G = \left\{ x\in H_1(X\times BO(2);\mathbb{Z}_2) \;\middle|\; \begin{matrix} c(x)=0 \text{ for all } c=(c_1,c_0)\in H^1(X\times BO(2);\mathbb{Z}_2)=H^1(X;\mathbb{Z}_2)\oplus\mathbb{Z}_2 \\ \text{satisfying } w_2(\phi)\cdot c_1+w_3(\phi)\cdot c_0 = 0 . \end{matrix} \right\}$$

If $w_1(\phi) = 0$ and $w_2(\phi) = 0$, then $\ker f_2' \cong H_1(X;\mathbb{Z}_2) \oplus \mathbb{Z}_4$. ∎

Given a low dimension n, this theorem allows sometimes to compute $\ker \sigma\circ j_n = \text{image } f_n$ and $\operatorname{coker} \sigma\circ j_{n+1} = \ker f_n$ and hence $\Omega_n(X;\phi)$ which is built up from these groups. Note that the homomorphisms ρ (cf. 7.12) are uninteresting in this context since they vanish here.

Observe also that for $n \leq 3$ the homomorphism f_n in the theorem has finite kernel and cokernel.

Before proving theorem 9.3 we collect a few facts which we will need both for the proof and for applications.

Fact 9.4 (see Conner-Floyd [13], chapter I and II, and Olk [50], p.0.12). The Hurewicz homomorphism μ from $\mathfrak{N}_*(X)$ to $H_*(X;\mathbb{Z}_2)$ is onto. Any dimension preserving right inverse of it determines a bijective $\mathfrak{N}_*$-homomorphism $\mathfrak{N}_* \otimes H_*(X;\mathbb{Z}_2) \cong \mathfrak{N}_*(X)$.

In particular, $\mu : \mathfrak{N}_1(X) \cong H_1(X;\mathbb{Z}_2)$.

Similarly, we have an isomorphism $\mu : \Omega_1(X) \cong H_1(X;\mathbb{Z})$ (since both groups can be represented as a quotient of $\pi_1(X)$, the homology group being even universal among the abelian homomorphic images of $\pi_1(X)$).

More generally, the Hurewicz homomorphism

$$\mu \ : \ \Omega_i(X) \longrightarrow H_i(X;\mathbb{Z})$$

is bijective for $i \leq 3$, and so is the obvious map

$$(\mu,\text{const}_*) \ : \ \Omega_4(X) \longrightarrow H_4(X;\mathbb{Z}) \oplus \Omega_4 \ .$$

Fact 9.5 (see Wu [73]). Given a vector bundle η over a paracompact space, the Steenrod operations act on the Stiefel-Whitney classes of η as follows.

$$Sq^r(w_s(\eta)) = \sum_{t=0}^{r} \binom{s-r+t-1}{t} w_{r-t}(\eta)\cdot w_{s+t}(\eta) \qquad (s \geq r) \ ,$$

where $\binom{a}{b} := 1$ for $b = 0$, $\binom{a}{b} := 0$ for $a < b$, $b \neq 0$. E.g. for all $s \geq 0$ we have

$$Sq^1(w_s(\eta)) = w_1(\eta)\cdot w_s(\eta) + (s+1)\cdot w_{s+1}(\eta)$$

and

$$Sq^2(w_s(\eta)) = w_2(\eta)w_s(\eta) + sw_1(\eta)\cdot w_{s+1}(\eta) + \frac{(s-1)(s-2)}{2}\, w_{s+2}(\eta)\ .$$

Fact 9.6 (see Wu [72]). Given a closed n-dimensional manifold N, the Steenrod operation

$$Sq^r \ :\ H^{n-r}(N;\mathbb{Z}_2) \longrightarrow H^n(N;\mathbb{Z}_2)$$

is just multiplication by the Wu class $U_r \in H^r(N;\mathbb{Z}_2)$ of N, where e.g. $U_1 = w_1(N)$, $U_2 = w_2(N)+w_1(N)^2$ and $U_3 = w_1(N)\cdot w_2(N)$.

Fact 9.7. Let N be a closed smooth manifold and $S \subset N$ a closed smooth submanifold. Let $[N]$, $[S]$ denote the corresponding fundamental classes, and let $\mathcal{D}(S)$ be the cohomology class on N derived from the Thom class of the normal bundle $\nu(S,N)$ in the obvious way. Then for all $c \in H^*(N)$ we have

$$(c|S)[S] \;=\; \pm\, c\cdot\mathcal{D}(S)[N]\ .$$

(Coefficients are in $\mathbb{Z}$ if both N and S are oriented, and in $\mathbb{Z}_2$ otherwise).

This follows, via excision, from the well-known local characterization of Thom classes and fundamental classes and from basic properties of the (external) cup product.

Fact 9.8. Given two vectorbundles λ and τ over a paracompact space, we can compute the Stiefel-Whitney class of the tensor product $\lambda \otimes \tau$ as if λ and τ were direct sums of line bundles (see the splitting principle in [27], 4.4.3.). E.g. if $\dim \lambda = 1$ and $\dim \tau = r$, we can write (formally)

$$w(\lambda) = (1+x)$$

$$w(\tau) = (1+y_1)(1+y_2)\cdot\ldots\cdot(1+y_r),$$

so that $w_s(\tau)$ is the s-th elementary symmetric polynomial in the one-dimensional classes $y_1,\ldots y_r$. Then

$$w(\lambda\otimes\tau) = (1+x+y_1)\cdot\ldots\cdot(1+x+y_r),$$

and hence for $s \geq 1$

$$(9.9) \qquad w_s(\lambda\otimes\tau) = \sum_{1\leq j_1<j_2<\ldots j_s\leq r}(x+y_{j_1})(x+y_{j_2})\ldots(x+y_{j_s})$$

$$= \sum x^s + x^{s-1}(y_{j_1}+\ldots+y_{j_s})+\ldots+x^{s-i}(y_{j_1}\cdot y_{j_2}\cdots y_{j_i}+\ldots+y_{j_{s-i+1}}\cdots y_{j_s})+\ldots+y_{j_1}\cdots y_{j_s}$$

$$= \binom{r}{s}w_1(\lambda)^s+\binom{r-1}{s-1}w_1(\lambda)^{s-1}w_1(\tau)+\ldots+\binom{r-i}{s-i}w_1(\lambda)^{s-i}w_i(\tau)+\ldots+w_s(\tau).$$

For later use we note the first Stiefel-Whitney classes of the virtual bundle $\lambda\otimes\tau-\tau$, that is, the homogeneous components of the quotient

$$\frac{w(\lambda\otimes\tau)}{w(\tau)} = w(\lambda\otimes\tau)\cdot(1+w_1(\tau)+(w_1^2(\tau)+w_2(\tau))+(w_1^3(\tau)+w_3(\tau))+(w_1^4(\tau)+w_1^2(\tau)w_2(\tau)+w_2^2(\tau)+w_4(\tau))+\ldots)$$

<u>Lemma 9.10.</u> <u>If λ and τ are vectorbundles of dimensions 1 and r resp. over a paracompact space, we have for $w_i = w_i(\lambda\otimes\tau-\tau)$ (where we put $x = w_1(\lambda)$):</u>

$$w_1 = rx$$
$$w_2 = \binom{r}{2}x^2+xw_1(\tau)$$
$$w_3 = \binom{r}{3}x^3+(r-1)x^2w_1(\tau)+xw_1(\tau)^2$$
$$w_4 = \binom{r}{4}x^4+\binom{r-1}{2}x^3w_1(\tau)+x^2(w_2(\tau)+(r-1)w_1(\tau)^2)+x(w_3(\tau)+w_1(\tau)w_2(\tau)+w_1(\tau)^3).$$

<u>Fact 9.11.</u> Let α, β be smooth vector bundles of dimension a, b over a closed manifold N, and let $S\subset N$ be the singularity of a non-degenerate (a-1)-morphism (resp. (b-1)-morphism) from α to β, where a-1 (resp. b-1) lies between 0 and min(a,b). Then any

characteristic number of S of the form

$$(w^r(S)\cdot w^s(\underset{\sim}{\text{Coker}})\cdot w^t(\underset{\sim}{\text{Ker}})\cdot y|S)\,[S] \quad ,$$

where r, s, t are multi-indices and $y \in H^*(N;\mathbb{Z}_2)$, can be expressed as a Whitney number of N by the following procedure. First eliminate $w^r(S)$, using the identity

$$(w(S)\cdot w(\underset{\sim}{\text{Ker}} \otimes \underset{\sim}{\text{Coker}}) = w(N)\mid S$$

derived from (1.6). Then use

$$w(\underset{\sim}{\text{Ker}}) = w(\alpha - \beta)\mid S\cdot w(\underset{\sim}{\text{Coker}})$$

(cf. (1.7) and (1.8)) to substitute the Stiefel-Whitney classes of $\underset{\sim}{\text{Coker}}$ (resp. of $\underset{\sim}{\text{Ker}}$) and to write our characteristic number as a sum of expressions of the form

$$(w_1\ (\underset{\sim}{\text{Ker}})^i\cdot(z|S))[S] = (w_{b-a+i+1}(\beta-\alpha)\cdot z)[N]$$

$$(\text{resp. } (w_1(\underset{\sim}{\text{Coker}})^i\ (z|S))[S] = (w_{a-b+i+1}(\alpha-\beta)\cdot z)[N])$$

where $i \geq 0$ and $z \in H^*(N;\mathbb{Z}_2)$.

To obtain the last identity e.g. in the first case (i.e. when $\underset{\sim}{\text{Ker}}$ is one-dimensional), apply fact 9.7 and proposition 5.3. twice. Observe that the zero set S' of a nondegenerate 0-morphism from $\underset{\sim}{\text{Ker}}$ to $S\times\mathbb{R}^i$ has dual class $w_1(\underset{\sim}{\text{Ker}})^i$ in S; but S' is also the singularity of a non-degenerate (a-1)-morphism from α to $\beta \oplus \underset{\sim}{\mathbb{R}}^i$ over N, and hence has dual class $w_{b-a+i+1}(\beta-\alpha)$ in N. ∎

<u>Proof of theorem 9.3.</u> It follows from fact 9.8 that $w(\gamma_2\otimes\xi_{\gamma_2}) = w(\gamma_2)$ and hence

$$(9.12) \qquad w(\Gamma) = 1 + w_1(\gamma_2)\ .$$

In particular, $w_1(\phi+\Gamma) \neq 0$.

The horizontal sequence in the statement of our theorem is essentially a piece of the singularity sequence for the set of data $(X,(\phi^+\oplus \mathbb{R},\phi^-),\ \xi_\phi,\ b^- -1)$. It can be extended to contain also the isomorphism $\Omega_o(X,\phi) \cong \overline{\Omega}_o(X,\phi)$. It is an easy exercise to identify these groups with $\mathbb{Z}$ or $\mathbb{Z}_2$. (Note: $w_1(\phi) = 0$ iff every pullback of ξ_ϕ to the unit circle is trivial, iff ξ_ϕ is trivial over X). Similarly $\Omega_o(X\times BO(2);\phi+\Gamma) = \mathbb{Z}_2$.

Next consider an element $[N,g,\overline{g},or]$ of $\Omega_n(X,(\phi^+\oplus\mathbb{R},\phi^-),\xi_\phi,b^- -1)$, and assume $\overline{g}$ is nondegenerate and gives rise to singularity data $(S,h=(h_1,h_2) : S \longrightarrow X\times BO(2),\overline{h})$. Then, given $x\in H^{n-2}(X;\mathbb{Z}_2)$, we have, by facts 9.7 and 9.11,

$$\begin{aligned} x\,(h_{1*}[S]) &= (g^*(x)|S)[S] \\ &= g^*(x)\cdot w_2(TN + g^*(\phi))[N] \\ &= (w_2(N) + w_1(N)\cdot g^*(w_1(\phi)) + g^*(w_2(\phi)))(g^*(x))[N] \\ &= (Sq^2(g^*(x)) + g^*(w_2(\phi))\cdot g^*(x))[N] \\ &= ((Sq^2 + w_2(\phi)\cdot)(x))(g_*[N]) \quad , \end{aligned}$$

since $g^*(w_1(\phi)) = w_1(N)$ by the existence of $or : \xi_N \cong g^*(\xi_\phi)$, and $w_2(N) + w_1(N)^2$ is the second Wu class of N (see fact 9.6). It follows that the diagram

$$\begin{array}{ccc} \Omega_n(X,(\phi^+\oplus\mathbb{R},\phi^-),\xi_\phi,b^- -1) & \xrightarrow{\ \sigma\circ j\ } & \Omega_{n-2}(X\times BO(2);\phi+\Gamma) \\ \big\downarrow \mu & & \big\downarrow proj_*\circ\mu \\ H_n(X;\mathbb{Z}_2) & \xrightarrow[\ Sq^2+w_2(\phi)\cdot\ ]{\text{adjoint of}} & H_{n-2}(X;\mathbb{Z}_2) \end{array}$$

is commutative. This implies our claim on how $\sigma\circ j_2$, $f_1'\circ\sigma\circ j_3$ and the first component of $f_2'\circ\sigma\circ j_4$ look like.

Similarly, given $[N,g,or] \in \overline{\Omega}_4(X,\phi)$, let a nondegenerate morphism $\overline{g} : TN \oplus g^*(\phi^+\oplus\mathbb{R}) \longrightarrow g^*(\phi^-)$ have a surface S as its singular set. Then $w_1(N) = g^*(w_1(\phi))$ and $w_1(\mathbf{Ker}) = w_1(\mathbf{Coker})$. Applying fact 9.11, we get

$$w_1(S)^2[S] = (g^*(w_1(\phi))^2|S)[S] + w_1(\underset{\sim}{\text{Coker}})^2[S]$$
$$= (w_2(N) + w_1(N)^2 + g^*(w_2(\phi)))g^*(w_1(\phi)^2[N] + w_4(TN+g^*(\phi))[N].$$

Now use the facts 9.5 and 9.6; note e.g. that

$$w_1(N)^2\cdot w_2(N) = Sq^1(w_1(N)w_2(N)) = 2\cdot w_1(N)^2\cdot w_2(N) + w_1(N)w_3(N)$$

and

$$w_2(N)\cdot g^*(w_1(\phi)^2+w_2(\phi)) = (Sq^2 + g^*(w_1(\phi))^2\cdot)\ g^*(w_1(\phi)^2 + w_2(\phi))$$
$$= g^*(w_2(\phi)^2 + w_1(\phi)^2\cdot w_2(\phi)) .$$

It follows easily that

$$w_1(S)^2[S] = w_4(N)[N] + g^*(w_1(\phi)^4 + w_2(\phi)^2 + w_1(\phi)w_3(\phi) + w_4(\phi))[N].$$

Moreover, a nondegenerate 0-morphism from $\underset{\sim}{\text{Ker}}$ to $S\times\mathbb{R}$ gives rise to a non-degenerate $(b^- -1)$-morphism $(\bar{g},\tilde{g}) : TN \oplus g^*(\phi^+ \oplus \underset{\sim}{\mathbb{R}}) \longrightarrow g^*(\phi^-) \oplus \underset{\sim}{\mathbb{R}}$ over all of N with the same singularity. Therefore proposition 5.3 shows that $W_2(\sigma\circ j_4[N,g,or])$ is $\pm p_1(TN+\phi)[N]$ or $w_2(TN+\phi)^2[N]$ according to whether $w_1(\phi) = 0$ or not. This completes our calculation of $f_2^!\circ\sigma\circ j_4$.

In order to obtain the vertical exact sequences in the statement of theorem 9.3, we apply the preceding discussion to $X\times BO(2)$ (or rather, to the product of X with a suitable (compact) skeleton; this will again be paracompact and of the right homotopy type). After obvious identifications, we obtain the following commuting diagram with an exact horizontal sequence

$$\Omega_2(X\times BO(2);\phi+\Gamma) \xrightarrow{f_2'} \overline{\Omega}_2(X\times BO(2);\phi+\Gamma) \xrightarrow{\sigma\circ j_2'} \mathbb{Z}_2 \xrightarrow{\delta_1'} \Omega_1(X\times BO(2);\phi+\Gamma) \xrightarrow{f_1'} \overline{\Omega}_1(X\times BO(2);\phi+\Gamma) \to 0$$

$$(\mathrm{proj}_*, W_2)\colon \overline{\Omega}_2(X\times BO(2);\phi+\Gamma) \xrightarrow{\ \cong\ } (H_2(X;\mathbb{Z}_2)\oplus \mathfrak{N}_2)\oplus \begin{cases} \mathbb{Z} & \text{if } w_1(\phi)=0 \\ \mathbb{Z}_2 & \text{if } w_1(\phi)\neq 0 \end{cases} \xrightarrow{w_2(\phi)\circ \mathrm{proj}_1} \mathbb{Z}_2$$

$$\overline{\Omega}_1(X\times BO(2);\phi+\Gamma) \cong \mathfrak{N}_1(X) \overset{\mu}{\cong} H_1(X;\mathbb{Z}_2)$$

Here the vertical arrows are defined either in the obvious way or as in the statement of 9.3; they are all isomorphisms. Consider e.g. the left hand arrow. Given $[S,h_1] \in \mathfrak{N}_2(X) \cong H_2(X;\mathbb{Z}_2) \oplus \mathfrak{N}_2$ (cf. fact 9.4), pick a classifying map h_2 of the planebundle $(\xi_S \otimes h_1^*(\xi_\phi)) \oplus \underset{\sim}{\mathbb{R}}$ over S and an obvious isomorphism $or : \xi_S \cong h_1^*(\xi_\phi) \otimes h_2^*(\xi_{\gamma_2})$; then $(\mathrm{proj}_*,W_2)[S,(h,h_2),or] = ([S,h_1],0)$. Also note that the complex projective line $\mathbb{C}P_1$, together with a constant map into X and a classifying map of the canonical complex line bundle, gives rise to a bordism class which gets mapped to $(0,\pm 1)$ under (proj_*,W_2). Thus this homomorphism is onto. To see that it is also injective, let $[S,(h_1,h_2),or]$ be in its kernel. By lemma 7.16 we may assume that S is connected. Hence we may push all the zeroes of a nondegenerate section of $h_2^*(\gamma_2)$ into a small ball in S and cancel them (possibly after first pushing some zeroes around closed arcs in S along which $h_1^*(\xi_\phi)$ is non-trivial; here we may have to attach a handle). Therefore, the bundle $h_2^*(\gamma_2)$ is isomorphic to $(\xi_S \otimes h_1^*(\xi_\phi)) \oplus \underset{\sim}{\mathbb{R}}$ and can easily be extended over a zero bordism of (S,h_1), and so can h_2 and or. Thus $[S,(h_1,h_2),or] = 0$.

Moreover, for any $[S,h,or] \in \overline{\Omega}_2(X\times BO(2);\phi+\Gamma)$ we have $w_1(S) = h_1^*(w_1(\phi)) + h_2^*(w_1(\gamma_2))$ (see 9.12), and hence

$$\begin{aligned} \sigma\circ j_2'\,[S,h,or] &= (h_1^*(w_2(\phi)) + h_1^*(w_1(\phi))^2 + w_1(s)h_1^*(w_1(\phi)))[S] \\ &= h_1^*(w_2(\phi))[S] \\ &= w_2(\phi)(h_{1*}[S]) \end{aligned}$$

(see also fact 9.6). This gives the desired commutativity of the triangle in the diagram above. It follows that $w_2(\phi) \neq 0$ if and only if $\sigma\circ j_2'$ is onto, or, equivalently, if and only if f_1' is an isomorphism. Also, we can identify the domain and range of $(proj_*, W_2)$, and then the image of f_2' is as claimed in the statement of our theorem.

In order to obtain our description of δ_1 and δ_1', consider the relative bordism class given by a constant map on the compact disk D^2, together with a suitable stabilization of the non-degenerate 0-morphism $TD^2 = D^2 \times \mathbb{R}^2 \longrightarrow \mathbb{R}$ which takes (x,v) to the inner product of x with $r(v)$ (r = rotation by 270^0 degrees). Since the singularity consists of a single point, the bordism class of singularity data corresponds to $1 \in \mathbb{Z}_2$. On the other hand, restriction to the boundary circle gives $\delta_1(1)$ and $\delta_1'(1)$ as described in the statement of the theorem.

Finally, we study the kernel of f_2'. Let $\tilde{\mathfrak{N}}_3(X\times BSO(3))$ denote the bordism group of tripels (N, g, i) where N is a closed 3-manifold, $g = (g_1, g_2) : N \longrightarrow X \times BSO(3)$ is continuous, and $i : \xi_N \otimes g_1^*(\xi_\phi) \hookrightarrow g_2^*(\gamma_3)$ is a vector bundle monomorphism. Associated to such a tripel, we have an element $[N, h = (h_1, h_2), or] \in \overline{\Omega}_3(X\times BO(2);\phi+\Gamma)$ obtained as follows: $h_1 = g_1$; h_2 classifies the complement of the image of i; its orientation bundle is canonically isomorphic to $\xi_N \otimes g_1^*(\xi_\phi)$ (since $g_2^*(\gamma_3)$ is oriented), and this leads to or.

In this way we obtain the isomorphism in the top line of the following diagram.

$$\begin{array}{ccccc} \tilde{\mathfrak{N}}_3(X\times BSO(3)) & \xrightarrow[\cong]{is} & \overline{\Omega}_3(X\times BO(2);\phi+\Gamma) & \xrightarrow{\sigma\circ j_3'} & \Omega_1(X\times BO(2)^2;\phi+\Gamma+\Gamma) \\ \downarrow{\scriptstyle \tilde{f}} & & \downarrow{\scriptstyle \mu\circ\tilde{f}\cdot is^{-1}} & & \downarrow{\scriptstyle proj_{1*}\circ\mu} \\ \mathfrak{N}_3(X\times BSO(3)) & \xrightarrow{\mu} & H_3(X\times BSO(3);\mathbb{Z}_2) & \overset{\ell^*}{\dashrightarrow} & H_1(X\times BO(2);\mathbb{Z}_2) \end{array} \tag{9.13}$$

Here $\tilde{f}$ forgets about the monomorphism i. Now, given $[N,g] \in \mathfrak{N}_3(X\times BSO(3))$, we may assume that N is connected and, if $w_1(\phi) \neq 0$, that also $g_1^*(w_1(\phi)) \neq 0$ (just attach suitable handles). According to theorem 3.7 there exists a monomorphism $i : \xi_N \otimes g_1^*(\xi_\phi) \hookrightarrow g_2^*(\gamma_3)$ if and only if the obstruction

$$\omega_1(\xi_N \otimes g_1^*(\xi_\phi), g_2^*(\gamma_3)) \in \Omega_o(N; \xi_N \otimes g_1^*(\xi_\phi) \otimes g_2^*(\gamma_3) - TN) \cong \begin{cases} \mathbb{Z} & \text{if } w_1(\phi) = 0 \\ \mathbb{Z}_2 & \text{if } w_1(\phi) \neq 0 \end{cases}$$

vanishes. But it is not hard to show that this obstruction gives rise to a welldefined homomorphism on $\mathfrak{N}_3(X\times BSO(3))$. If $w_1(\phi) = 0$, this homomorphism is trivial (the domain consists only of torsion!), and hence $\tilde{f}$ is onto. If $w_1(\phi) \neq 0$, the obstruction above is equal to the following, by facts 9.11, 9.6 and 9.5. (Here we do not indicate pullbacks in our notations):

$$\begin{aligned} w_3(\gamma_3 - \xi_N \otimes \xi_\phi)[N] &= (w_3(\gamma_3) + (w_1(N) + w_1(\phi))w_2(\gamma_3) + (w_1(N) + w_1(\phi))^3)[N] \\ &= (\cancel{w_3(\gamma_3) + Sq^1(w_2(\gamma_3))} + w_1(\phi)w_2(\gamma_3) + \cancel{Sq^1(w_1(N)^2 + w_1(\phi)^2)} + w_1(N)^2 w_1(\phi) + w_1(\phi)^3)[N] \\ &= w_1(\phi)(w_2(\gamma_3) + w_1(N)^2 + w_1(\phi)^2)[N] \; . \end{aligned} \tag{9.14}$$

As an example consider $z = [S^1 \times P^2, e \times (\text{constant})] \in \mathfrak{N}_3(X\times BSO(3))$ with $e^*(w_1(\phi)) \neq 0$. Here our obstruction equals

$$(e^*(w_1(\phi)) \times w_1(P^2)^2)\,[S^1 \times P^2] = 1 \; ;$$

on the other hand, $\mu(z) = 0$. Now recall that μ is onto (cf. 9.4). Thus for any $x \in H_3(X\times BSO(3);\mathbb{Z}_2)$ there is a $y \in \mathfrak{N}_3(X\times BSO(3))$ such that $\mu(y) = x$. Adding a suitable multiple of z, we may actually assume that our obstruction vanishes on y, and hence that y is in the image of $\tilde{f}$. It follows that $\mu \circ \tilde{f} \circ is^{-1}$ is always onto.

Next we apply previous results of this proof to $(X\times BO(2), \phi + \Gamma)$ instead of (X,ϕ). Thus the composed homomorphism $(proj_{1*} \circ \mu) \circ (\sigma \circ j_3')$ in our diagram (9.13) coincides with the composite

$$\overline{\Omega}_3(X\times BO(2);\phi+\Gamma) \xrightarrow{\mu} H_3(X\times BO(2);\mathbb{Z}_2) \xrightarrow[(w_2(\phi)+w_1(\phi)w_1(\gamma_2))\cdot]{\text{adjoint of}} H_1(X\times BO(2);\mathbb{Z}_2)$$

Given $u = [N,g,i] \in \tilde{\mathcal{N}}_3(X\times BSO(3))$ as above, then for all $c \in H^1(X\times BO(2);\mathbb{Z}_2)$ we have

$$\begin{aligned} & c((\mathrm{proj}_{1*}\cdot\mu\circ\sigma\circ j_3')\circ \mathrm{is}\ (u)) \\ (9.15)\qquad &= (c\cdot(w_2(\phi) + w_1(\phi)w_1(\gamma_2)))\ (\mu[N,\ h,\ \mathrm{or}]) \\ &= h^*\ (c\cdot w_2(\phi) + c\cdot w_1(\phi)w_1(\gamma_2))[N] \end{aligned}$$

If $c \in H^1(X;\mathbb{Z}_2)$, this expression becomes

$$\begin{aligned} &(g^*(c\cdot w_2(\phi)) + g^*(c\cdot w_1(\phi))\cdot(w_1(N) + g^*(w_1(\phi)))[N] \\ &= g^*(c\cdot w_2(\phi) \quad + c^2 w_1(\phi))[N] \end{aligned}$$

since $h^*(w_1(\gamma_2)) = w_1(\mathrm{Im}\ i^\perp) = w_1(\xi_N \otimes g_1^*(\xi_\phi)) = w_1(N) + g^*(w_1(\phi))$ and $w_1(N)\cdot v = Sq^1(v)$ for all $v \in H^2(N;\mathbb{Z}_2)$.

If $c = w_1(\gamma_2)$, the expression (9.15) equals

$$\begin{aligned} &((w_1(N) + g^*(w_1(\phi)))g^*(w_2(\phi))) + (w_1(N)^2 + g^*(w_1(\phi))^2)g^*(w_1(\phi)))[N] \\ &= g^*(w_3(\phi) + w_1(\phi)w_2(\gamma_3))[N] \end{aligned}$$

since $Sq^1 w_2 = w_3 + w_1 w_2$ and since, by 9.14, $g^*(w_1(\phi))(g^*(w_2(\gamma_3)) + w_1(N)^2 + g^*(w_1(\phi)^2))[N] = 0$.

Thus the diagram 9.13 becomes commutative if we define

$$\ell\ :\ H^1(X\times BO(2);\mathbb{Z}_2) \longrightarrow H^3(X\times BSO(3);\mathbb{Z}_2)$$

by

$$\ell(c) = (w_2(\phi)\cdot + (w_1(\phi)\cdot)\circ Sq^1)(c), \qquad c \in H^1(X;\mathbb{Z}_2),$$

and $\ell(w_1(\gamma_2)) = w_3(\phi) + w_1(\phi)\cdot w_2(\gamma_3)$.

If $w_1(\phi) \neq 0$, then $\ell(w_1(\gamma_2)) \notin H^3(X;\mathbb{Z}_2) \supset \ell(H^1(X;\mathbb{Z}_2))$, and hence $H_1(BO(2);\mathbb{Z}_2)$ lies in the image of ℓ^*.

Now recall from the exact sequence of $(X\times BO, \phi+\Gamma)$ that

Ker f_2' = Im δ_2' = Coker $\sigma\circ j_3'$. If $w_1(\phi) \neq 0$ or $w_2(\phi) \neq 0$, $\mathrm{proj}_{1*}\circ\mu$ is bijective and hence induces an isomorphism from Coker $\sigma\circ j_3'$ onto Coker ℓ^*. Most of the last statement of the theorem follows.

It remains to consider ker f_2' when $w_1(\phi) = 0$ and $w_2(\phi) = 0$. Then also $w_2(\phi+\Gamma) = 0$, and according to an observation of C. Olk (see the footnote of our theorem), we conclude that δ_2' is isomorphic in the following diagram with exact column

$$\begin{array}{ccc}
 & 0 & \\
 & \downarrow & \\
 & \mathbb{Z}_2 & \\
 & \downarrow{\scriptstyle \delta_1''} & \\
A(X) := & \Omega_1(X\times BO(2)\times BO(2);\pi_1^*(\phi)+\pi_2^*(\Gamma)+\pi_3^*(\Gamma)) & \xrightarrow[\cong]{\delta_2'} \ker f_2' \\
 & \downarrow{\scriptstyle f_1''} & \\
 & H_1(X\times BO(2);\mathbb{Z}_2) & \\
 & \downarrow & \\
 & 0 &
\end{array}$$

Denote the middle group by $A(X)$ for short. Clearly we have the exact sequence

$$0 \longrightarrow A(\text{point}) \longrightarrow A(X) \xrightarrow{\pi_{1*}\circ f_1''} H_1(X;\mathbb{Z}_2) \longrightarrow 0 ,$$

and $A(\text{point})$ is a group of order 4, generated by $\delta_1''(1)$ and $[S^1;\text{Möbius}\oplus\mathbb{R}, \text{Möbius}\oplus\mathbb{R};$ suitable isomorphism$]$. These elements come already from the generators

$$x = [S^1;\ \text{constant};\ \text{stabilization of } TS^1 \cong \mathbb{R}]$$

and

$$y = [S^1;\ \text{Möbius band } \lambda;\ TS^1 \oplus (\lambda\oplus\lambda) \cong \mathbb{R} \oplus \mathbb{R}^2]$$

of the group $\Omega_1(P^\infty;2\lambda)$ under the obvious homomorphism induced by the map

$$P^\infty = BO(1) \longrightarrow BO(2) \xrightarrow{\text{diag}} \text{point} \times BO(2) \times BO(2) .$$

A simple-minded bordism involving the punctured annulus shows that $2y = x$. Hence $\Omega_1(P^\infty;2\lambda)$ and $A(\text{point})$ must be isomorphic to $\mathbb{Z}_4$, and $\ker f_2' \cong A(X) \cong H_1(X;\mathbb{Z}_2) \oplus \mathbb{Z}_4$. This completes the proof of theorem 9.3. ∎

Example 9.16 (compare with exercise 7.19). If X is a point and ϕ is trivial, then $\Omega_n(X;\phi)$ is canonically isomorphic to the stable homotopy group π_n^S. From theorem 9.3 we obtain the wellknown isomorphisms

$$\mathbb{Z}_2 \xrightarrow[\cong]{\delta_1} \Omega_1(X,\phi) \cong \pi_1^S$$

and

$$\mathbb{Z}_2 \xrightarrow[\cong]{\delta_2 \cdot \delta_1'} \Omega_2(X;\phi) \cong \pi_2^S \quad .$$

It is clear from the construction of these isomorphisms that the generator of $\Omega_n(X;\phi)$, $n = 1,2$, can be represented by the invariantly framed n-torus $(S^1)^n$.

Exercise 9.17 (see also § 11). Use theorem 9.3 to show $\pi_3^S \cong \mathbb{Z}_{24}$. How does the factor 3 in the order of this group enter the picture? (Warning: the complete calculation of π_3^S is more intricate than the computations in the last example).

When using theorem 9.3 we often need more information on the auxiliary groups $\overline{\Omega}_n(X;\phi)$. For many future applications, the following result will suffice.

Proposition 9.18. Let Y be a pathconnected, locally pathconnected and paracompact space, $q \geq 1$ a natural number and ϕ a virtual bundle over $P^{q-1} \times Y$ such that the orientation bundle ξ_ϕ has the form $\gamma \otimes \eta$ where γ is the (pullback of the) canonical line bundle over the projective space P^{q-1} and η is any line bundle over Y. Then we have the exact sequence

$$0 \longrightarrow \ker \pi_{2*} \longrightarrow \overline{\Omega}_{q-1}(P^{q-1} \times Y;\ \phi) \xrightarrow{\pi_{2*}} \mathfrak{N}_{q-1}(Y) \longrightarrow 0$$

with

$$\ker \pi_{2*} \cong \begin{cases} \mathbb{Z} & \text{if } q \equiv 1(2) \text{ and } \eta \text{ is trivial}, \\ 0 & \text{otherwise}. \end{cases}$$

Moreover, $\ker \pi_{2*}$ is generated by $2 \cdot [P^{q-1}, (\mathrm{id}, \mathrm{constant}), \mathrm{or}] \in \overline{\Omega}_{q-1}(P^{q-1} \times Y; \phi)$. Thus $\overline{\Omega}_{q-1}(P^{q-1} \times Y; \phi)$ is isomorphic to the direct sum of $\mathfrak{N}_{q-1}(Y)/\mathbb{Z}_2$ and $\mathbb{Z}$ or $\mathbb{Z}_2$ respectively.

Proof. Due to the special form of ξ_ϕ, we can identify $\overline{\Omega}_{q-1}(P^{q-1} \times Y; \phi)$ with the bordism group of tripels $(N^{q-1},\ g : N \longrightarrow Y,\ \overline{g} : \xi_N \otimes g^*(\eta) \hookrightarrow N \times \mathbb{R}^q)$. Then the homomorphism π_{2*} just forgets about $\overline{g}$ and hence fits into an exact sequence very similar to (7.2):

$$(9.19) \qquad \begin{array}{l} \cdots \to \mathfrak{N}_q(Y) \xrightarrow{j} \mathfrak{N}_q^{rel}(Y) \xrightarrow{\partial} \overline{\Omega}_{q-1}(P^{q-1} \times Y; \phi) \xrightarrow{\pi_{2*}} \mathfrak{N}_{q-1}(Y) \longrightarrow 0 \\ \qquad\qquad\qquad \searrow^{\sigma} \begin{cases} \mathbb{Z} & \text{if } q \equiv 1(2) \text{ and } w_1(\eta) = 0, \\ \mathbb{Z}_2 & \text{otherwise}. \end{cases} \end{array}$$

Here $\mathfrak{N}_q^{rel}(Y)$ denotes the (relative) bordism group of compact q-manifolds M with a continuous map $g : M \to Y$ and a monomorphism $\overline{g}^\partial : \xi_M \otimes g^*(\eta)|\partial M \hookrightarrow \partial M \times \mathbb{R}^q$ over the boundary. $\sigma[M, g, \overline{g}^\partial]$ is the number of singular points of any non-degenerate 0-morphism $\overline{g}$ which extends $\overline{g}^\partial$ over all of M. If q is odd and η is trivial, one can count such a singular point with multiplicity $+1$ or -1 according to whether the tangent map of $s_{\overline{g}}$ (cf. 1.4) is "orientation preserving" there or not; hence one obtains an integer invariant in this case. It is not hard to see that σ is always an isomorphism. E.g., in order to see injectivity, attach suitable handles to "seal off" the singular points, while keeping track of the orientation behavior of M and $g^*(\mu)$.

When $\sigma \circ j$ is integer-valued, this composite homomorphism must vanish since its domain is a torsion group. When it takes values in $\mathbb{Z}_2$, we have

$$\sigma \circ j\,[N, g] = (w_1(N) + g^*(w_1(\eta)))^q[N]$$

(see fact 9.11), and hence $\sigma \circ j$ is nontrivial: just evaluate it on $[P^q,\text{constant}] \in \mathfrak{N}_q(Y)$ if q is even, or else on the class of $P^{q-1} \times S^1 \longrightarrow S^1 \xrightarrow{\ell} Y$ *) where ℓ is a loop such that $\ell^*(\eta)$ is not trivial. This proves the first statement in 9.18 concerning $\ker \pi_{2*}$.

Next assume that q is odd. Let v be a vectorfield on P^{q-1} with just one (non-degenerate) zero. Then the vectorfield $\tilde{v}$ on $M = P^{q-1} \times [-1,1]$, defined by

$$\tilde{v}(x, t) = (v(x), t) \in T_{(x,t)}(P^{q-1} \times [-1, 1]) = T_x(P^{q-1}) \times \mathbb{R} ,$$

also has precisely one zero. Now we can identify $TM = \pi_1^*(TP^{q-1}) \oplus \underset{\sim}{\mathbb{R}} \cong \pi_1^*(\gamma^q)$ with $\underset{\sim}{\mathrm{Hom}}(\xi_M, \underset{\sim}{\mathbb{R}}^q)$. Hence $\alpha = [(P^{q-1} \times [-1,1], \text{constant map}, \tilde{v}|P^{q-1} \times \{-1,1\})]$ can be interpreted as a generator of $\mathfrak{N}_q^{rel}(Y)$. Therefore $\partial(\alpha) = 2\,[P^{q-1},\text{id},\text{constant},\text{or}]$ generates $\operatorname{Im} \partial = \ker \pi_{2*}$. Our proposition follows. ∎

Finally we present one more tool to get information on the groups $\Omega_n(X;\phi)$ and $\overline{\Omega}_n(X;\phi)$, namely the <u>Gysin sequence.</u>

<u>Proposition 9.20</u> <u>(see Salomonsen [57], 5.3). Let $S(\eta)$ be the spherebundle of a q-plane bundle η over a space X. Then for any virtual bundle ϕ over X there is a long exact sequence</u>

$$\ldots \to \Omega_n(S(\eta);\phi) \xrightarrow{\text{proj}_*} \Omega_n(X;\phi) \xrightarrow{\Delta} \Omega_{n-q}(X;\eta+\phi) \xrightarrow{d} \Omega_{n-1}(S(\eta);\phi)$$

<u>(Here $\Delta[M,g,\overline{g}]$ is represented by the zero set Z of a non-degenerate</u>

*) This example (and its usefulness here) were brought to my attention by Christof Olk.

section of $g^*(\eta)$ together with $g|Z$ and the isomorphism

$$TZ \oplus g^*(\eta)|Z \oplus g^*(\phi^+)|Z \cong TM|Z \oplus g^*(\phi^+)|Z \xrightarrow{\bar{g}|} g^*(\phi^-)|Z .$$

$d[N,h,\bar{h}]$ is represented by the map $S(h^*(\eta)) \longrightarrow S(\eta)$ together with an obvious isomorphism).

Exercise 9.21. Let ϕ be a virtual bundle over a paracompact space X.

(i) Identify $\bar{\Omega}_n(X,\phi)$ with $\Omega_n(X\times BSO(m);\xi_\phi-\gamma_m)$ for large m.

(ii) Show that the Gysin sequence of the line bundle $\pi_1^*(\xi_\phi)$ over $X\times BSO(m)$ takes the following form

$$\ldots \to \Omega_n(\tilde{X}) \xrightarrow{proj_*} \bar{\Omega}_n(X,\phi) \xrightarrow{\Delta} \Omega_{n-1}(X) \xrightarrow{d} \Omega_{n-1}(\tilde{X}) \longrightarrow \ldots$$

where $\tilde{X} = S(\xi_\phi)$ and d corresponds to taking double covers.

(iii) Compare this sequence with a similar sequence in integer homology and obtain in particular a canonical isomorphism

$$\mu : \bar{\Omega}_1(X;\xi) \xrightarrow{\cong} H_1(X;\tilde{\mathbb{Z}})$$

(where the twisted coefficients $\tilde{\mathbb{Z}}$ are associated to ξ_ϕ).

(iv) Identify $\bar{\Omega}_n(X,\phi)$ with $\Omega_n(S(\xi_\phi\otimes\xi_\gamma);-\gamma_m)$, where $S(\xi_\phi\otimes\xi_\gamma)$ is the indicated double cover of $X\times BO(m)$, m large. Obtain a long exact Gysin sequence

$$\ldots \to \bar{\Omega}_n(X;\phi) \xrightarrow{f} \mathfrak{N}_n(X) \xrightarrow{\Delta} \bar{\Omega}_{n-1}(X\times BO(1);\phi) \xrightarrow{d} \bar{\Omega}_{n-1}(X;\phi) \longrightarrow \ldots$$

and compare it to the homology sequence induced by $0 \longrightarrow \tilde{\mathbb{Z}} \xrightarrow{\cdot 2} \tilde{\mathbb{Z}} \longrightarrow \mathbb{Z}_2 \longrightarrow 0$.

Exercise 9.22. Show that

$$\Omega_n(X;\phi) = \bigoplus_{j\in J} \Omega_n(X_j;\phi) ,$$

where $(X_j)_{j\in J}$ are the pathconnected components of X. (Thus the connectedness assumption on X in this section is not really restrictive).

Exercise 9.23. Recall that we can identify $\Omega_n(X;\phi)$ also with $\Omega_n(X,(\phi^+,\phi^-\oplus\mathbb{R}),\xi_\phi,b^-)$. Would this alternate approach change the final statement of theorem 9.3?

§10. Bordism of immersions.

In this section we study the bordism groups

$$\mathfrak{I}\mathfrak{N}_n(m) := \mathfrak{N}_n(m,n) \qquad (m>n)$$

of (unoriented) closed n-manifolds smoothly immersed into $\mathbb{R}^m$.

Our main tool will be the exact sequence (7.23). First note that we have isomorphisms

$$\mathfrak{N}_n(m,n-1) \xrightarrow{\cong} \mathfrak{N}_n(m,n-2) \xrightarrow{\cong} \cdots \xrightarrow{\cong} \mathfrak{N}_n(m,-1) \cong \mathfrak{N}_n$$

when $2m>3(n-1)$; indeed, in this dimension range each of the partial arrows is sandwiched between negative-dimensional (and hence vanishing) normal bordism groups in the singularity sequence (7.23). Thus, equating n and k in (7.23), we obtain the following exact sequence, provided $m>\frac{3}{2}n-1$ and $m>n$.

(10.1)

$$\mathfrak{N}_{n+1} \xrightarrow{\sigma\circ j} \Omega_{2n-m}(BO(1)\times BO(q);\phi_{1,q}) \underset{\rho}{\overset{\delta}{\rightleftarrows}} \mathfrak{I}\mathfrak{N}_n(m) \xrightarrow{f} \mathfrak{N}_n \xrightarrow{\sigma\circ j}$$

$$\xrightarrow{\sigma\circ j} \Omega_{2n-m-1}(BO(1)\times BO(q);\phi_{1,q}) \longrightarrow \cdots$$

(with $\mathfrak{I}\mathfrak{N}_{n+1}(m-1) \xrightarrow{f}$ preceding $\mathfrak{N}_{n+1}$)

Here

(10.2) $$\phi_{1,q} = \gamma_1 \otimes \gamma_q + \gamma_q - \gamma_1 ,$$

$q=m-n+1$, and we have $\delta\circ\rho=2\cdot\mathrm{Id}$ on $\mathfrak{I}\mathfrak{N}_n(m)$.

It will be of prime importance to determine the image of the obvious forgetful homomorphism f. Fortunately, in some cases this problem has already been solved.

<u>Theorem 10.3 (see R. Brown [11]). Let $\alpha(n)$ be the number of ones in the binary expansion of n. If $\alpha(n)\geq 2n-m$, then</u>

$$f : \mathfrak{I}\mathfrak{N}_n(m) \longrightarrow \mathfrak{N}_n$$

<u>is onto.</u>

If $\alpha(n) > 2n-m$, then f has even a right inverse. Indeed, under this stronger assumption the composite of f with the forgetful map $f': \mathfrak{J}\mathfrak{N}_n(m-1) \longrightarrow \mathfrak{J}\mathfrak{N}_n(m)$ is already onto; on the other hand, the image of f' consists of elements of order 2 since the extra (m-th) dimension can be used to deform an immersion into its (additive) inverse. Hence, given a basis $z_1,\ldots z_r$ of the $\mathbb{Z}_2$-vectorspace $\mathfrak{N}_n$, we can pick elements $y_1,\ldots,y_r$ of order 2 in $\mathfrak{J}\mathfrak{N}_n(m)$ such that $f(y_i)=z_i$ for $1\leq i\leq r$. This determines the required half-inverse of f.

Moreover, for $k\geq 0$ we have

$$\alpha(2k) = \alpha(k) \leq k$$

$$\alpha(2k+1) = \alpha(k)+1 \leq k+1 \;;$$

hence

$$2\alpha(n) \leq n+1$$

and

$$\tfrac{3}{2}n-1 < 2n-\alpha(n)$$

for all $n\geq 0$. Thus if $\alpha(n)\geq 2n-m$, we are also in the right dimension range for the exact sequence (10.1). In particular, we obtain

<u>Theorem 10.4.</u> <u>If $m>n$ are nonnegative integers such that $\alpha(n)$ and $\alpha(n+1)$ are both strictly larger than $2n-m$, then</u>

$$\mathfrak{J}\mathfrak{N}_n(m) \cong \mathfrak{N}_n \oplus \Omega_{2n-m}(BO(1)\times BO(m-n+1);\phi_{1,m-n+1}).$$

We will now use the computational techniques of §9 to calculate the normal bordism groups in (10.1) and 10.4 when m is close to $2n$. For $m\geq 2n-2$ this will lead to a nearly complete determination of $\mathfrak{J}\mathfrak{N}_n(m)$ even in those situations where the conditions on $\alpha(n)$ and $\alpha(n+1)$ above are not satisfied.

<u>Proposition 10.5.</u> <u>The low-dimensional normal bordism groups of $BO(1)\times BO(q)$ with coefficients in</u> $\phi_{1,q}$ <u>(cf. 10.2) are given by the following table.</u>

	$\Omega_0(BO(1)\times BO(n+1);\phi_{1,n+1})$ $(n\geq -1)$	$\Omega_1(BO(1)\times BO(n);\phi_{1,n})$ $n\geq +1)$	$\Omega_2(BO(1)\times BO(n-1);\phi_{1,n-1})$ $(n\geq 3)$
$n\equiv 1(4)$	$\mathbb{Z}_2$	$\mathbb{Z}_2\oplus\mathbb{Z}_2$	$\mathbb{Z}_2\oplus\mathbb{Z}_2$
$n\equiv 2(4)$	$\mathbb{Z}$	$\mathbb{Z}_4$	group of order 4
$n\equiv 3(4)$	$\mathbb{Z}_2$	$\mathbb{Z}_2\oplus\mathbb{Z}_2$	$\mathbb{Z}_2\oplus\mathbb{Z}_2\oplus\mathbb{Z}_2$
$n\equiv 4(4)$	$\mathbb{Z}$	$\mathbb{Z}_2$	$\mathbb{Z}_2$

Proof. Write x for (the pullback of) $w_1(\gamma_1)\in H^1(BO(1),\mathbb{Z}_2)$. Since $\phi=\phi_{1,q}=(\gamma_1\otimes\gamma_q-\gamma_q)+2\gamma_q-\gamma_1$, we can apply lemma 9.10 and we obtain

$$\begin{aligned} w(\phi_{1,q}) &= w(\gamma_1\otimes\gamma_q-\gamma_q)\cdot w(\gamma_q)^2\cdot w(\gamma_1)^{-1} \\ &= (1+qx+(\tbinom{q}{2}x^2+xw_1(\gamma_q))+..)(1+w_1(\gamma_q)^2+..)(1+x+x^2). \end{aligned}$$

Therefore

$$\begin{aligned} w_1(\phi_{1,q}) &= (q+1)x\ , \\ w_2(\phi_{1,q}) &= w_1(\gamma_q)^2+xw_1(\gamma_q)+\tbinom{q-1}{2}x^2 \neq 0 \end{aligned}$$

for all $q\geq 0$. According to theorem 9.3, this implies already the statement about the 0-dimensional bordism groups. Also we have the exact sequence

$$\overline{\Omega}_2(BO(1)\times BO(q);\phi)\xrightarrow{\sigma\circ j_2}\mathbb{Z}_2\xrightarrow{\delta_1}\Omega_1(BO(1)\times BO(q);\phi)\xrightarrow{f_1}\overline{\Omega}_1(BO(1)\times BO(q);\phi)\longrightarrow 0.$$

For all $i\in\mathbb{Z}$ the obvious isomorphisms allow us to identify

$$(10.6)\qquad \overline{\Omega}_i(BO(1)\times BO(q);\phi_{1,q}) = \begin{cases} \mathfrak{N}_i(BO(q)) & q \text{ even} \\ \Omega_i(BO(1)\times BO(q)) & q \text{ odd,} \end{cases}$$

since the orientation bundle of $\phi_{1,q}$ is γ_1 or trivial according to the parity of q. If q is even

$$\sigma\circ j_2([N,\lambda,\gamma,or]) = \tbinom{q-1}{2}w_1(N)^2[N]$$

since $w_1(\lambda)w_1(\gamma) = w_1(N)\cdot w_1(\gamma) = Sq^1(w_1(\gamma))$. If q is odd,

$$\sigma\circ j_2([N,\lambda,\gamma,or]) = w_1(\lambda)w_1(\gamma)[N] ,$$

since $Sq^1 = w_1(N)\cdot$ vanishes on any closed orientable surface N. (Here we confuse vector bundles with their classifying maps). Easy examples involving the projective plane and the torus show that $\sigma\circ j_2$ is onto (and hence f_1 is isomorphic) if and only if $q \not\equiv 2(4)$. For odd q, we get the isomorphism

$$\begin{array}{ccc} \Omega_1(BO(1)\times BO(q);\phi_{1,q}) & \xrightarrow{\cong} & \mathbb{Z}_2 \oplus \mathbb{Z}_2 \\ [N,\lambda,\gamma,is] & \longrightarrow & (w_1(\gamma)[N],w_1(\lambda)[N]) \end{array} ;$$

for $q \equiv 0(4)$, $q>0$

$$\begin{array}{ccc} \Omega_1(BO(1)\times BO(q);\phi_{1,q}) & \xrightarrow{\cong} & \mathbb{Z}_2 \\ [N,\lambda,\gamma,is] & \longrightarrow & w_1(\gamma)[N] . \end{array}$$

Finally, if $q \equiv 2(4)$, we get the exact sequence

$$\begin{array}{l} 0 \longrightarrow \mathbb{Z}_2 \xrightarrow{\delta_1} \Omega_1(BO(1)\times BO(q);\phi_{1,q}) \longrightarrow \mathbb{Z}_2 \longrightarrow 0; \\ \qquad\qquad\qquad [N,\lambda,\gamma,is] \longrightarrow w_1(\gamma)[N] \end{array}$$

moreover, by deleting a small disk from $S^1\times I$ and identifying one of the outer boundary circles via an orientation reversing map, it is easy to find a bordism which shows that

$$\begin{aligned} 2[S^1, \underset{\sim}{\mathbb{R}}, \text{Möbius}\oplus \underset{\sim}{\mathbb{R}}^{q-1}, \text{suitable isom.}] &= [S^1; \underset{\sim}{\mathbb{R}}, \underset{\sim}{\mathbb{R}}^q, TS=\mathbb{R}] \\ &= \delta_1(1) \neq 0 ; \end{aligned}$$

thus $\Omega_1(BO(1)\times BO(q);\phi_{1,q}) \cong \mathbb{Z}_4$ in this case.

Next we study our normal bordism groups in dimension 2. From theorem 9.3 we obtain the exact sequence

$$\overline{\Omega}_3(BO(1)\times BO(q);\phi) \xrightarrow{\sigma\circ j_3} \mathbb{Z}_2\oplus\mathbb{Z}_2 \xrightarrow{\delta_2} \Omega_2(BO(1)\times BO(q);\phi) \xrightarrow{f_2} \overline{\Omega}_2(BO(1)\times BO(q);\phi) \xrightarrow{\sigma\circ j_2} \mathbb{Z}_2 .$$

Here we have identified $H_1(BO(1)xBO(q);\mathbb{Z}_2)$ with $\mathbb{Z}_2 \oplus \mathbb{Z}_2$ via the isomorphism which applies the cohomology classes $(x, w_1(\gamma_q))$. Given

$$z = [N, \lambda, \gamma; \xi_N \cong \underbrace{\lambda \otimes \lambda \otimes .. \otimes \lambda}_{q+1 \text{ copies}}] \in \overline{\Omega}_3(BO(1)xBO(q);\phi) \quad ,$$

the first component of $\sigma \circ j_3(z)$ equals

$$w_2(\phi) \circ x\ (\mu(z)) = (w_1(\lambda)w_1(\gamma)^2 + w_1(\lambda)^2 w_1(\gamma) + (\tfrac{q-1}{2})w_1(\lambda)^3)[N]$$

$$= \begin{cases} w_1(N)^2 w_1(\gamma)[N] & \text{if } q \text{ is even} \\ (\tfrac{q-1}{2})w_1(\lambda)^3[N] & \text{if } q \text{ is odd} \end{cases}$$

(recall that $Sq^1 : H^2(N,\mathbb{Z}_2) \longrightarrow H^3(N;\mathbb{Z}_2)$ is multiplication with $w_1(N)$ (cf. 9.6); thus e.g. $w_1(N)w_1(\lambda)w_1(\gamma_q) = w_1(\lambda)^2 w_1(\gamma) + w_1(\lambda)w_1(\gamma)^2$). Similarly, the second component of $\sigma \circ j_3(z)$ equals

$$w_2(\phi) \cdot w_1(\gamma_q)(\mu(z)) = (w_1(\gamma)^3 + w_1(\lambda)w_1(\gamma)^2 + (\tfrac{q-1}{2})w_1(\lambda)^2 w_1(\gamma))[N]$$

$$= \begin{cases} (w_1(\gamma)^3 + (\tfrac{q-1}{2})w_1(N)^2 w_1(\gamma))[N] & \text{if } q \text{ is even} \\ (w_1(\gamma)^3 + (1 + (\tfrac{q-1}{2}))w_1(\lambda)^2 w_1(\gamma))[N] & \text{if } q \text{ is odd.} \end{cases}$$

In particular, for all $q \geq 1$

$$\sigma \cdot j_3\ [P^3, \underset{\sim}{\mathbb{R}}, (\gamma_1 | P^3) \oplus \underset{\sim}{\mathbb{R}}^{q-1}, \xi_P \cong \underset{\sim}{\mathbb{R}}] = (0,1) .$$

Moreover, for even q

$$\sigma \cdot j_3[N = P^2 x S^1, \xi_N, \pi_2^*(\text{Möbius} \oplus \underset{\sim}{\mathbb{R}}^{q-1}), \text{id}] = (1, (\tfrac{q-1}{2})) ;$$

for $q \equiv 3(4)$ $\quad \sigma \cdot j[P^3, \gamma_1 | P^3, \mathbb{R}^q, \text{or}] = (1,0)$.

Thus $\sigma \cdot j_3$ is onto (and hence f_2 is injective) if and only if $q \not\equiv 1(4)$. If $q \equiv 1(4)$, then the image of $\sigma \cdot j_3$ is $\{0\} x \mathbb{Z}_2$.

Furthermore recall from our previous calculations that $\sigma \cdot j_2$ vanishes (and hence f_2 is onto) if and only if $q \equiv 2(4)$. Also, we can use (10.6) and the work of Conner and Floyd [13] to get isomorphisms

$$\overline{\Omega}_2(BO(1)\times BO(q);\phi_{1,q}) \cong \mathfrak{N}_2(BO(q)) \cong \mathbb{Z}_2\oplus\mathbb{Z}_2\oplus\mathbb{Z}_2$$
$$[N,\ \lambda,\ \gamma,\ or] \longrightarrow (w_2(\gamma)[N],\ w_1(\gamma)^2[N],\ w_1(N)^2[N])$$

for q even, $q>0$; and for odd q:

$$\overline{\Omega}_2(BO(1)\times BO(q);\phi_{1,q}) \cong H_2(BO(1)\times BO(q);\mathbb{Z}) \cong \mathbb{Z}_2\oplus\begin{cases}\mathbb{Z}_2 & q>1\\ 0 & q=1\end{cases}$$
$$[N,\ \lambda,\ \gamma,\ or] \longrightarrow (w_1(\lambda)w_1(\gamma)[N], w_2(\gamma)[N]) .$$

(In order to see that the last map is isomorphic, first prove surjectivity by evaluating on $[S^1\times S^1, \pi_1^*(\text{Möbius}), \pi_2^*(\text{Möbius}\oplus \underset{\sim}{\mathbb{R}}^{q-1})]$ and on $[\mathbb{C}P(1), \underset{\sim}{\mathbb{R}}, \gamma_1(\mathbb{C})\oplus \underset{\sim}{\mathbb{R}}^{q-2}]$; then apply the Künneth theorems and [13], p. 23 and 42, to compute the orders of our groups).

It follows now from our exact sequence that the composite of the map

(10.7)
$$\tau : \Omega_2(BO(1)\times BO(q);\phi_{1,q}) \longrightarrow \overbrace{\underbrace{\mathbb{Z}_2}_{q\equiv 3(4)} \oplus \mathbb{Z}_2}^{q\equiv 0(4)} \oplus \mathbb{Z}_2 \quad (\text{all three: } q\equiv 2(4))$$
$$[N,\ \lambda,\ \gamma,\ is] \longrightarrow (w_2(\gamma)[N],\ w_1(\gamma)^2[N],\ w_1(N)^2[N]),$$

with the projection onto the direct sum of the indicated $\mathbb{Z}_2$-factors is bijective for $q\not\equiv 1(4)$. If $q\equiv 1(4)$, $q>1$, we get the exact sequence

$$0\longrightarrow \mathbb{Z}_2\oplus\{0\} \xrightarrow{\ \delta_2|\ } \Omega_2(BO(1)\times BO(q);\phi_{1,q}) \xrightarrow{\ w_2(\gamma_q)\cdot\mu\ } \mathbb{Z}_2 \longrightarrow 0 .$$

This finishes the proof of proposition 10.5. ∎

Now we apply our calculations to study bordism groups of immersions. The following theorem extends a result of R. Wells [68] who determined the groups $\mathfrak{I}\mathfrak{N}_n(2n)$ and $\mathfrak{I}\mathfrak{N}_{4n}(8n-1)$ for all n, as well as all $\mathfrak{I}\mathfrak{N}_n(m)$ with $m>n$, $n\leq 4$.

<u>Theorem 10.8.</u> <u>For</u> $n>0$ $\mathfrak{I}\mathfrak{N}_n(2n) \cong \mathfrak{N}_n\oplus\mathbb{Z}$ <u>if n is even</u>, $\mathfrak{I}\mathfrak{N}_n(2n) \cong \mathfrak{N}_n\oplus\mathbb{Z}_2$ <u>if n is odd. Moreover, for</u> $n>2$, <u>we have the following table listing unoriented bordism groups of immersions.</u>

	$\mathfrak{J}\mathfrak{N}_n(2n-1)$	$\mathfrak{J}\mathfrak{N}_n(2n-2)$
$n \equiv 1(4)$	$\mathfrak{N}_n \oplus \mathbb{Z}_2 \oplus \mathbb{Z}_2$	$\mathfrak{N}_n \oplus (\mathbb{Z}_2)^{\beta(n-1)}$
$n \equiv 2(4)$	$\mathfrak{N}_n \oplus \mathbb{Z}_4$	extension of $\mathfrak{N}_n$ by a group of order 4 *)
$n \equiv 3(4)$	$\mathfrak{N}_n \oplus (\mathbb{Z}_2)^{\beta(n+1)}$	$\mathfrak{N}_n \oplus (\mathbb{Z}_2)^{\gamma(n+1)}$
$n = 4(4)$ and: $\alpha(n)=1$	$\mathfrak{N}_n/\mathbb{Z}_2 \oplus \mathbb{Z}_4$	$\mathfrak{N}_n/\mathbb{Z}_2$
$\alpha(n)=2$	$\mathfrak{N}_n \oplus \mathbb{Z}_2$	$\mathfrak{N}_n/\mathbb{Z}_2 \oplus Z_4$
$\alpha(n)\geq 3$	$\mathfrak{N}_n \oplus \mathbb{Z}_2$	$\mathfrak{N}_n \oplus \mathbb{Z}_2$

(Here $\alpha(q)$ denotes the number of ones in the binary expansion of a natural number q, $\beta(q)=\min(\alpha(q),2)$, $\gamma(q)=\min(\alpha(q),3)$). Finally, $\mathfrak{J}\mathfrak{N}_2(3) \cong \mathbb{Z}_8$.

Proof. We start from the exact sequence (10.1). Most of the time we will be able to express the homomorphisms $\sigma \cdot j$ and ρ by Stiefel-Whitney numbers. Therefore, computing the images of $\sigma \cdot j$ and ρ amounts on one hand, to finding relations between Stiefel-Whitney numbers, and, on the other hand, to finding manifolds which show that certain relations do not hold. The relevant relations can often be deduced easily from the Wu formulas (see facts 9.5 and 9.6; for interesting consequences concerning the normal Stiefel-Whitney classes $\bar{w}_i(M)$ of a closed manifold M, see especially [44]). Since we have Brown's theorem 10.3 at our disposal here, we can often shortcut the whole process.

The first statement in our theorem follows immediately from theorem 10.4 which also settles the case when $m=2n-1$ and n, $n+1$ are both no power of 2.

Now assume that $n+1$ is a power of 2. Then $n \equiv 3(4)$, $\alpha(n)>1$, and (10.1) takes the following form

*) According to Christof Olk, $\mathfrak{J}\mathfrak{N}_n(2n-2) \cong \mathfrak{N}_n \oplus \mathbb{Z}_2 \oplus \mathbb{Z}_2$ for $n \equiv 2(4)$.

$$(10.9)\quad \begin{array}{ccccccccc} \cdots \to \mathfrak{N}_{n+1} & \xrightarrow{\sigma\circ j} & \mathbb{Z}_2 \oplus \mathbb{Z}_2 & \longrightarrow & \mathfrak{I}\mathfrak{N}_n(2n-1) & \xrightarrow{f} & \mathfrak{N}_n & \to 0. \\ [M] & \longrightarrow & (\bar{w}_{n+1}(M)[M], w_1\bar{w}_n(M)[M]) & & & & & \end{array}$$

(Here we have identified $[N,\lambda,\gamma,is] \in \Omega_1(BO(1)xBO(n);\phi_{1,n})$ with $(w_1(\lambda)[N],(w_1(\lambda)+w_1(\gamma))[N]) \in \mathbb{Z}_2 \oplus \mathbb{Z}_2$, and we computed $\sigma\circ j$ by 9.11). It is wellknown, e.g. already from Whitney's work [71], that $\bar{w}_{n+1}(M)[M]$ vanishes always. Also, recall that here we have

$$(10.10)\quad \begin{aligned} w(P^{n+1}) &= (1+x)^{n+2} = (1+x^{n+1})(1+x) , \\ \bar{w}(P^{n+1}) &= (1+x^{n+1})(1+x+x^2+..+x^{n+1}) \\ &= 1+x+x^2+..x^n , \end{aligned}$$

and hence

$$w_1\bar{w}_n(P^{n+1})[P^{n+1}] = x^{n+1}[P^{n+1}] = 1$$

(x denotes the generator of $H^1(P^{n+1};\mathbb{Z}_2)$). Thus the image of $\sigma\circ j$ is $\{0\}x\mathbb{Z}_2$, and $\mathfrak{I}\mathfrak{N}_n(2n-1) \cong \mathfrak{N}_n \oplus \mathbb{Z}_2$.

Next, let n be a power of 2, $n \geq 2$. Then we get the exact sequences

$$\begin{array}{ccccccccc} & & \mathbb{Z}_2 & & & & & & \\ & & \downarrow{\scriptstyle \delta_1} & & & & & & \\ 0 & \longrightarrow & \Omega_1(BO(1)xBO(n);\phi_{1,n}) & \underset{\rho}{\overset{\delta}{\rightleftarrows}} & \mathfrak{I}\mathfrak{N}_n(2n-1) & \xrightarrow{f} & \mathfrak{N}_n & \longrightarrow & 0 \\ & & \downarrow{\scriptstyle w_1(\gamma_n)\circ\mu} & & & & & & \\ & & \mathbb{Z}_2 & & & & & & \\ & & \downarrow & & & & & & \\ & & 0 & & & & & & \end{array} ,$$

and $w_1(\gamma_n)\circ\mu$ is an isomorphism iff $n \neq 2$. Given an immersion $i:M^n \looparrowright \mathbb{R}^{2n-1}$, let $N \subset M$ be the zero set of a nondegenerate section of the normal bundle $\nu(i)$ of i. Then

$$(10.11)\quad \rho[i] = [N, \underset{\sim}{\mathbb{R}}, (\nu(i)\oplus \underset{\sim}{\mathbb{R}})N, \text{ suitable } is]$$

and, by 9.11.,

$$(w_1(\gamma_n)\circ\mu)\rho[i] = w_1(M)\bar{w}_{n-1}(M)[M].$$

As we saw above, this characteristic number equals 1 for $M=P^n$. Therefore $\mathfrak{I}\mathfrak{N}_n(2n-1)$ is the direct sum of a complement of $\mathbb{Z}_2\cdot[P^n]$ in $\mathfrak{N}_n$, and a cyclic factor (of order 4 (when $n>2$) or 8 (when $n=2$)) generated by any immersion of P^n into $\mathbb{R}^{2n-1}$. Note in particular how $\mathfrak{I}\mathfrak{N}_2(3)\cong\mathbb{Z}_8$ is built up from three copies of $\mathbb{Z}_2$, which are increasingly hard to detect by characteristic numbers.

Finally, we consider the case $m=2n-2$, $n>2$. From (10.1) we get the exact line

$$\begin{array}{l} \mathfrak{N}_{n+1} \xrightarrow{\sigma\circ j} \Omega_2(BO(1)\times BO(n-1);\phi_{1,n-1}) \underset{\rho}{\overset{\delta}{\rightleftarrows}} \mathfrak{I}\mathfrak{N}_n(2n-2) \xrightarrow{f} \mathfrak{N}_n \xrightarrow{\overline{w}_{n-1}w_1} \mathbb{Z}_2 \\ \qquad\qquad\qquad \downarrow \tau=(\tau_1,\tau_2,\tau_3) \\ \qquad\qquad\qquad \mathbb{Z}_2\oplus\mathbb{Z}_2\oplus\mathbb{Z}_2 \quad . \end{array} \tag{10.12}$$

f is onto if and only if n is not a power of 2, by 10.3, 10.9 and 10.10, and f has a right inverse when $\alpha(n)\geq 3$ (cf. proof of 10.4). Also, δ is injective when $\alpha(n+1)\geq 3$, since then the forgetful map $\mathfrak{I}\mathfrak{N}_{n+1}(2n-1)\longrightarrow\mathfrak{N}_{n+1}$ is onto by 10.3.

In particular, if $n\equiv 2(4)$, then $\alpha(n)\geq 2$, $\alpha(n+1)\geq 3$, and $\mathfrak{I}\mathfrak{N}_n(2n-2)$ is an extension of $\mathfrak{N}_n$ by the group $\Omega_2(BO(1)\times BO(n-1);\phi_{1,n-1})$ which has four elements.

If $n\equiv 2(4)$, the injective map τ, defined by evaluating $w_2(\gamma_{n-1})$, $w_1(\gamma_{n-1})^2$, $w_1(S)^2$, determines $\Omega_2(BO(1)\times BO(n-1);\phi_{1,n-1})$ (see 10.7). We obtain from 9.11 that

$$\tau\circ\sigma\circ j[M^{n+1}] = ((\overline{w}_2\overline{w}_{n-1}(M)+w_1\overline{w}_n(M))[M]\ ,\ nw_1\overline{w}_n(M)\ [M]\ ,\ 0)\quad ,$$

since $w_1^2\overline{w}_{n-1}(M) = Sq^1(w_1\cdot\overline{w}_{n-1}(M)) = n\cdot w_1\overline{w}_n(M)$, and $\overline{w}_{n+1}(M)=0$. Similarly, we compute for an immersion class $[i:N^n \looparrowright \mathbb{R}^{2n-2}]\in\mathfrak{I}\mathfrak{N}_n(2n-2)$ that

$$\tau\circ\rho\ [i] = (\overline{w}_2(N)\overline{w}_{n-2}(N)[N],\ 0,\ 0)\ .$$

E.g. if $n+1$ is a power of 2, then we get from (10.10)

$$\tau\circ\sigma\circ j\ [P^{n+1}] = (0,\ 1,\ 0)\ .$$

If $n+1$ is the sum of two (possibly equal) powers of 2, say $r=2^p$ and $s=2^q$, such that $r,s\geq 2$, then

$$\text{(10.13)}\qquad \tau\circ\sigma\circ j\ [P^r \times P^s] = (1,\ 0,\ 0)\ ,$$

since $\overline{w}(P^r \times P^s) = (1+x_1+x_1^2+..+x_1^{r-1})(1+x_2+x_2^2+...x_2^{s-1})$ where x_1, x_2 generate $H^1(P^r;\mathbb{Z}_2)$ and $H^1(P^s;\mathbb{Z}_2)$ respectively.

Now consider the special case $n \equiv 3(4)$. Here τ is an isomorphism and f has a right inverse. The examples above show that the image of $\tau\circ\sigma\circ j$ is $\mathbb{Z}_2\oplus\mathbb{Z}_2\oplus\{0\}$, $\mathbb{Z}_2\oplus\{0\}\oplus\{0\}$ or $\{0\}$ according to whether $\alpha(n+1)$ is 1, 2 or greater. Correspondingly, $\mathfrak{I}\mathfrak{N}_n(2n-2)$ is the direct sum of $\mathfrak{N}_n$ and $\mathbb{Z}_2$, $\mathbb{Z}_2\oplus\mathbb{Z}_2$ or $\mathbb{Z}_2\oplus\mathbb{Z}_2\oplus\mathbb{Z}_2$.

Next we study the case $n \equiv 0(4)$. According to (10.7) τ_1 is an isomorphism here. If n is not a power of 2, then $\alpha(n+1)\geq 3$, and we get the exact sequence (which splits provided $\alpha(n)\geq 3$)

$$0 \longrightarrow \mathbb{Z}_2 \underset{\tau_1\circ\rho}{\overset{\delta\circ\tau_1^{-1}}{\rightleftarrows}} \mathfrak{I}\mathfrak{N}_n(2n-2) \xrightarrow{\ f\ } \mathfrak{N}_n \longrightarrow 0\ .$$

If in addition $n=r+s$, where $r=2^p$, $s=2^q$, then there exists an immersion $i:N^n \looparrowright \mathbb{R}^{2n-2}$ such that N^n is bordant to $P^r \times P^s$, and hence

$$\tau_1\circ\rho\ [i] = \overline{w}_2(P^r \times P^s)\overline{w}_{n-2}(P^r \times P^s)[P^r \times P^s] = 1$$

(compare with the computation in the examples above); it follows that $\mathfrak{I}\mathfrak{N}_n(2n-2)$ is isomorphic to the direct sum of the cyclic group of order 4 generated by $[i]$, and of a complement of $\mathbb{Z}_2\cdot[N]$ in $\mathfrak{N}_n$. Finally, if n is a power of 2, then $\delta \equiv 0$, and $\mathfrak{I}\mathfrak{N}_n(2n-2)$ is the kernel of the $\mathbb{Z}_2$-valued epimorphism on $\mathfrak{N}_n$ defined by the characteristic number $w_1(N)\overline{w}_{n-1}(N)[N]$; here we have used the fact, to be proved later, that there is an (n+1)-manifold M such that

$$\overline{w}_2(M)\overline{w}_{n-1}(M)[M] = 1 .$$

Such a manifold still exists if $n-1$ is a power of 2 (take e.g. $P^{n-1}xP^2$, cf. (10.13)). Thus, for $n \equiv 1(4)$, the isomorphism (τ_1,τ_2) maps image $(\sigma\circ j)$ = image (ρ) onto $\mathbb{Z}_2\oplus\{0\}$ or $\{0\}$, according to whether $\alpha(n-1)$ is 1 or greater; correspondingly, $\mathcal{I}\mathfrak{N}_n(2n-2)$ is the direct sum of $\mathfrak{N}_n$ and $\mathbb{Z}_2$ or $\mathbb{Z}_2\oplus\mathbb{Z}_2$.

It remains to exhibit closed manifolds M in dimensions $=2^t+1$, $t>1$, such that $\overline{w}_2(M)\overline{w}_{s-2}(M)[M] = 1$. This will be done in the following example.

<u>Example 10.14</u> <u>(Stong's manifolds</u>, see [61], 3.4). Given a k-tupel $r=(r_1,r_2,..,r_k)$ of nonnegative integers, $k \geq 1$, let λ_j denote the canonical line bundle over the projective space P^{r_j} and define P^r to be the (total space of the) projective space bundle of the bundle $\lambda_1\oplus.....\oplus\lambda_k$ over $P^{r_1} x..xP^{r_k}$,

$$P^r = G_1(\lambda_1\oplus\lambda_2\oplus..\oplus\lambda_k) ,$$

(cf. § 1). By the theorem of Leray-Hirsch, we have

$$H^*(P^r;\mathbb{Z}_2) = \mathbb{Z}_2[x_1,x_2,..x_k;c] \Big/ \begin{array}{l} x_j^{r_j+1} = 0 \text{ for } i = i,..,k; \\ c^k = w_1c^{k-1}+..w_ic^{k-i}+..+w_k \end{array}$$

where $x_1,..x_k$ and c are the first Stiefel-Whitney classes of $\lambda_1,..\lambda_k$ and of the canonical line bundle λ over P^r; w_i is the ith Stiefel-Whitney class of $\lambda_1\oplus..\oplus\lambda_k$. The last relation can be obtained as follows. For the tangent bundle along the fibers of the projection we have from (1.2)

$$\begin{aligned} TF(P^r)\oplus \mathbb{R} &\cong \underset{\sim}{\mathrm{Hom}}(\lambda,\lambda^\perp) \oplus \underset{\sim}{\mathrm{Hom}}(\lambda,\lambda) \\ &\cong \underset{\sim}{\mathrm{Hom}}(\lambda,p^*(\lambda_1\oplus\lambda_2\oplus...\oplus\lambda_k)) , \end{aligned}$$

eorem 11.1.

Compare the composite homomorphism $\mathfrak{I}\mathfrak{N}_2(3) \longrightarrow \pi_3^S$ in

e Kahn-Priddy (or "figure-8") homomorphisms

$\to \pi_n^S$ in [39].

and the k-dimensional Stiefel-Whitney class of this bundle must vanish. This allows us also to determine the Stiefel-Whitney class of the closed, smooth, $(r_1+..+r_k+k-1)$-dimensional manifold P^r as follows

$$w(P^r) = \prod_{j=1}^{k} (1+x_j)^{r_j+1} \cdot \prod_{j=1}^{k} (1+x_j+c) .$$

The family of manifolds P^r, $r=(r_1,..r_k)$, provides us with a rich reservoir of examples with possibly interesting Stiefel-Whitney numbers. As an illustration, consider the special case when k is a power of 2, $k \geq 4$, and r is the k-tupel $(2,0,.,0)$. Then P^r is a $(k+1)$-manifold such that

$$\begin{aligned} w(P^r) &= (1+x_1)^{-1}(1+x_1+c)(1+c)^{k-1} \\ \bar{w}(P^r) &= (1+x_1)(1+x_1+c)^{-1}(1+c)(1+c^k) \\ &= (1+(x_1+c)+x_1c)(1+(x_1+c)+..+(x_1+c)^i+..)(1+c^k) . \end{aligned}$$

Hence

$$\begin{aligned} \bar{w}_2\bar{w}_{k-1}(P^r)[P^r] &= (x_1c)\cdot x_1c(x_1+c)^{k-3}[P^r] \\ &= x_1^2c^2(c^{k-3}+x_1(c^{k-4}+..))[P^r] \\ &= 1 , \end{aligned}$$

since $x_1^3=0$, and $x_1^2c^{k-1}$ is the generator of $H^{k+1}(P^r;\mathbb{Z}_2)$. This example completes the proof of theorem 10.8. ∎

§11. An example of odd torsion: calculation of π_3^S.

So far, we encountered mainly 2-primary torsion in explicitly computed normal bordism groups. However, the interaction of free groups in our exact sequences can give rise to odd torsion, too. As an illustration, we reprove the following classical result (cf. [66], p. 186).

Theorem 11.1. The stable homotopy group π_3^S is isomorphic to $\mathbb{Z}_{24}$.

Proof. Let X be a point and ϕ be trivial. Then, by theorem 9.3, $\Omega_3(X;\phi) \cong \pi_3^S$ fits into the exact sequence

$$(11.2) \qquad \Omega_4 \xrightarrow{\sigma\circ j_4} \Omega_2(BO(2);\Gamma) \xrightarrow{\delta_3} \pi_3^S \longrightarrow \Omega_3 \ ,$$

where the oriented bordism group $\Omega_4 \cong \mathbb{Z}$ is generated by $\mathbb{C}P(2)$, and $\Omega_3 = 0$.

On the other hand, we have the Gysin sequence (cf. proposition 9.20)

$$(11.3) \qquad \ldots \to \Omega_2(S(\gamma_2);\pi^*(\Gamma)) \xrightarrow{\pi_*} \Omega_2(BO(2);\Gamma) \xrightarrow{\Delta = W_2} \mathbb{Z} \ .$$

Here Δ coincides with the homomorphism W_2 in 9.3, and hence is onto. Since the pullback bundle $\pi^*(\gamma_2)$ over the spherebundle $S(\gamma_2)$ has a section, it follows from (9.2) that $\pi^*(\Gamma)$ is stably isomorphic to the pulled-back orientation bundle of γ_2.

Therefore, $\Omega_2(S(\gamma_2);\pi^*(\Gamma))$ can be identified with the bordism group of tripels

$$(\text{surface } S, \text{ line bundle } \lambda \text{ over } S, \ is{:}TS\oplus\lambda \cong \mathbb{R}^3) \ ,$$

or, equivalently, with the immersion group $\mathfrak{Im}_2(3) \cong \mathbb{Z}_8$ (cf. theorem 10.8). Finally, we claim that π_* in (11.3) is injective. To see this, consider the diagram

$$\Omega_3(BO(2);\Gamma) \xrightarrow{\Delta} \ldots$$

Here Δ belongs to the extension of ou… left. As an easy consequence of (9.3) an… an isomorphism. Now observe that there is… $\bar{h}$ so that the triple

$$(P^1 \times P^2, \ \eta = (\gamma_1|P^1) \otimes (\gamma_1| \ldots$$

defines an element $x \in \Omega_3(BO(2);\Gamma)$; indeed, η… is trivial. We have by fact 9.11 that

$$(w_1(\gamma_2)\circ\mu\circ f_1)\circ\Delta(x) = w_2(\eta)w_1(\eta)[P \ldots$$

Hence Δ is onto, and π_* in (11.3) is injective.

It follows that (11.2) and (11.3) give rise to … of exact sequences.

$$(11.4) \qquad \begin{array}{ccccc} & & 0 & & \\ & & \downarrow & & \\ & & \mathbb{Z}_8 \cong \mathfrak{Im}_2(3) & & \\ & & \downarrow & & \\ \Omega_4 & \xrightarrow{\sigma\circ j_4} & \Omega_2(BO(2);\Gamma) & \xrightarrow{\delta_3} & \pi_3^S \longrightarrow \\ & & \downarrow W_2 & & \\ & & \mathbb{Z} & & \\ & & \downarrow & & \\ & & 0 & & \end{array}$$

By 9.3, we have

$$W_2\circ\sigma\circ j_4[\mathbb{C}P(2)] = \pm\, p_1(\mathbb{C}P(2))[\mathbb{C}P(2)] = \pm\, 3 \ .$$

This implies th…

Exercise 11.5. … (11.4) with th… $\mathfrak{Im}_{n-1}(n)$ —…

§12. Bordism of framefields.

In this section we study some of the bordism groups $\Omega_n(k,k)$ and $\mathfrak{N}_n(k,k)$ of manifolds with k-framefields (compare 6.1).

Already the case $k=0$ is very interesting. Here we are dealing with manifolds without any extra structure, but the bordisms are required to carry a nonsingular vector field. The following result is essentially due to Bruce Reinhart [55].

Theorem 12.1. For any dimension $n \geq 0$ we have isomorphisms

$$\mathfrak{N}_n(0,0) \xrightarrow{(\mathrm{proj}\circ f,\chi)} \mathfrak{N}_n/\mathbb{Z}_2[P^n] \oplus \mathbb{Z} \qquad \text{if } n \text{ is even},$$

$$\mathfrak{N}_n(0,0) \xrightarrow{f} \mathfrak{N}_n \qquad \text{if } n \text{ is odd};$$

and

$$\Omega_n(0,0) \xrightarrow{(f,\frac{1}{2}(\chi-\tau))} \Omega_n \oplus \mathbb{Z} \qquad \text{if } n \equiv 0(2),$$

$$\Omega_n(0,0) \xrightarrow{f} \Omega_n \qquad \text{if } n \equiv 3(4),$$

$$\Omega_n(0,0) \xrightarrow{(f,k)} \Omega_n \oplus \mathbb{Z}_2 \qquad \text{if } n \equiv 1(4).$$

Here χ, τ and k denote Euler number, signature and real Kervaire semicharacteristic respectively ($k(M) = \sum_i \dim H^{2i}(M;\mathbb{R}) \bmod 2$), and f is the obvious forgetful homomorphism.

Proof. We have the exact sequence (cf. 7.23 and also 9.3)

$$\begin{array}{ccccccccc}
\mathfrak{N}_{n+1}(1,0) & \xrightarrow{\sigma\circ j} & \Omega_0(BO(n+1)\times BO(1);\psi) & \underset{\rho}{\overset{\delta}{\rightleftarrows}} & \mathfrak{N}_n(0,0) & \xrightarrow{f} & \mathfrak{N}_n(0,-1) & \longrightarrow & 0 \\
\wr\wr & & \wr\wr \begin{cases} \mathbb{Z} & n \text{ even} \\ \mathbb{Z}_2 & n \text{ odd} \end{cases} & & & & \wr\wr & & \\
\mathfrak{N}_{n+1} & & & & & & \mathfrak{N}_n & &
\end{array},$$

where $\psi=\gamma_{n+1}\otimes\gamma_1+\gamma_1-\gamma_{n+1}$ is orientable iff n is even. $\sigma\circ j$ and ρ both assign to each manifold its (mod 2) Euler number, and $\delta(1)$ is the class of the n-sphere. When n is odd, $\sigma\circ j$ is onto, and f is an isomorphism. When n is even, then $\rho\circ\delta=2\cdot\mathrm{Id}$ and $\chi(P^n)=1$; hence any element in the kernel of $(\mathrm{proj}\circ f,\chi)$ lies also in the kernel of f (and of $\rho=\chi$) and thus has to vanish; it is easily seen that $(\mathrm{proj}\circ f,\chi)$

is also onto.

Similarly, we have the exact sequence

$$\Omega_{n+1} \xrightarrow{\ \chi\ } \left\{\begin{matrix} \mathbb{Z} & n \text{ even} \\ \mathbb{Z}_2 & n \text{ odd} \end{matrix}\right\} \underset{\chi}{\overset{\delta}{\rightleftarrows}} \Omega_n(0,0) \xrightarrow{\ f\ } \Omega_n \xrightarrow{\ f\ } 0.$$

If n is even, then $\chi(\Omega_{n+1})=0$, and the welldefined homomorphism $\frac{1}{2}(\chi-\tau)$ is a left inverse of δ. On the other hand, $\chi(\Omega_{n+1})$ is $\mathbb{Z}_2$ or 0 according to whether $n \equiv 3(4)$ or $n \equiv 1(4)$. In the latter case, $\mathfrak{k}$ gives rise to a welldefined left inverse of δ; this claim, which completes the proof of theorem 12.1, follows easily from the theorem of Poincaré-Hopf and from the following observation.

<u>Proposition 12.2.</u> <u>Let M be a compact orientable smooth manifold of dimension $n=2k$, k odd, with boundary M. Then</u>

$$\underline{\mathfrak{k}(\partial M) = \text{Euler number of } M, \text{ mod } 2.}$$

<u>Proof.</u> Consider the commuting diagram

$$\begin{array}{ccccccccc} \cdots \xrightarrow{j_{*q+1}} & H_{q+1}(M,\partial M) & \xrightarrow{\partial_{q+1}} & H_q(\partial M) & \xrightarrow{i_{*q}} & H_q(M) & \xrightarrow{j_{*q}} & H_q(M,\partial M) & \longrightarrow \cdots \\ & \updownarrow & & \updownarrow & & \updownarrow & & \updownarrow & \\ \cdots \longrightarrow & H^{n-q-1}(M) & \longrightarrow & H^{n-q-1}(\partial M) & \longrightarrow & H^{n-q}(M,\partial M) & \xrightarrow{j^*_{n-q}} & H^{n-q}(M) & \longrightarrow \cdots \end{array} \tag{12.3}$$

of exact sequences, where the vertical arrows are Lefschetz (or Poincaré) duality isomorphisms. Coefficients are taken in $\mathbb{R}$. For a suitable space X, we write

$$b_q(X) = \dim H_q(X) = \dim H^q(X).$$

Then we have

$$\begin{aligned} b_q(\partial M) &= \dim(\operatorname{im} i_{*q}) + \dim(\operatorname{im} \partial_{q+1}) \\ &= b_q(M) - \dim(\operatorname{im} j_{*q}) + b_{n-q-1}(M) - \dim(\operatorname{im} j_{*q+1}), \end{aligned}$$

and hence

$$\sum_{s\in\mathbb{Z}} b_{2s}(\partial M) = \sum_{r\in\mathbb{Z}} b_r(M) - \sum_{r\in\mathbb{Z}} \dim(\mathrm{im} j_{*r})$$
$$= \chi(M) - 2\sum_{r \text{ odd}} b_r(M) - 2\sum_{r=0}^{k-1} \dim(\mathrm{im} j_{*r}) - \dim(\mathrm{im} j_{*k})$$

Here we have used the commutativity of the last square above to deduce that $\dim(\mathrm{im} j_{*q}) = \dim(\mathrm{im} j_{*n-q})$. Moreover observe that the cup product on the middle dimensional cohomology space $H^k(M, \partial M)$ induces a bilinear antisymmetric map from $\mathrm{im} j_{*k} \times \mathrm{im} j_{*k}$ to $\mathbb{R}$, which is also nondegerate by Lefschetz duality. Now it is an easy exercise in linear algebra to show that $\dim(\mathrm{im} j_{*k})$ is even. The proposition follows.∎

Before we calculate more bordism groups of framefields, we have to establish the following relations (cf. [32], 2.12).

<u>Lemma 12.4.</u> <u>Let M be a closed smooth n-dimensional manifold. If n is odd, we have</u>

$$0=w_1(M)^n=\ldots=w_1(M)^{n-i}w_i(M)=\ldots=w_1(M)^2w_{n-2}(M)=w_1(M)w_{n-1}(M)=w_n(M).$$

<u>If n is even and</u> $0 \le i < n$, <u>we have</u>

$$w_1(M)^{n-i}w_i(M)=w_1(M)^{n-i-1}w_{i+1}(M) \qquad \text{for } i \equiv 0(2)$$

<u>and</u>

$$\binom{n}{2}w_1(M)^{n-i}w_i(M)=w_1(M)^{n-i-2}w_{i+2}(M) \qquad \text{for } i \equiv 0(4).$$

<u>Proof.</u> Write w_i for $w_i(M)$. By the Wu formulas (see facts 9.5 and 9.6) we have for $0 \le i < n$

$$w_1{}^{n-i}w_i = Sq^1(w_1{}^{n-i-1}w_i)$$
$$= (n-i-1)w_1{}^{n-i}w_i+w_1{}^{n-i-1}(w_1w_i+(i+1)w_{i+1}),$$

and hence

$$(n-i-1)w_1{}^{n-i}w_i = (i+1)w_1{}^{n-i-1}w_{i+1} .$$

Similarly, for $0 \le i < n-1$

$$
\begin{aligned}
w_2 w_1^{n-i-2} w_i + w_1^{n-i} w_i \\
&= Sq^2(w_1^{n-i-2} w_i) \\
&= \binom{n-i-2}{2} w_1^{n-i} w_i + (n-i) w_1^{n-i-1}(w_1 w_i + (i+1)(w_{i+1}) \\
&\quad + w_1^{n-i-2}(w_2 w_i + i w_1 w_{i+1} + \binom{i-1}{2} w_{i+2})
\end{aligned}
$$

The claim for odd n follows from applying both formulas to odd i (or to $i=n-1$). The case $n\equiv 0(2)$ is immediate. ∎

Next represent an element $x\in\mathfrak{N}_n(k,k)$ by an n-manifold M with a k-field $u:\underline{\mathbb{R}}^k \hookrightarrow TM$. Let η denote a complement of the image of u, and let $S\subset M$ be the zero set of a generic section of η. Then S is a closed k-manifold naturally equipped with an isomorphism

$$TS\oplus\eta|S \cong TM|S = (\underline{\mathbb{R}}^k\oplus\eta)|S$$

and hence with a stable parallelization $\overline{h}$. The homomorphisms

$$
(12.5)\qquad
\begin{aligned}
&\chi' : \mathfrak{N}_n(k,k) \longrightarrow \Omega_k(BO(n-k);\text{trivial}) , \qquad \text{and} \\
&\chi'' : \mathfrak{N}_n(k,k) \longrightarrow \Omega_k(\text{point};\text{trivial})
\end{aligned}
$$

defined by $\chi'(x)=[S,\eta|S,\overline{h}]$ and $\chi''(x)=[S,\overline{h}]$, will be studied in greater detail in §19. Note here, however, that the diagram

$$
(12.6)\qquad
\begin{array}{ccc}
\mathfrak{N}_n(k,k) & \xrightarrow{\ \chi'\ } & \Omega_k(BO(n-k);\text{trivial}) \\
 & \searrow^{\rho} & \downarrow \text{incl}_* \\
 & & \Omega_k(BO(n-k+1)\times BO(1);\gamma_{n-k+1}\otimes\gamma_1+\gamma_1-\gamma_{n-k+1})
\end{array}
$$

commutes; so χ' gives sharper information than the homomorphism ρ in (7.23) does.

The following result will be needed in the next theorem.

<u>Lemma 12.7.</u> <u>Denote the nontrivial line bundle over the circle by "Möbius". Let u be a "horizontal" vector field on the (total space of the) sphere bundle S (Möbius $\oplus\,\underline{\mathbb{R}}^{n-1}$) (i.e. the bundle projection maps</u>

<u>u into a nowhere zero vector field on the base circle)</u>.

<u>Then $\chi''[S(\text{Möbius} \oplus \mathbb{R}^{n-1});u]$ is the generator of $\Omega_1(\text{point;trivial}) \cong \mathbb{Z}_2$, provided $n \geq 1$.</u>

<u>Similarly, $\chi''[S(\text{Möbius} \oplus \mathbb{R}^{n-2}) \times S^1;(u,\text{vectorfield along } S^1)]$ is the generator of $\Omega_2(\text{point;trivial}) \cong \mathbb{Z}_2$, if $n \geq 2$.</u>

<u>Proof.</u> We know from example 9.16 that for $k=1$ or 2 $\Omega_k(\text{point;trivial}) \cong \mathbb{Z}_2$ is generated by the invariantly framed k-torus. Thus the second statement follows from the first one.

Now note that

$$S(\text{Möbius} \oplus \mathbb{R}^{n-1}) = I \times S^{n-1} / (0;x_1,..x_n) \sim (1;-x_1,x_2,..x_n),$$

and that the tangent bundle along the S^{n-1}-component is complementary to u. The first statement of our lemma is proved as soon as we exhibit a nondegenerate vectorfield in this S^{n-1}-direction with a singularity consisting of a parallelized circle. According to example 4.16, for even n $\{0\} \times S^{n-1}$ and $\{1\} \times S^{n-1}$ carry vectorfields u_o and u_1 which correspond to one another under the idenfication above; the linear homotopy between u_o and u_1 then gives the global vectorfield with the required singularity. If n is odd, any nonzero vectorfield along the equator $S^{n-1} \cap (\{0\} \times \mathbb{R}^{n-1})$ extends symmetrically to the two halfspheres of S^{n-1} so that there are nondegenerate zeroes precisely at the poles $(\pm 1,0,..0)$. Applying this construction fiberwise, we get the desired vectorfield; its singularity is the circle $S(\text{Möbius} \oplus \{0\}) \subset S(\text{Möbius} \oplus \mathbb{R}^{n-1})$.■

<u>Theorem 12.8.</u> <u>For $k=1$ or 2 and $n \geq k$ we have the isomorphism</u>

$$\mathcal{H}_n(k,k) \xrightarrow[\cong]{(\chi'',f)} \mathbb{Z}_2 \oplus \ker(w_n:\mathcal{H}_n \longrightarrow \mathbb{Z}_2)$$

<u>where χ'' is the homomorphism into $\Omega_k(\text{point;trivial}) \cong \mathbb{Z}_2$ defined in (12.5), and f is the natural forgetful homomorphism.</u>

Proof. First let us compute the auxiliary groups $\Omega_i(Y_s;\psi_s)$ in (7.23); here $Y_s=BO(S)\times BO(1)$ and

$$\psi_s = \gamma_s\otimes\gamma_1+\gamma_1-\gamma_s = (\gamma_s\oplus\underset{\sim}{\mathbb{R}})\otimes\gamma_1-(\gamma_s\oplus\underset{\sim}{\mathbb{R}}) .$$

By lemma 9.10 we have for $s>0$

$$w_1(\psi_s) = (s+1)w_1(\gamma_1) ,$$
$$w_2(\psi_s) = \binom{s+1}{2}w_1(\gamma_1)^2 + w_1(\gamma_1)\cdot w_1(\gamma_s) \neq 0.$$

Therefore we get the exact sequence (see 9.3)

$$\begin{array}{ccccccccc}
\overline{\Omega}_3(Y_s;\psi_s) & \xrightarrow{\sigma\circ j_3} & \Omega_1(Y_s\times BO(2);\psi_s+\Gamma) & \xrightarrow{\delta_2} & \Omega_2(Y_s;\psi_s) & \xrightarrow{f_2} & \overline{\Omega}_2(Y_s;\psi_s) & \xrightarrow{\sigma\circ j_2} & \mathbb{Z}_2 \\
\downarrow\mu & & \cong\downarrow proj_*\circ\mu & & & & & & \\
H_3(Y_s;\mathbb{Z}_2) & \xrightarrow[\text{of } w_2(\psi_s)\cdot]{\text{adjoint}} & H_1(Y_s;\mathbb{Z}_2) & & & & & & \\
& & \cong\downarrow(w_1(\gamma_s),w_1(\gamma_1)) & & & & & & \\
& & \mathbb{Z}_2\oplus\mathbb{Z}_2 & & & & & &
\end{array}$$

Note that $\overline{\Omega}_i(Y_s;\psi_s)$ is canonically isomorphic to $\Omega_i(BO(S)\times BO(1))$ when s is odd, and to $\mathfrak{N}_i(BO(S))$ when s is even ($[N,\gamma]\in\mathfrak{N}_i(BO(S))$ corresponds to $[N,\gamma,\xi_N,id]\in\overline{\Omega}_i(Y_s;\psi_s)$).

The examples $[S^1\times S^1,\pi_1^*(\text{Möbius}\oplus\underset{\sim}{\mathbb{R}}^{s-1}),\pi_2^*(\text{Möbius})]$ and $[P^2,\gamma=\underset{\sim}{\mathbb{R}}^{s-1}\oplus\gamma_1|P^2$ or trivial$]$ in $\overline{\Omega}_2(Y_s;\psi_s)$ show that $\sigma\circ j_2$ is always onto. Hence, by 9.3, we have the isomorphism

$$f_1:\Omega_1(Y_s;\psi_s)\xrightarrow{\cong}\overline{\Omega}_1(Y_s;\psi_s)\cong\begin{cases}\mathbb{Z}_2\oplus\mathbb{Z}_2 & \text{if } s \text{ odd}\\ \mathbb{Z}_2 & \text{if } s \text{ even ,}\end{cases}$$

given by evaluating $w_1(\gamma_s)$(and $w_1(\gamma_1)$).

In order to compute $\sigma\circ j_3$, consider an element $y=[N,\gamma,\lambda,or]$ in $\overline{\Omega}_3(Y_s;\psi_s)$. We have

$$w_1(\gamma_s)(proj_*\circ\mu\circ\sigma\circ j_3(Y)) = (w_1(\lambda)w_1(\gamma)^2+\binom{s+1}{2}w_1(\lambda)^2w_1(\gamma))[N] ,$$
$$w_1(\gamma_1)(proj_*\circ\mu\circ\sigma\circ j_3(Y)) = (w_1(\lambda)^2w_1(\gamma)+\binom{s+1}{2}w_1(\lambda)^3)[N].$$

If s is odd, N is orientable and

$$0 = w_1(N)w_1(\lambda)w_1(\gamma) = Sq^1(w_1(\lambda)w_1(\gamma)) = w_1(\lambda)^2 w_1(\gamma) + w_1(\lambda)w_1(\gamma)^2;$$

the example $y = [P^3, \gamma = \mathbb{R}^{s-1} \oplus \gamma_1|P^3$ or trivial, $\lambda = \gamma_1|P^3]$ then shows that $\mathrm{im}(proj_* \circ \mu \circ \sigma \circ j_3 = \mathbb{Z}_2 (\binom{s+1}{2}+1, 1)$. If s is even, then $w_1(\lambda)\cdot = w_1(N)\cdot = Sq^1$ annihilates the squares of $w_1(\gamma)$ and $w_1(\lambda)$; Therefore, the example $y = [P^2 \times S^1, \text{Möbius} \oplus \mathbb{R}^{s-1}]$ shows that the image of $proj_* \circ \mu \circ \sigma \circ j_3$ equals $\mathbb{Z}_2(\binom{s+1}{2}, 1)$. Thus in both cases $\mathbb{Z}_2 \oplus \{0\}$ is a complement of this image. It follows that $\ker f_2 \cong \mathbb{Z}_2$, and it is not hard to describe a generator of this kernel (recall the construction of δ_2!).

For odd $s>1$, elements $[N,\gamma,\lambda]$ in $\overline{\Omega}_2(Y_s;\psi) = \Omega_2(BO(S) \times BO(1))$ are detected by the characteristic numbers $w_2(\gamma)[N]$ and $w_1(\lambda)w_1(\gamma)[N] = \sigma \circ j_2[N,\gamma,\lambda]$ (cf. the arguments immediately prior to 10.7). When s is even, we have the complete invariants $w_2(\gamma)[N]$, $w_1(N)^2[N]$ and $(w_1(\gamma)^2 + \binom{s+1}{2} w_1(N)^2[N] = \sigma \circ j_2[N,\gamma]$ for $[N,\gamma] \in \overline{\Omega}_2(Y_s;\psi_s) = \mathfrak{N}_2(BO(S)) \cong \mathfrak{N}_2 \oplus H_2(BO(S);\mathbb{Z}_2)$. Hence for $s>1$ we obtain the exact sequence

$$0 \longrightarrow \mathbb{Z}_2 \xrightarrow{\delta_2|} \Omega_2(Y_s;\psi_s) \xrightarrow{f_2} \begin{Bmatrix} \mathbb{Z}_2 & s \text{ odd} \\ \mathbb{Z}_2 \oplus \mathbb{Z}_2 & s \text{ even} \end{Bmatrix} \longrightarrow 0$$

where $f_2[N,\gamma,\lambda,\overline{h}]$ is given by $w_2(\gamma)[N]$ (and $w_1(N)^2[N]$).

Next note that the forgetful homomorphism $\mathfrak{N}_n(k,k-1) \longrightarrow \mathfrak{N}_n$ is bijective if $n>2k-3$, by an argument involving the codimension of singularities. Thus, for $n \geq 1$ and $m=k$, the exact sequence 7.23 contains the following piece (for an analysis of the adjoining piece to the right, see the proof of 12.1).

$$\ldots \longrightarrow \mathfrak{N}_{n+1} \xrightarrow{\sigma \circ j} \Omega_1(BO(n) \times BO(1);\psi_n) \xrightarrow{\delta} \mathfrak{N}_n(1,1) \xrightarrow{f} \mathfrak{N}_n \xrightarrow{w_n} \mathbb{Z}_2$$

(12.9)

$$\Omega_1(BO(n) \times BO(1);\psi_n) \xrightarrow[\cong]{w_1(\gamma_n)(,w_1(\gamma_1))} \underbrace{\underbrace{\mathbb{Z}_2}_{n \text{ even}} (\oplus \mathbb{Z}_2)}_{n \text{ odd}}$$

Given an $(n+1)$-manifold N, let $u: TN \longrightarrow \mathbb{R}^2$ be a nondegenerate

1-morphism with singularity S. Then, by the procedure 9.11 and by the relations 12.4, we have

$$\begin{aligned} w_1(\gamma_n)\sigma\circ j[N] &= w_1(\mathrm{Ker})[S] = (w_1(N)|S)[S] + w_1(\mathrm{Coker})[S] \\ &= w_n(N)w_1(N)[N] + w_{n+1}(N)[N] \\ &= \begin{cases} 0 & \text{if } n \not\equiv 3(4) \\ w_{n+1}(N)[N] & \text{if } n \equiv 3(4) \end{cases} \end{aligned}$$

and

$$\begin{aligned} w_1(\gamma_1)\sigma\circ j[N] &= w_1(\mathrm{Coker})[S] \\ &= w_{n+1}(N)[N]. \end{aligned}$$

Therefore, $\mathbb{Z}_2 (\oplus 0)$ corresponds to a complement of the image of $\sigma\circ j$. We get the exact sequence

$$0 \longrightarrow \mathbb{Z}_2 \xrightarrow{\ \delta|\ } \mathfrak{N}_n(1,1) \xrightarrow{\ f\ } \ker(w_n : \mathfrak{N}_n \to \mathbb{Z}_2) \longrightarrow 0 ,$$

where $\delta|(1)$ can be represented by a horizontal vectorfield on the (total space of the) sphere bundle $S(\text{Möbius} \oplus \mathbb{R}^{n-1})$ over S^1. Hence, by lemma 12.7, χ'' is a left inverse of $\delta|$, and the theorem follows for $k=1$.

The calculation above shows that the homomorphisms $\sigma\circ j$ and w_{n+1} have the same kernel in $\mathfrak{N}_{n+1}$ (and similarly in $\mathfrak{N}_n$). So the next piece of our exact singularity sequence 7.23 takes the form (for $n \geq 3$)

(12.10)

$$\begin{array}{ccccccccccc} & & & & 0 & & & & & & \\ & & & & \downarrow & & & & & & \\ & & & & \mathbb{Z}_2 & & & & & & \\ & & & & \downarrow{\scriptstyle \delta_2|} & & & & & & \\ \mathfrak{N}_{n+1}(3,3) & \xrightarrow{f} & \mathfrak{N}_{n+1}(3,2) & \xrightarrow{\sigma\circ j} & \Omega_2(BO(n-1)\times BO(1);\psi_{n-1}) & \xrightarrow{\delta} & \mathfrak{N}_n(2,2) & \xrightarrow{f} & \mathfrak{N}_n & \xrightarrow{w_n} & \mathbb{Z}_2 \\ & & \downarrow{\scriptstyle \cong} & & \downarrow{\scriptstyle f_2} & & & & & & \\ & & \mathfrak{N}_{n+1} & & \underbrace{\underbrace{\mathbb{Z}_2}_{n \text{ even}} (\oplus \mathbb{Z}_2)}_{n \text{ odd}} & & & & & & \end{array}$$

where $f_2[S,\gamma,\lambda,\overline{h}]$ is given by $w_2(\gamma)[S]$ (and $w_1(S)^2[S]$).

Consider an (n+1)-manifold N and a nondegenerate 2-morphism $TN \longrightarrow \underset{\sim}{\mathbb{R}}^3$ with singularity S. We know again from 9.11 that $f_2 \circ \sigma \circ j[N]$ is given by

$$w_2(\underset{\sim}{Ker})[S] = (w_2(N)|S)[S] + (w_1(N)|S)w_1(\underset{\sim}{Coker})[S]$$
$$= w_2(N)w_{n-1}(N)[N] + w_1(N)w_n(N)[N]$$

(and by

$$w_1(S)^2[S] = nw_1(\underset{\sim}{Coker})^2[S]$$
$$= nw_{n+1}(N)[N] \qquad).$$

We claim that $f_2 \circ \sigma \circ j$ is always onto. To see this, it suffices to exhibit an m-manifold N for every dimension $m \geq 4$ such that

$$w_2(N)w_{m-2}(N)[N] = 1 \quad \text{and} \quad w_m(N)[N] = 0$$

(and hence $w_1(N)w_{m-1}(N)[N]=0$ by lemma 12.4). For $m=4$ the manifold $N=P^2 \times P^2 \amalg P^4$ will do. If $m=5$, the nontrivial element $[N]$ in $\Omega_5 = \mathbb{Z}_2$ must necessarily be detected by the characteristic number $w_2(N)w_3(N)[N]$. On the other hand, if an arbitrary m-manifold N satisfies the condition above, then so does the (m+2r)-manifold $N \times P$, $P=P^{2r}$, for $r \geq 0$. Indeed,

$$w_2(N\times P)w_{m+2r-2}(N\times P)[N\times P]$$
$$= (w_2(N)+w_1(N)w_1(P)+w_2(P))(w_{m-2}(N)w_{2r}(P)+w_{m-1}(N)w_{2r-1}(P)+w_m(N)w_{2r-2}(P))[N\times P]$$
$$= (w_2w_{m-2}(N)w_{2r}(P)+w_1w_{m-1}(N)w_1w_{2r-1}(P)+w_m(N)w_2\,w_{2r-2}(P))[N\times P]$$
$$= w_2(N)w_{m-2}(N)[N]\cdot w_{2r}(P)[P]$$
$$= 1$$

So the required manifold exists in every dimension $m \geq 4$.

It follows that the sequence

$$\mathbb{Z}_2 \xrightarrow{\delta\circ\delta_2|} \mathfrak{N}_n(2,2) \xrightarrow{f} \ker(w_n : \mathfrak{N}_n \longrightarrow \mathbb{Z}_2) \longrightarrow 0$$

is exact for $n \geq 2$ (if $n=2$, see also 7.21). Here $\delta\circ\delta_2|(1)$ can be represented by $S(\text{Möbius} \oplus \underset{\sim}{\mathbb{R}}^{n-2}) \times S^1$, together with a horizontal vectorfield on the spherebundle $S(\text{Möbius} \oplus \underset{\sim}{\mathbb{R}}^{n-2})$ over the unit circle, and the vectorfield along the second factor S^1. Hence we know from lemma 12.7 that χ'' is a left inverse of $\delta\circ\delta_2|$. This implies the theorem also for $k=2$. Moreover, since $\delta\circ\delta_2|$ is injective, we conclude that

$$(12.11) \qquad f(\mathfrak{N}_{n+1}(3,3)) = \ker(f_2\circ\sigma\circ j)$$

in the diagram (12.10). ∎

Given integers n and k, consider the following two conditions concerning a bordism class $x = [M] \in \mathfrak{N}_n$.

$B_k(x)$: x lies in the image of the forgetful map $f:\mathfrak{N}_n(k,k)\longrightarrow\mathfrak{N}_n$, i.e. M is bordant to a manifold M' which allows a k-framefield $\underset{\sim}{\mathbb{R}}^k \hookrightarrow TM'$.

$C_k(x)$: all Stiefel-Whitney numbers of M which contain some $w_i(M)$, $i>n-k$, as a factor, vanish.

Clearly $B_k(x)$ implies $C_k(x)$ since $w_i(M') = 0$ for $i > n-k$.

Definition 12.12. A pair (n,k) of integers is called $\mathfrak{N}$-pleasant if the conditions $B_k(x)$ and $C_k(x)$ are equivalent for all $x\in\mathfrak{N}_n$, i.e. if

$$f(\mathfrak{N}_n(k,k)) = \{x \quad \mathfrak{N} \mid C_k(x) \text{ is satisfied}\}.$$

It is important to know whether there is a dimension combination (n,k) which fails to be $\mathfrak{N}$-pleasant; such a pair would have to satisfy $0<k<n$. The following partial answer will be very useful in the next chapter.

Theorem 12.13. *For $k\leq 4$ and arbitrary $n\in\mathbb{Z}$ the pair (n,k) is $\mathfrak{N}$-pleasant.*

Proof. For $k\leq 3$, this claim follows from the results (12.9), (12.10) and (12.11) in the last proof. The full statement of the theorem can be deduced from the work of Stong, who shows e.g. in [61], proposition 7.2., that any bordism class $x \in \mathfrak{N}_n$ satisfying condition $C_4(x)$ contains a manifold M' which fibers over the torus $(S^1)^4$ (and hence carries a "horizontal" 4-frame field). ∎

CHAPTER III. FRAMEFIELDS.

§13. The complete obstructions $\omega_k(M)$ and $\omega'_k(M)$.

Throughout this section M will be a closed smooth n-dimensional manifold, and k is a natural number.

Definition 13.1. A k-field on M is a vectorbundle homomorphism $u : M\times \mathbb{R}^k \longrightarrow TM$ (or, equivalently, a system of k tangential vectorfields $v_1,..v_k$ on M, given by $v_i(x)=u(x,e_i)$ for $x\in M$, where $e_i=(0,..0,1,0,.)$ is the i-th standard basis vector in $\mathbb{R}^k$, $1 \leq i \leq k$).

The singularity S of a k-field u is defined by

$$S = \{ x\in M \mid \operatorname{rank}(u_x) < k \}$$

(i.e., S is the locus of points x in M where the vectors $v_1(x),\dots,v_k(x)$ fail to be linearly independent in the tangent space $T_x(M)$).

The span of the manifold M is the maximum number k such that M has a k-field without singularities (i.e. the maximum number of linearly independent vector fields on M).

If $n\geq 2k-3$, there exists a smooth nondegenerate (k-1)-morphism $u : M\times\mathbb{R}^k \longrightarrow TM$ (see definition 1.4). Its singularity is a (k-1)-dimensional smooth submanifold of M, equipped with vectorbundles $\underset{\sim}{Ker}$, $\underset{\sim}{Coker}$ and $\underset{\sim}{Im}$ of dimension 1, n-k+1 and k-1 respectively.

In particular, the line bundle $\underset{\sim}{Ker} \subset S\times\mathbb{R}^k$ and the inclusion $S \subset M$ define a map

$$g : S \longrightarrow P^{k-1} \times M .$$

Moreover, we have isomorphisms (which are canonical up to homotopy)

$$S\times\mathbb{R}^k \cong \underset{\sim}{Ker} \oplus \underset{\sim}{Im} ,$$

$$TS \oplus (\underset{\sim}{Ker}\otimes\underset{\sim}{Coker}) \cong TM|S \cong \underset{\sim}{Coker} \oplus \underset{\sim}{Im} ,$$

and therefore, adding $\underset{\sim}{Ker} \otimes\underset{\sim}{Im}\oplus \underset{\sim}{\mathbb{R}} \cong \underset{\sim}{Ker}\otimes \underset{\sim}{\mathbb{R}}^k$,

$$\overline{g} : TS \oplus (\underset{\sim}{Ker} \otimes TM|S) \oplus \underset{\sim}{\mathbb{R}} \cong TM|S \oplus \underset{\sim}{Ker} \otimes \underset{\sim}{\mathbb{R}}^k.$$

Thus we can define

$$\omega_k(M) = [S , g , \overline{g}] \in \Omega_{k-1}(P^{k-1}xM;\phi_M)$$

and

$$\omega_k'(M) = [S,(Idx\tau)\circ g,\overline{g}] \in \Omega_{k-1}(P^{k-1}xB(S)O(n);\phi).$$

Here we are treating two parallel theories simultaneously. If M is oriented, then $B(S)O(n)$ stands for the classifying space $BSO(n)$ of oriented n-plane bundles; if an orientation of M is to be neglected or nonexistent, then $B(S)O(n)$ stands for $BO(n)$. In both cases, $\tau:M \longrightarrow B(S)O(n)$ denotes a classifying map of the tangent bundle of M. The coefficient bundles ϕ_M and ϕ over $P^{k-1}xM$ and $P^{k-1}xB(S)O(n)$ are defined by

$$\text{(13.2)} \qquad \begin{aligned} \phi_M &= \lambda \otimes TM - k\lambda - TM \quad \text{and} \\ \phi &= \lambda \otimes \gamma - k\lambda - \gamma , \end{aligned}$$

where λ and γ are the canonical bundles over P^{k-1} and $B(S)O(n)$. We will see presently that the invariants above can be defined even when our condition on k is not satisfied.

<u>Theorem 13.3.</u> <u>For all</u> $1 \leq k \leq n$ $\omega_k(M)$ <u>is a welldefined invariant of M.</u> <u>Moreover, for</u> $k < \frac{n}{2}$ <u>we have:</u>

$$\underline{\omega_k(M) = 0 \quad \text{if and only if} \quad span(M) \geq k.}$$

<u>Proof.</u> $\omega_k(M)$ coincides with the invariant $\omega_k(\underset{\sim}{\mathbb{R}}^k,TM,\emptyset)$ defined in (2.12); for $n < 2k-3$, see also proposition 2.15. If $k < \frac{n}{2}$, then by theorem 3.7 this is the precise obstruction to the existence of a vector bundle monomorphism $\underset{\sim}{\mathbb{R}}^k \hookrightarrow TM$. ∎

While we may think of $\omega_k(M)$ as a <u>characteristic class</u> living in a group which depends on M, the following result shows that $\omega_k'(M)$ plays very much the rôle of a <u>characteristic number.</u>

Theorem 13.4. For all $1 \le k \le n$ $\omega_k'(M)$ is welldefined and depends only on the bordism class of M in the "fine" (or vector field) bordism group $\mathfrak{N}_n(0,0)$, or $\Omega_n(0,0)$ resp., (cf. 12.1); consequently, $\omega_k'(M)$ depends only on the following classical invariants of M: in the unoriented theory on the Stiefel-Whitney numbers and the Euler number of M, and in the oriented theory in addition on the Pontrjagin numbers and, for $n \equiv 1(4)$, on the (real) Kervaire semicharacteristic $\mathfrak{k}(M)$ of M.

Moreover, for $n > 2k-2$, we have: $\omega_k'(M)$ vanishes if and only if M is bordant (in $\mathfrak{N}_n(0,0)$ or $\Omega_n(0,0)$ resp.) to a manifold M' with $\operatorname{span}(M') \ge k$.

Proof. Two n-manifolds M and M' are bordant in the "fine" sense mentioned above if a bordism B from M to M' allows a vectorfield without zeroes which points inwards at M and outwards at M'. Such a vectorfield determines a complementary n-plane bundle η in TB which extends $TM \cup TM'$ over all of B. Now recall from 2.15 that $\omega_k'(M)$ may always be formed from the singularity data of a nondegenerate section of the n-plane bundle $\gamma \otimes TM$ over $P^{k-1} \times M$. Applying the same procedure to the n-plane bundle $\lambda \otimes \eta$ over $P^{k-1} \times B$, we get a bordism which shows that $\omega_k'(M) = \omega_k'(M')$.

It follows from theorem 12.1. and from classical bordism theory that the characteristic numbers listed above determine the fine bordism class of $[M]$, and hence $\omega_k'(M)$. Moreover, note that ω_k' defines a homomorphism on $\mathfrak{N}_n(0,0)$ and $\Omega_n(0,0)$.

If M is bordant in the "fine" sense to a manifold with $\operatorname{span} \ge k$, then clearly $\omega_k'(M)=0$.

Conversely, assume that $n > 2k-2$ and that $\omega_k'(M)=0$. Then, given a nondegenerate (k-1)-morphism $u: \underset{\sim}{\mathbb{R}}^k \longrightarrow TM$ with singularity data $(S,(\underset{\sim}{Ker}, TM|S), \bar{g})$, there is a bordism $\mathcal{S}$ of S with vectorbundles $\widetilde{\underset{\sim}{Ker}}$ and η of dimensions 1 and n, extending $\underset{\sim}{Ker}$ and $TM|S$, and there is also an appropriate extension $\overline{G}$ of $\overline{g}$. Then we have also splittings $\mathcal{S} \times \mathbb{R}^k = \underset{\sim}{\widetilde{Ker}} \oplus \underset{\sim}{\widetilde{Coim}}$ and $\eta = \underset{\sim}{\widetilde{Coker}} \oplus \underset{\sim}{\widetilde{Im}}$ which restrict to the canonical splittings at S; just define $\underset{\sim}{\widetilde{Coim}}$ as the complement of $\widetilde{Ker}$ in $\mathcal{S} \times \mathbb{R}^k$, and let $\underset{\sim}{\widetilde{Im}}$ be the image of any monomorphism $\tilde{u}_2 : \underset{\sim}{\widetilde{Coim}} \hookrightarrow \eta$ extending $u|S$. ($\tilde{u}_2$ exists because $n-k+2>k$). Finally, we can destabilize $\overline{G}$, and we obtain a vector bundle isomorphism $T\mathcal{S} \oplus (\underset{\sim}{\widetilde{Ker}} \otimes \underset{\sim}{\widetilde{Coker}}) \cong \eta \oplus \underset{\sim}{\mathbb{R}}$ with an appropriate behaviour at $S = \partial \mathcal{S}$.

Now we can apply the model construction of §1 and the structure lemma 1.10. We fit $M \times I$ together with the total space of the disk bundle $D(\zeta)$ of $\zeta = \underset{\sim}{Hom}(\underset{\sim}{\widetilde{Ker}}, \underset{\sim}{\widetilde{Coker}})$ to get a bordism B from $M = M \times \{0\}$ to $M' = (M - D(\zeta|S)) \times \{1\} \cup S(\zeta)$. B carries a nonvanishing vectorfield v which points inwards at M and which allows k linearly independent complementary vectorfields at M'; however, v might not point outward at M', so that (B, v) does not give us a fine bordism. This difficulty can be remedied by stabilizing.

Consider the commutative diagram

$$\begin{array}{ccc}
\mathcal{N}_n(k,k) & \xrightarrow[\cong]{St} & \Omega_n(BO(1), (\underset{\sim}{\mathbb{R}}, \underset{\sim}{\mathbb{R}}^{k+1}), \gamma_1, k+1) \\
\downarrow f & & \downarrow f \\
\mathcal{N}_n(0,0) & \xrightarrow[\cong]{St} & \Omega_n(BO(1), (\underset{\sim}{\mathbb{R}}, \underset{\sim}{\mathbb{R}}), \gamma_1, 1)
\end{array}$$

or its oriented analogue; e.g. the right hand lower group can be identified with the fine bordism group of n-manifolds N with a monomorphism $\underset{\sim}{\mathbb{R}} \hookrightarrow TN \oplus \underset{\sim}{\mathbb{R}}$. Recall from proposition 7.7 that the stabilizing map St is always an isomorphism.

Now return to the vectorfield v on B. Stabilizing provides the extra space which allows us to rotate v into the outward pointing vectorfield at M', and there is also a monomorphism $\underset{\sim}{\mathbb{R}}^{k+1} \longrightarrow TM' \oplus \underset{\sim}{\mathbb{R}}$. This shows that $St[M]$ lies in the image of the right hand forgetful map f, and hence $[M] \in f(\mathcal{N}_n(k,k))$. Thus M is bordant in $\mathcal{N}_n(0,0)$

to a manifold with span $\geq k$. ■

Example. We indicate how one can extract the Euler number $\chi(M)$ of M from the invariant $\omega_k(M)$ or $\omega'_k(M)$.

Given an element $z = [S , (g_1,g_2):S \longrightarrow P^{k-1}xB(S)O(n) , \overline{g}]$ of $\Omega_{k-1}(P^{k-1}xB(S)O(n);\phi)$, we define the degree $\deg(z)$ as follows. If n is odd, $\deg(z)$ is just the mod 2 mapping degree of $g_1:S \longrightarrow P^{k-1}$ (S is a (k-1)-dimensional closed manifold!). If n is even, then $\overline{g}$ can be brought into the following form (at least stably)

$$\overline{g} : TS \oplus g_1^*(\lambda) \otimes g_2^*(\gamma) \cong g_1^*(TP^{k-1}) \oplus g_2^*(\gamma) ,$$

and the orientations of $g_1^*(\lambda) \otimes g_2^*(\gamma)$ and of $g_2^*(\gamma)$ correspond to one another canonically; hence, so do the orientations of TS and $g_1^*(TP^{k-1})$. Now pick a regular value y of the map g_1 (which we may assume to be smooth), and define $\deg(z)$ to be the algebraic number of points in $g_1^{-1}(y)$; a point $x \in g_1^{-1}(y)$ is counted as +1 or -1 according to whether or not the tangent isomorphism $Tg_1 : T_xS \longrightarrow T_y(P^{k-1})$ relates orientations in the same way as $\overline{g}$ does.

This way we obtain a welldefined homomorphism deg from $\Omega_{k-1}(P^{k-1}xB(S)O(n);\phi)$ into $\mathbb{Z}$ (when n is even), resp. into $\mathbb{Z}_2$ (when n is odd). In the same fashion, deg can also be defined on $\Omega_{k-1}(P^{k-1}xM;\phi_M)$.

Proposition 13.5. Let $1 \leq k \leq n$. If n is even, then the integers $\deg(\omega_k(M))$, $\deg(\omega'_k(M))$ and the Euler number $\chi(M)$ are all equal. If n is odd, then the mod 2 numbers $\deg(\omega_k(M)) = \deg(\omega'_k(M))$ vanish.

Proof. Pick a nondegenerate 0-morphism $v : Mx\mathbb{R} \longrightarrow TM$ with zeroes $x_1,\ldots,x_r \in M$, and pick a trivialization $t : \lambda|U \cong Ux\mathbb{R}$ over a suitable neighborhood U of some point y in P^{k-1}. Then there is a nondegenerate 0-morphism $u : \pi_1^*(\lambda) \longrightarrow \pi_2^*(TM)$ over all of $P^{k-1}xM$ which restricts to

$$u| = \pi_2^*(v)\cdot\pi_1^*(t) : \pi_1^*(\lambda|U) \xrightarrow{\pi_1^*(t)} U \times M \times \mathbb{R} \xrightarrow{\pi_2^*(v)} \pi_2^*(TM)$$

over $U\times M$. The singularity S of u, together with the inclusion $g : S \subset P^{k-1}\times M$ and the obvious isomorphism

$$\bar{g} : TS \oplus (\pi_1^*(\lambda)\otimes\pi_2^*(TM))|S \cong (\pi_1^*(TP^{k-1}) \oplus \pi_2^*(TM))|S ,$$

represents $\omega_k(M)$.

On the other hand, $S \cap U\times M = U\times\{x_1,..x_i,..x_r\}$. Therefore y is a regular value of $g_1=\pi_1|S$, and $g_1^{-1}(y)=\{(y,x_1),..,(y,x_i),..,(y,x_r)\}$. If n is even and if we form $\deg(\omega_k(M))$, the point (y,x_i) is counted as $+1$ if and only if the automorphism of $T_{x_i}(M)$ induced by v is orientation preserving, i.e. x_i contributes $+1$ to the index sum of v. Thus $\deg(\omega_k(M))=\deg(\omega_k'(M))$ is the index sum of v which in turn equals $\chi(M)$ by the theorem of Poincaré-Hopf. If n is odd, $\deg(\omega_k(M))=\deg(\omega_k'(M))$ is the mod 2 Euler number which vanishes.∎

Next consider a k-field $u:M\times\mathbb{R}^k \longrightarrow TM$ with finite singularities, i.e. the singularity S of u consists of finitely many points. (Thus, in particular, u fails to be a nondegenerate (k-1)-morphism unless $k=1$ or $S=\emptyset$). If $x\in S$, pick an orientation or_x of T_xM and a chart $\phi : U \xrightarrow{\cong} \mathbb{R}^n$ such that $U\cap S = \{x\}$, $\phi(x) = 0$, and or_x corresponds to the standard orientation of $\mathbb{R}^n$ under the tangent map of ϕ. Then ϕ carries u into a k-field $\phi_*(u)$ on $\mathbb{R}^n$, which, when restricted to the unit sphere $S^{n-1}\subset\mathbb{R}^n$, determines a map $h:S^{n-1}\longrightarrow W^k(\mathbb{R}^k,\mathbb{R}^n)\sim V_{n,k}$. The resulting class $\iota(u,x,or_x) = [h] \in [S^{n-1},V_{n,k}]$ ($\approx \pi_{n-1}(V_{n,k})$ for $k < n$) in the homotopy of the Stiefel manifold $V_{n,k}$ is independent of the choice of the chart ϕ (compare [49], p. 33-35). The sum (defined when $k < n$)

$$\mathrm{Index}(u,or) = \sum_{x\in S} \iota(u,x,or_x) \in \pi_{n-1}(V_{n,k})$$

is called the <u>index of u</u> with respect to the system of orientations $or = \{or_x\}_{x\in S}$.

Proposition 13.6. Let M be a connected n-manifold which is either oriented or nonorientable, and let u be a k-field with finite singularity set S, $1 \leq k \leq n$. Pick a point $x_0 \in M$, and let or_x, $x \in S \cup \{x_0\}$, be the given, respectively, an arbitrary, orientation of $T_x M$. Then there exists a k-field u' on M with just one singular point x_0 such that

$$\iota(u', x_0, or_{x_0}) = \sum_{x \in S} \iota(u, x, or_x) = \mathrm{Index}(u, or).$$

In particular, if $\mathrm{Index}(u,or) = 0$, then $\mathrm{span}(M) \geq k$.

Proof. Using vectorfields with support around embedded arcs, we can isotop all points $x \in S$ into the interior of a disc D around x_0. We may assume that all orientations are compatible; otherwise we perform some additional isotopies along orientation reversing loops on M.

u gets deformed correspondingly into a k-field u'' which is nonsingular outside of $D - \partial D$. Using just $u''|\partial D$, multiplied with the radius function on D, we can modify u'' in D and we obtain the desired k-field u' with singularity $\{x_0\}$. Then $\iota(u', x_0, or_{x_0})$ is represented by $u'|\partial D = u''|\partial D$, and so is $\mathrm{Index}(u'', or) = \mathrm{Index}(u, or)$ (analyse the definition of sums in $\pi_{n-1}(V_{n,k})$!). The identity above follows.

If $\mathrm{Index}(u,or)$ vanishes, then $u''|\partial D$ can be extended to a nonsingular k-field over all of D; this fits together with $u''|M-D$ to define a nonsingular k-field over M. ∎

Finally we relate the main invariants of this section with one another. Consider the commutative diagram

(13.7)

$$\begin{array}{ccccc}
 & & \pi_{n-1}(V_{n,k}) & & \\
 & & \Big\downarrow \sigma \quad (\cong \text{ for } n>2k) & & \\
 & & \Omega_{k-1}(P^{k-1};(n-k)\lambda) & & \\
 & \swarrow \Theta_M & & \Theta \searrow & \\
\Omega_{k-1}(P^{k-1}\times M;\phi_M) & & \xrightarrow{(id\times\tau)_*} & & \Omega_{k-1}(P^{k-1}\times B(S)O(n);\phi) \\
\cup & & & & \cup \\
\omega_k(M) & & & & \omega_k'(M)
\end{array}$$

Here $(id\times\tau)_*$ is derived in the obvious fashion from any classifying map τ of the tangent bundle of M. Θ_M is deduced from the inclusion $P^{k-1}\times\{x_o\}\subset P^{k-1}\times M$ and from an isomorphism $is:T_{x_o}M \cong \mathbb{R}^n$ (orientation preserving in the oriented theory). This definition is independent of the choice of the point x_o in M and of is if M is connected (and non-orientable in the unoriented theory). In a similar fashion, Θ is defined canonically. The singularity map σ (cf. 5.1) is homomorphic unless $(n,k)=(1,1)$, and bijective for $n>2k$ (see proposition 5.4).

<u>Proposition 13.8.</u> <u>Let $1 \le k \le n$. Then we have</u>

$$(id \times \tau)_*(\omega_k(M)) = \omega_k'(M)$$

<u>Moreover, if M is connected and has a k-field u with finite singularity S and if $k < n$, then</u>

$$\Theta_M \circ \sigma(\mathrm{Index}(u,or)) = \omega_k(M) \qquad \text{and}$$
$$\Theta \circ \sigma(\mathrm{Index}(u,or)) = \omega_k'(M)$$

<u>(where the system or of orientations is arbitrary, if M is non-orientable; and compatible with the choice of $is:T_{x_o}M \cong \mathbb{R}^n$ in the definition of Θ_M, if M is orientable. In particular, in the oriented theory or is given by the orientation of M).</u>

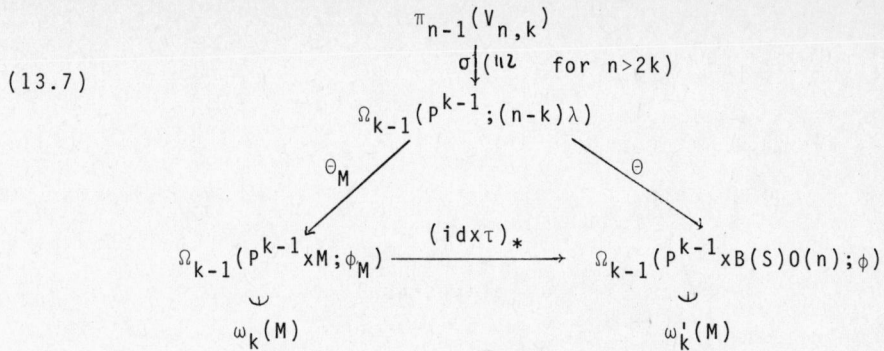

<u>Proof.</u> The first identity follows directly from the definitions of $\omega_k(M)$ and $\omega'_k(M)$. The rest follows from proposition 13.6 and the definitions of σ and Θ_M. ∎

There are also interesting relations between $\omega_k(M)$ and $\omega_{k+1}(M)$, $\omega'_k(M)$ and $\omega'_{k+1}(M)$, etc. In order to study them, let Y stand for {point}, M or $B(S)O(n)$, and let $\phi_{Y,k}$ denote the virtual bundle $(n-k)\lambda$, ϕ_M or ϕ over $P^{k-1}xY$ (defined as in 13.2 with respect to the number k). Thus e.g. $(\lambda+\phi_{Y,k+1})|P^{k-1}xY = \phi_{Y,k}$. Note that (1.2) induces a canonical isomorphism which identifies the normal bundle ν of the inclusion $P^{k-1}xY \subset P^kxY$ with the canonical line bundle λ over $P^{k-1}xY$:

$$j: \ \nu \cong \left(\underset{\sim}{\mathrm{Hom}}(\lambda, \underset{\sim}{\mathbb{R}}^{k+1}/\lambda) \big/ \underset{\sim}{\mathrm{Hom}}(\lambda, \underset{\sim}{\mathbb{R}}^{k}/\lambda)\right) \cong \underset{\sim}{\mathrm{Hom}}(\lambda, \underset{\sim}{\mathbb{R}}) \cong \lambda .$$

We can use the resulting Thom isomorphism (cf. [57], sections 3 and 5) to simplify the normal bordism sequence of the pair

$$P = (P^kxY, (P^k-P^{k-1}) \times Y) .$$

We obtain the commuting diagram of long exact sequences

$$\begin{array}{ccccccccc}
\cdots \to \Omega_{i+1}(P;\phi_{Y,k+1}) & \xrightarrow{\pm\partial} & \Omega_i((P^k-P^{k-1})xY;\phi_{Y,k+1}) & \to & \Omega_i(P^kxY;\phi_{Y,k+1}) & \to & \Omega_i(P;\phi_{Y,k+1}) & \longrightarrow & \cdots \\
\text{Thom}\ \updownarrow\ \cong & & \text{incl}*\ \uparrow\ \cong & & \| & & \text{Thom}\ \updownarrow\ \cong & & \\
\cdots \to \Omega_i(P^{k-1}xY;\phi_{Yk}) & \xrightarrow{d} & \Omega_i(Y;\text{trivial}) & \xrightarrow{\text{incl}*} & \Omega_i(P^kxY;\phi_{Y,k+1}) & \xrightarrow{\Delta} & \Omega_{i-1}(P^{k-1}xY;\phi_{Y,k}) & \to &
\end{array} \tag{13.9}$$

Here the homomorphisms incl_* are induced by the obvious inclusion

$$\begin{array}{ccc}
\mathbb{R} & \subset & \lambda| \\
\downarrow & & \downarrow \\
\{[(0,0,..0,1)]\} & \xrightarrow[\sim]{\subset} & P^k-P^{k-1}
\end{array}$$

$\Delta[\ N\ ,\ g=(g_1,g_2):N \longrightarrow P^kxY\ ,\ \bar{g}\ ]$ is represented by the submanifold $Z = g_1^{-1}(P^{k-1}) \subset N$ (we may assume that g_1 is smooth and transverse to P^{k-1}) together with $g|Z$ and the isomorphism

$$TZ \oplus g|^*(\phi^+_{Y,k}) = TZ \oplus g|^*(\lambda\oplus\phi^+_{Y,k+1}) \cong TN|Z \oplus g^*(\phi^+_{Y,k+1})| \xrightarrow{\overline{g}|Z} g^*(\phi^-_{Y,k(+1)})|$$

(where the isomorphism $g|^*(\lambda) \cong \nu(Z,N)$ is induced from the idenfication j above).

Finally, $d[L\ ,\ h=(h_1,h_2):L \to P^{k-1}xY\ ,\ \overline{h}]$ is represented by the double cover manifold $\tilde{L}=S(h_1^*(\lambda))$ (associated with the line bundle $h_1^*(\lambda)$), together with the obvious map $\tilde{L} \xrightarrow{\pi} L \xrightarrow{h_2} Y$ and the isomorphism

$$T\tilde{L}\oplus\pi^*(h^*(\eta)) \cong \pi^*(TL\oplus h^*(\lambda\otimes\eta)) \xrightarrow{\pi^*(\overline{h})} \pi^*h^*(\eta)(\oplus\pi^*h^*(k\lambda))$$

(we omit trivial factors); here we use the obvious trivialisation of the linebundle $\pi^*(h_1^*(\lambda))$ over $\tilde{L}$, as well as the fact that $\phi_{Y,k}$ is of the form $\lambda\otimes\eta-\eta-k\lambda$.

Before applying the exact sequence defined above, we first define one more invariant. Given a k-framefield $v: \mathbb{R}^k \hookrightarrow TM$ on M, let ζ denote a complement of the image of v, and let $S \subset M$ be the zero set of a generic section of ζ. Then S is a closed k-manifold naturally equipped with the inclusion map $j:S \subset M$ and with the stable parallelization

$$\overline{h}\ :\ TS\oplus\zeta|S \cong TM|S \cong \mathbb{R}^k\oplus\zeta|S\ .$$

These data give rise to the welldefined invariant

(13.10) $\qquad \chi(M,v) = [S,j,\overline{h}] \in \Omega_k(M;\text{trivial})$

which depends only on the (regular) homotopy class of v and which is closely related to the bordism invariants

$$\chi'(M,v) = [S,\zeta|S,\overline{h}] \in \Omega_k(B(S)O(n-k);\text{trivial})$$

and

$$\chi''(M,v) = [S,\ \overline{h}] \qquad \in \Omega_k(\text{point};\text{trivial})$$

defined in (12.5). In the extreme case when $k=0$ (and M is connected) all three invariants coincide with the Euler number $\chi(M) \in \mathbb{Z}$; this motivates our notation.

Theorem 13.11. Let $1 \leq k < n$. Then the diagram

$$
\begin{array}{ccccccc}
\cdots\to\pi_n(V_{n,k}) & \xrightarrow{\partial} & \pi_{n-1}(S^{n-k-1}) & \xrightarrow{incl_*} & \pi_{n-1}(V_{n,k+1}) & \xrightarrow{proj_*} & \pi_{n-1}(V_{n,k}) \to\cdots \\
 & & & & & & \ni \ \mathrm{Index}(u,or) \\
\downarrow{\scriptstyle \pm\sigma} & & \downarrow{\scriptstyle (-1)^k\sigma} & & \downarrow{\scriptstyle \sigma} & & \downarrow{\scriptstyle \sigma} \\
\cdots\to\Omega_k(P^{k-1};(n-k)\lambda) & \xrightarrow{d} & \Omega_k(\mathrm{point};\mathrm{trivial}) & \xrightarrow{incl_*} & \Omega_k(P^k;(n-k-1)\lambda) & \xrightarrow{\Delta} & \Omega_{k-1}(P^{k-1};(n-k)\lambda)\to\cdots \\
\downarrow{\scriptstyle \Theta_M} & & \downarrow{\scriptstyle \Theta_M} & & \downarrow{\scriptstyle \Theta_M} & & \downarrow{\scriptstyle \Theta_M} \\
\cdots\to\Omega_k(P^{k-1}\times M;\phi_{M,k}) & \xrightarrow{d} & \Omega_k(M;\mathrm{trivial}) & \xrightarrow{incl_*} & \Omega_k(P^k\times M;\phi_{M,k+1}) & \xrightarrow{\Delta} & \Omega_{k-1}(P^{k-1}\times M;\phi_{M,k})\to\cdots \\
\ni\ d_k(v_o,v_1) & & \ni\ \chi(M,v) & & \ni\ \omega_{k+1}(M) & & \ni\ \omega_k(M) \\
\downarrow{\scriptstyle (id\times\tau)_*} & & \downarrow{\scriptstyle \tau_*} & & \downarrow{\scriptstyle (id\times\tau)_*} & & \downarrow{\scriptstyle (id\times\tau)_*} \\
\cdots\to\Omega_k(P^{k-1}\times B(S)O(n),\phi_k) & \xrightarrow{d} & \Omega_k(B(S)O(n);\mathrm{trivial}) & \xrightarrow{incl_*} & \Omega_k(P^k\times B(S)O(n),\phi_{k+1}) & \xrightarrow{\Delta} & \Omega_{k-1}(P^{k-1}\times B(S)O(n),\phi_k)\to\cdots \\
 & & \ni\ i_*(\chi'(M,v)) & & \ni\ \omega'_{k+1}(M) & & \ni\ \omega'_k(M) \\
 & & \downarrow{\scriptstyle const_*} & & & & \\
 & & \Omega_k(\mathrm{point};\mathrm{trivial}) & & & & \\
 & & \ni\ \chi''(M,v) & & & &
\end{array}
$$

commutes (here the top line is the exact homotopy sequence of the fibration $proj: V_{n,k+1} \longrightarrow V_{n,k}$ which drops the last vector of a $(k+1)$-frame; the other lines are the exact sequences described in (13.9); the homomorphisms σ, Θ_M, $(id\times\tau)_*$ are defined as in (5.1) and (13.7), and i_* is induced by the canonical map $i: B(S)O(n-k) \longrightarrow B(S)O(n)$; note that $const_* \circ \tau_* \circ \Theta_M$ is the identity on $\Omega_k(\mathrm{point};\mathrm{trivial})$).

Moreover, we have

$$\omega_k(M) = \Delta(\omega_{k+1}(M)) ,$$

and hence

$$\omega'_k(M) = \Delta(\omega'_{k+1}(M)) .$$

If M has a k-framefield v, then $i_*(\chi'(M,v)) = \tau_*(\chi(M,v))$, $\chi''(M,v) = const_*(\chi'(M,v))$ and

$$\omega_{k+1}(M) = (-1)^k \mathrm{incl}_*(\chi(M,v)) \ ,$$

and hence

$$\omega'_{k+1}(M) = (-1)^k \mathrm{incl}_* i_*(\chi'(M,v)) \ .$$

Finally, for any two k-framefields v_o, v_1 on M we have (cf.4.2)

$$(-1)^{k+1} d(d_k(v_o,v_1)) = \chi(M,v_1) - \chi(M,v_o) \ .$$

Proof. Note that the equation $(\omega_1(M)=)\chi(M) = \deg(\omega_k(M))$ in 13.5 is closely related with the identity $\omega_k(M) = \Delta(\omega_{k+1}(M))$. The proofs of this identity and of the commutativity relation $\Delta\circ\sigma = \sigma\circ\mathrm{proj}_*$ involve the same arguments, so we need only do the latter.

Given $z\in\pi_{n-1}(V_{n,k+1})$, interprete a representative as a (k+1)-morphism $u^\partial : S^{n-1}\times \mathbb{R}^{k+1} \hookrightarrow S^{n-1}\times \mathbb{R}^n$, and let $p^{**}(u^\partial)$ be the corresponding section of the bundle $\underset{\sim}{\mathrm{Hom}}(\lambda,\underset{\sim}{\mathbb{R}}^n)$ over $P^k\times S^{n-1}$, as in (2.14). Then $\sigma(z)$ is represented by the singularity data of an extension of $p^{**}(u^\partial)$ over all of $P^k\times D^n$. Similarly, $\sigma(\mathrm{proj}_*(z))$ is represented by the singularity data of an extension of $p^{**}(u^\partial)|P^{k-1}\times S^{n-1}$ over all of $P^{k-1}\times D^n$. The two extensions can be chosen compatibly, and one just reads off the desired relation between σ, Δ and proj_*.

The last identity in 13.11 and the commutativity of the lefthand upper square will be shown simultaneously in 19.12.

The remaining claims are checked in a straight forward manner. ∎

Definition 13.12. The stable span of the manifold M is the maximum number ℓ such that $TM\oplus\underset{\sim}{\mathbb{R}}$ has $\ell+1$ linearly independent sections.

Example. For even n, $\mathrm{span}(S^n) = 0$, but stable $\mathrm{span}(S^n) = n$.

Exercise 13.13.

a.) Fix any integer $m\geq 1$. Show that the stable span of M is the maximum number ℓ such that $TM\oplus\underset{\sim}{\mathbb{R}}^m$ has $\ell+m$ linearly independent sections.

b.) Let P be a nonempty parallelizable manifold of dimension $p > 0$. Prove:

$$\begin{aligned} \text{stable span}(M) &= \text{span}(M\times P) - p \\ &= \text{stable span}(M\times P) - p \,. \end{aligned}$$

<u>Exercise 13.14.</u> Let $S \subset M$ be the singularity of a non-degenerate $(k-1)$-morphism from $\underset{\sim}{\mathbb{R}}^k$ into the tangent bundle TM of a closed n-manifold M, where $0<k\leq n$. Clearly, the Stiefel-Whitney class $w_{n-k+1}(M)$ equals the cohomology class $D(S)$ constructed in the obvious way from the Thom class of the normal bundle of S in M (see proposition 5.3). Conclude that

$$w_{n-k+1}(M) = w_{n-k+2}(M) = \ldots = w_{n-1}(M) = 0$$

for any closed n-manifold which allows a k-field with finite singularities (at least if $n>2k-4$).

<u>Exercise 13.15.</u> Let Y be a pathconnected paracompact space, let ϕ be a virtual vectorbundle over $P^q\times Y$, $q\geq 1$, and assume $i\leq 4$. Along the lines of the formula following (13.9), define homomorphisms Δ which make the following diagram commute

$$\begin{array}{ccccccc}
\Omega_{i-1}(P^q\times Y\times BO(2);\phi+\Gamma) & \xrightarrow{\delta_i} & \Omega_i(P^q\times Y;\phi) & \xrightarrow{f_i} & \overline{\Omega}_i(P^q\times Y;\phi) & \xrightarrow{\sigma\circ j_i} & \Omega_{i-2}(P^q\times Y\times BO(2);\phi+\Gamma) \\
\downarrow{\scriptstyle \pm\Delta} & & \downarrow{\scriptstyle \Delta} & & \downarrow{\scriptstyle \Delta} & & \downarrow{\scriptstyle \Delta} \\
\Omega_{i-2}(P^{q-1}\times Y\times BO(2);\lambda+\phi+\Gamma) & \xrightarrow{\delta_{i-1}} & \Omega_{i-1}(P^{q-1}\times Y;\lambda+\phi) & \xrightarrow{f_{i-1}} & \overline{\Omega}_{i-1}(P^{q-1}\times Y;\lambda+\phi) & \xrightarrow{\sigma\circ j_{i-1}} & \Omega_{i-3}(P^{q-1}\times Y\times BO(2);\lambda+\phi+\Gamma)
\end{array}$$

Here the horizontal exact sequences are given by theorem 9.3.

<u>Exercise 13.16.</u> Assume M is connected and $n>2k>0$. Prove the following strengthening of part of proposition 13.8 :

M allows a k-field with finite singularities if and only if $\omega_k(M) = \Theta_M\circ\sigma\,(v)$ for some $v\in\pi_{n-1}(V_{n,k})$.

(If u is a k-field with index v, defined on a little ball in M, show that u can be extended over all of M without additional singularities.)

Example 13.17. For $q \geq 0$ consider S^{4q+1} as the unit sphere of $\mathbb{C} \oplus \mathbb{H}^q$, and let u_o be the standard vectorfield on S^{4q+1} given by complex multiplication with $i \in \mathbb{C}$. Then

$$\chi''(S^{4q+1}, u_o) \in \Omega_1(\text{point;trivial}) \cong \mathbb{Z}_2$$

is nontrivial.

Indeed, the vectorfield v on S^{4q+1} given by quaternionic multiplication with $j \in \mathbb{H}$, $v(z,h_1,\ldots h_q) := (0,jh_1,\ldots jh_q)$, defines a nondegenerate section of the complement η of $\mathbb{R}\cdot v$ in TS^{4q+1}; the singularity of v is the (invariantly framed) circle $S^1 \subset \mathbb{C} \times 0 \times 0$.

Exercise 13.18. Use theorems 13.11, 9.3 and 13.3 to show that span $S^{4q+1} = 1$. Conclude that in general span $S^n \geq 2$ if and only if $n \equiv 3(4)$.

Exercise 13.19. Let β be an n-plane bundle over a closed connected n-manifold, $n \geq 2$, such that $w_1(\beta) \neq w_1(M)$ and $w_n(\beta) = 0$.

(i) Show that a nondegenerate section s of β must have an even number of zeroes.

(ii) Use suitable isotopies to deform s until all zeroes lie in a ball $B \subset M$. Use trivializations $TB \cong \underset{\sim}{\mathbb{R}}^n \cong \beta|B$ to define an "index" of s.

(iii) Show that there is an embedded circle $S^1 \subset M$ along which the orientation bundles of β and of TM are not isomorphic. Isotop enough zeroes of s around S^1 until the "index" of s vanishes (i.e. until the positively counted zeroes cancel the negatively counted ones).

(iv) Conclude that β allows a nowhere vanishing section.

This result, which extends proposition 3.10 (i) also to the case of surfaces, will be needed in the next section.

§14. Existence of two or three independent vectorfields.

As a first example we study now the invariants $\omega_k(M)$ and $\omega'_k(M)$ for $k=2$ or 3, and we deduce criteria for the span of a manifold M to be $\geq k$. For orientable M this question has been treated in detail by Thomas [65], Frank [22], Atiyah-Dupont [6] and Dupont [18], so we may concentrate on the nonorientable case. Actually, it turns out that our approach is more suited to deal with nonorientable manifolds anyway. In particular, the case of 2-fields on even-dimensional non-orientable manifolds, considered quite hard by Thomas (see [65], p. 648, problem 2), is the easiest case in our framework.

Throughout this section M will denote a closed connected smooth n-manifold. In view of the commutative diagram (13.7) we will consider the cases $Y=$point, M, $B(S)O(n)$ simultaneously, and the n-plane-bundle η over Y will stand for $\mathbb{R}^n$, TM or γ respectively, so that the coefficient bundle

$$\phi_Y = \lambda \otimes \eta - k\lambda - \eta$$

over $P^{k-1}\times Y$ is just $(n-k)\lambda$, ϕ_M or ϕ respectively (cf. 13.2 and 13.7), where k is a fixed natural number. Then we know from lemma 9.10 that

$$(14.1)\qquad \begin{aligned} w_1(\phi_Y) &= (n-k)w_1(\lambda) \\ w_2(\phi_Y) &= \binom{n-k}{2} w_1(\lambda)^2 + w_1(\lambda)w_1(\eta) \ . \end{aligned}$$

First, let $k=2$. By theorem 9.3, the relevant obstruction group fits into the following exact sequence

$$(14.2)\qquad \bar{\Omega}_2(P^1\times Y;\phi_Y) \xrightarrow{\sigma\circ j_2} \mathbb{Z}_2 \xrightarrow{\delta_1} \Omega_1(P^1\times Y;\phi_Y) \xrightarrow{f_1} \bar{\Omega}_1(P^1\times Y;\phi_Y) \longrightarrow 0$$

$$\bar{\Omega}_1(P^1\times Y;\phi_Y) \cong \begin{cases} H_1(P^1\times Y;\mathbb{Z}) & n \text{ even} \\ H_1(Y;\mathbb{Z}_2) & n \text{ odd} \ . \end{cases}$$

The isomorphism to the right is given by the obvious forgetful maps, see also proposition 9.18. Moreover, $\sigma\circ j_2$ maps a bordism class $x=[N \to P^1 x Y$, or: $\xi_N \cong \lambda^{\otimes n}]$ into $w_1(\lambda)w_1(\eta)[N]$. The behaviour of $\sigma\circ j_2$ decides whether f_1 is an isomorphism or whether $\mathbb{Z}_2$ gives rise to a nontrivial subgroup of $\Omega_1(P^1 x Y; \phi_Y)$ whose elements might be hard to detect.

If n is even, a suitable example of the form

$$x = [P^1 \times S^1 \xrightarrow{Id \times \ell} P^1 \times Y, or]$$

shows that $\sigma\circ j_2$ is onto (and hence f_1 is bijective) if and only if $w_1(\eta)\neq 0$. On the other hand, for $Y=M$ the homomorphism (bijective iff $w_1(M)\neq 0$)

$$\Omega_1(P^1 x M; \phi_M) \xrightarrow{f_1} H_1(P^1 x M; \mathbb{Z}) \cong \mathbb{Z}\oplus H^{n-1}(M; \tilde{\mathbb{Z}}_{TM})$$

maps our obstruction $\omega_2(M)$ to the pair $(\chi(M), W_{n-1}(M))$ of classical obstructions, by propositions 13.5 and 5.3. Thus we can apply theorem 13.3 and obtain

Theorem 14.3. Let M be a closed connected nonorientable manifold of even dimension $n\geq 6$. Then $\mathrm{span}(M)\geq 2$ if and only if the Euler number $\chi(M)$ and the (integer, twisted) Stiefel-Whitney class $W_{n-1}(M)$ of M vanish.

Remark: This result was obtained independently by D.Frank [23], and by the author in 1973/74 (see also [36]).

Now let n be odd. Then an element $x\in\overline{\Omega}_2(P^1 x Y; \phi_Y)$ can be given by a singular surface $g: N \longrightarrow Y$ together with a monomorphism of the orientation bundle ξ_N into $\mathbb{R}^2$ (representing ξ_N as a pullback from P^1). We have

$$\begin{aligned} \sigma\circ j_2(x) &= w_1(\xi_N)\cdot g^*(w_1(\eta))[N] \\ &= Sq^1(g^*(w_1(\eta)))[N] \\ &= g^*(w_1(\eta)^2)[N] \;. \end{aligned}$$

If $w_1(\eta)^2 \neq 0$ in $H^2(Y;\mathbb{Z}_2)$, then, by fact 9.4, there is a closed connected nonorientable zero-bordant surface N and a map $g: N \longrightarrow Y$ such that $g^*(w_1(\eta)^2)[N]=1$; moreover, by proposition 3.10 the homomorphism bundle $\underset{\sim}{\text{Hom}}(\xi_N, \underset{\sim}{\mathbb{R}}^2)$ over N allows a nowhere vanishing section, since

$$w_2(\underset{\sim}{\text{Hom}}(\xi_N, \underset{\sim}{\mathbb{R}}^2))[N] = w_1(N)^2[N] = 0 .$$

Thus for odd n $\sigma \circ j_2$ is onto (and hence f_1 is bijective) if and only if $w_1(\eta)^2 \neq 0$. On the other hand, for $Y=M$ the homomorphism (bijective iff $w_1(M)^2 \neq 0$)

$$\Omega_1(P^1 \times M;\phi_M) \xrightarrow{f_1} H_1(M;\mathbb{Z}_2) \cong H^{n-1}(M;\mathbb{Z}_2)$$

maps our obstruction $\omega_2(M)$ to the Stiefel-Whitney class $w_{n-1}(M)$ (see proposition 5.3). We obtain the following result of E.Thomas ([65], p.648, theorem 6).

Theorem 14.4. Let M be a closed connected manifold of odd dimension $n \geq 5$ such that $w_1(M)^2 \neq 0$. Then $\text{span}(M) \geq 2$ if and only if the Stiefel-Whitney class $w_{n-1}(M)$ vanishes.

Remark 14.5. From the (naturality of the) exact sequence (14.2) we get the isomorphisms

$$\pi_{n-1}(V_{n,2}) \underset{\substack{\cong \\ (\text{for } n>4)}}{\xrightarrow{\sigma}} \Omega_1(P^1;(n-2)\lambda) \underset{\cong}{\xrightarrow{\Theta}} \Omega_1(P^1 \times BSO(n);\phi) \cong \begin{cases} \mathbb{Z} \oplus \mathbb{Z}_2 & n \text{ even} \\ \mathbb{Z}_2 & n \text{ odd} \end{cases}$$

(see also (13.7)); the $\mathbb{Z}_2$-factor here is the kernel of f_1.

Now assume that M is orientable and $n>4$. Then a similar $\mathbb{Z}_2$-factor (detected by $(id \times \tau)_*$, see (13.7)) appears in the obstruction group $\Omega_1(P^1 \times M;\phi_M)$. Thus even if the appropriate (n-1)-dimensional Stiefel-Whitney class and the Euler number of M vanish, $\omega_2(M)=\omega_2'(M) \in \ker f_1=\mathbb{Z}_2$ may still be a nontrivial combination of the invariants listed in theorem 13.4. Results of Atiyah, Frank and Thomas (see [65]) imply that this additional $\mathbb{Z}_2$-invariant (which has to vanish if the span of M is to be ≥ 2) is $\frac{1}{2}(\chi(M)-\tau(M)) \bmod 2$ for $n \equiv 0(4)$, $k(M)$ for $n \equiv 1(4)$, and zero otherwise. In other words, precisely those invariants which distinguish the vector field bordism groups $\Omega_n(0,0)$ (see theorem 12.1) from ordinary oriented bordism, play a fundamental rôle as obstructions to the existence of two linearly independent vectorfields on a given oriented manifold.

If n is odd and we have only $w_1(M)^2=0$, there is a subtler $\mathbb{Z}_2$-invariant built into $\omega_2(M)$; for $n \equiv 1(4)$ this has been identified as a "twisted" semicharacteristic (see Atiyah-Dupont [6] or Dupont-Lusztig [19]).

Next we consider the case $k=3$. Then the orientation bundle of ϕ_Y is the $(n-1)$-fold tensor product of λ with itself, and we have that

$$w_2(\phi_Y) = \binom{n-3}{2} w_1(\lambda)^2 + w_1(\lambda)w_1(\eta) \tag{14.6}$$

is nonzero if and only if $w_1(\eta) \neq 0$ or $n \equiv 1,2,(4)$. By theorem 9.3 the relevant obstruction group fits into the following commuting diagram of exact sequences

(14.7)

$$\begin{array}{ccccccccc}
 & & \mathbb{Z}_2 & & & & & & \\
 & & \downarrow{\scriptstyle \delta_1'} & & & & & & \\
\overline{\Omega}_3(P^2 x Y; \phi_Y) & \xrightarrow{\sigma\circ j_3} & \Omega_1(P^2 x Y x BO(2); \phi_Y+\Gamma) & \xrightarrow{\delta_2} & \Omega_2(P^2 x Y; \phi_Y) & \xrightarrow{f_2} & \overline{\Omega}_2(P^2 x Y; \phi_Y) & \xrightarrow{\sigma\circ j_2} & \mathbb{Z}_2 \\
\downarrow{\scriptstyle \mu} & & \downarrow{\scriptstyle f_1' = proj_*\cdot\mu}\ (\cong \mathbb{Z} \text{ iff } w_1(\eta)\neq 0 \text{ or } n\equiv 1,2(4)) & & & & \cong \begin{cases} \mathbb{Z}\oplus H_2(Y;\mathbb{Z}_2) & n \text{ even} \\ H_1(Y;\mathbb{Z}_2)\oplus H_2(Y;\mathbb{Z}) & n \text{ odd} \end{cases} & & \\
H_3(P^2 x Y; \mathbb{Z}_2) & \xrightarrow{(w_2(\phi_Y)\cdot)_*} & H_1(P^2 x Y; \mathbb{Z}_2) & & & & & & \\
 & & \downarrow & & & & & & \\
 & & 0 & & & & & &
\end{array}$$

The isomorphism to the right is defined as follows. If n is even, its components are the map deg (as in 13.5) and the obvious map $\mu\circ\pi_{2*}$; this is bijective by proposition 9.18. If n is odd, we have the isomorphism (cf. fact 9.4)

$$\mathrm{is} : \Omega_2(P^2 \times Y) \xrightarrow{\mu} H_2(P^2 \times Y;\mathbb{Z}) \cong (H_1(P^2;\mathbb{Z}) \otimes H_1(Y;\mathbb{Z})) \oplus H_2(Y;\mathbb{Z})$$
$$\cong H_1(Y;\mathbb{Z}_2) \oplus H_2(Y;\mathbb{Z})$$

which maps any bordism class of the form $[P^1 \times S^1 \xrightarrow{\mathrm{incl} \times \ell} P^2 \times Y]$ to $(\mu[S^1 \xrightarrow{\ell} Y],0)$. Thus the first component is_1 can also be described as follows: given an oriented singular surface $g=(g_1,g_2):N \longrightarrow P^2 \times Y$, let $Z \subset N$ be the (one-dimensional) zero set of a nondegenerate section of $g_1^*(\lambda)$; then

$$\mathrm{is}_1[g:N \longrightarrow P^2 \times Y] = g_{2*}[Z] \in H_1(Y;\mathbb{Z}_2).$$

Moreover, the second component is_2 is again of the form $\mu \circ \pi_{2*}$. ∎

In the special case $Y=M$ we can use Poincaré duality and we obtain the following relation.

<u>Lemma 14.8.</u> <u>Assume $n \geq 3$. Then the homomorphism</u>

$$\Omega_2(P^2 \times M;\phi_M) \xrightarrow{f_2} \begin{cases} \mathbb{Z} \oplus H_2(M;\mathbb{Z}_2) \cong \mathbb{Z} \oplus H^{n-2}(M;\mathbb{Z}_2) & n \text{ even} \\ H_1(M;\mathbb{Z}_2) \oplus H_2(M;\mathbb{Z}) \cong H^{n-1}(M;\mathbb{Z}_2) \oplus H^{n-2}(M;\tilde{\mathbb{Z}}_{TM}) & n \text{ odd} \end{cases}$$

<u>maps our obstruction</u> $\omega_3(M)$ <u>to the pair</u> $(\chi(M),\ w_{n-2}(M))$ <u>(n even), resp. to the pair</u> $(w_{n-1}(M),\ W_{n-2}(M))$ <u>(n odd), of classical obstructions.</u>

<u>In particular, if</u> $f_2(\omega_3(M))$ <u>vanishes, then so does</u> $\omega_3'(M) \in \Omega_2(P^2 \times BO(n);\phi)$ <u>(here we take BO(n) even if M is orientable).</u>

<u>Proof.</u> The first statement follows from propositions 13.5 and 5.3. E.g. if n is odd, $\omega_3(M)$ is represented by the zero set $N \subset P^2 \times M$ of a nondegenerate section of the bundle $\underset{\sim}{\mathrm{Hom}}(\lambda,TM)$ over $P^2 \times M$, together with the inclusion and more data. We may assume that N lies transverse to $P^1 \times M$ and that $Z=N \cap (P^1 \times M)$, together with the inclusion into $P^1 \times M$ and additional data, represents $\omega_2(M)$. Notice that P^1 is the zero set of a nondegenerate section of the canonical bundle λ over P^2.

Thus, by the description of is_1 above, the first component of $f_2(\omega_3(M))$ is the Poincaré dual of $\mu[Z \subset P^1 xM \to M]$, i.e. the Stiefel-Whitney class $w_{n-1}(M)$ (see proposition 5.3).

If $f_2(\omega_3(M)) = 0$, then for all $n \geq 3$ we have by lemma 12.4

$$w_n(M) = w_1(M)w_{n-1}(M) = w_1(M)^2 w_{n-2}(M) = w_2(M)w_{n-2}(M) = 0.$$

(Exercise: try to deduce this also from fact 9.11 and from the equation $(\sigma_* j_2)_* f_2(\omega_3(M))=0$). Therefore, by theorem 12.13, M is (unoriented) bordant to a manifold M' with span≥ 3. Since clearly $\chi(M)=\chi(M')=0$, M and M' are bordant even in $\mathfrak{N}_n(0,0)$ (see theorem 12.1). Hence $\omega_3'(M)=0$ by theorem 13.4.∎

Recall that $\omega_3'(M)=(id x\tau)_*(\omega_3(M))$ (see 13.8). Hence the last lemma implies that $\omega_3(M)=0$ if $f_2(\omega_3(M))=0$ and if $(id x\tau)_*$ is injective on $\ker f_2 = \mathrm{Im}\,\delta_2$. So we will now study the lefthand half of the diagram 14.7 in order to see when this last injectivity condition holds. Again we treat the cases Y=point, M or $B(S)O(n)$ simultaneously.

Put $y_0=w_1(\eta)$, pick a $\mathbb{Z}_2$-base $y_1,\ldots,y_r$ of a complement of $\mathbb{Z}_2 \cdot w_1(\eta)$ in $H^1(Y;\mathbb{Z}_2)$, and identify any element $z \in H_1(P^2 xY;\mathbb{Z}_2)$ with the tupel

$$(14.9) \qquad (w_1(\lambda)(z); (y_0(z)); y_1(z),\ldots,y_r(z)) \in \mathbb{Z}_2 \oplus (\mathbb{Z}_2) \oplus \mathbb{Z}_2^r$$

of mod 2 integers (if $w_1(\eta)=0$, the second component is to be removed).

Given $x=[g:N^3 \longrightarrow P^2 xY, \xi_N \cong \lambda^{\otimes n-1}] \in \overline{\Omega}_3(P^2 xY;\phi_Y)$, we obtain from the commutativity of (14.7) and from the Wu relations 9.6 that

$$(14.10) \qquad w_1(\lambda)(f_1' \circ \sigma \circ j_3(x)) = w_1(\lambda)^2 w_1(\eta)[N] = \begin{cases} w_1(N)^2 w_1(\eta)[N] & n \text{ even}, \\ w_1(\lambda)w_1(\eta)^2[N] & n \text{ odd}, \end{cases}$$

and that for every $y \in H^1(Y;\mathbb{Z}_2)$

$$y(f_1^!\circ\sigma\circ j_3(x)) = (w_1(\lambda)w_1(\eta)y+\binom{n-3}{2}w_1(\lambda)^2y)[N]$$

(14.11)

$$=\begin{cases}(w_1(\eta)^2y+w_1(\eta)y^2+\binom{n-3}{2}w_1(N)^2y)[N] & n \text{ even,}\\ w_1(\lambda)(w_1(\eta)y+\binom{n-3}{2}y^2)[N] & n \text{ odd.}\end{cases}$$

If $n \equiv 2(4)$, pick loops $(\ell_o),\ell_1,..\ell_r : S^1 \to Y$ which correspond to the base of $H_1(Y;\mathbb{Z}_2)$ dual to $(y_o),y_1,..y_r$. Put $x_i = [id\times\ell_i : P^2\times S^1 \longrightarrow P^2\times Y, or] \in \overline{\Omega}_3(P^2\times Y;\phi_Y)$. Then

$$f_1^!\circ\sigma\circ j_3(x_i) = (w_1(\eta)(\ell_{i*}[S^1]);0,..,\underbrace{1}_{i^{th}},..0), \quad i=(0),1,..,r,$$

and it is not hard to see that the image of $f_1^!\circ\sigma\circ j_3$ is spanned by these elements and therefore is a complement of the first factor $\mathbb{Z}_2$ in $H_1(P^2\times Y;\mathbb{Z}_2)$. Hence for all three choices of Y, both in the oriented and unoriented theory, $\ker f_2 \cong \mathbb{Z}_2$ is generated by $\delta_2\circ f_1^{!-1}(1;0,0,..,0)$, and the maps $(id\times\tau)_*$, Θ_M and Θ of (13.7) restrict to isomorphisms between these kernels of f_2. Thus theorem 13.3 and lemma 14.8 lead to the following generalization of a result of Atiyah and Dupont who treated the case when M is orientable (see [6],[18]).

Theorem 14.12. Let M be a connected closed smooth manifold (orientable or not) of dimension $n > 6$, $n \equiv 2(4)$. Then:

span(M) ≥ 3 if and only if $\chi(M) = 0$ and $w_{n-2}(M) = 0$. ∎

Next consider the case $n \equiv 0(4)$. Assume that $w_1(\eta)\neq 0$, and that the kernel of the homomorphism

$$\overline{e} : H^1(Y;\mathbb{Z}_2) \longrightarrow H^3(Y;\mathbb{Z}_2)$$
$$y \longmapsto w_1(\eta)^2y+w_1(\eta)y^2$$

consists only of $\mathbb{Z}_2\cdot w_1(\eta)$. Thus $\overline{e}(y_1),...,\overline{e}(y_r)$ form a partial base of $H^3(Y;\mathbb{Z}_2)$, and we can find dual elements $x_1^!,...,x_r^! \in H_3(Y;\mathbb{Z}_2)$, i.e. $(\overline{e}(y_j))(x_i^!)=\delta_{ij}$. Now note that the composite homomorphism

$$\bar{\Omega}_3(P^2 \times Y;\phi_Y) \xrightarrow{\pi_{2*}} \mathfrak{N}_3(Y) \xrightarrow{\mu} H_3(Y;Z_2)$$

is onto since μ is onto by fact 9.4, and π_{2*} is onto by 3.10 indeed, any element in $\bar{\Omega}_3(P^2 \times Y;\phi_Y)$ can be given by a singular manifold $g:N^3 \longrightarrow Y$, equipped with a monomorphism $\xi_N \hookrightarrow N \times \mathbb{R}^3$, or, equivalently, with a nowhere vanishing section of $\xi_N^{dual} \oplus \xi_N^{dual} \oplus \xi_N^{dual}$, and such a section exists always. So we may pick $x_i \in \bar{\Omega}_3(P^2 \times Y;\phi_Y)$, $i=1,..r$, such that $\mu \circ \pi_{2*}(x_i) = x_i'$ and therefore $y_j(f_1' \circ \sigma \circ j_3(x_i)) = \bar{e}(y_j)(x_i') = \delta_{ij}$ (see 14.11). Hence for some $a_i \in \mathbb{Z}_2$

$$f_1' \circ \sigma \circ j_3(x_i) = (a_i; \underbrace{0;}_{0^{th}}\ 0,..,\underbrace{1,}_{i^{th}}.., 0)\ .$$

Moreover, for a suitable element of $\bar{\Omega}_3(P^2 \times Y;\phi_Y)$ of the form $\bar{x} = [id \times \ell : P^2 \times S^1 \longrightarrow P^2 \times Y, or]$ we have

$$f_1' \circ \sigma \circ j_3(\bar{x}) = (1;\ 0;\ 0,..,\ 0,..0)\ .$$

Thus clearly the image of $f_1' \circ \sigma \circ j_3$ equals the subgroup $\mathbb{Z}_2 \oplus \{0\} \oplus \mathbb{Z}_2^r$ of $H_1(P^2 \times Y;\mathbb{Z}_2)$ (cf. 14.9), and $\ker f_2 = \mathbb{Z}_2$ is generated by $\delta_2 \circ f_1'^{-1}(0;1;0,..0)$.

In particular, if M is nonorientable, the last example shows that the homomorphism Θ_M (see 13.7 and 13.8) maps $\ker f_2 \subset \Omega_2(P^2;(n-3)\lambda)$ to zero in $\Omega_2(P^2 \times M;\phi_M)$. If in addition M satisfies the condition above that $\ker(\bar{e}) = \mathbb{Z}_2 \cdot w_1(M)$, then the previous discussion applies to $Y=M$ and to $Y=BO(n)$, and it shows that $(id \times \tau)_*$ is injective on $\ker f_2$. Thus we obtain from 13.3, 13.8 and 14.8:

<u>Theorem 14.13. Let M be a nonorientable connected closed smooth manifold of dimension $n \geq 8$, $n \equiv 0(4)$, such that $\chi(M) = 0$.</u>

<u>If (i) $w_{n-2}(M)=0$ and $\mathbb{Z}_2 \cdot w_1(M) = \{y \in H^1(M;\mathbb{Z}_2) \mid w_1(M)^2 y = w_1(M) y^2\}$ or if (ii) M has a 3-field with finite singularities,</u>

<u>then: span $(M) \geq 3$.</u>

Remark. If (ii) holds, this is closely related to the work of D. Frank [23].

Finally, let n be odd. Given any element $v\in H_2(Y;\mathbb{Z}_2)$, there is a bordism class $x=[g:N^3 \longrightarrow P^2\times Y]$ in $\Omega_3(P^2\times Y)$ such that $g^*(w_1(\lambda)\, w)[N^3]=w(v)$ for all $w\in H^2(Y;\mathbb{Z}_2)$. Indeed, by fact 9.4 we may choose a nonorientable connected zero-bordant surface S with a continuous map g_2' into Y such that $g_{2*}'[S]=v$. As usual, let ξ_S denote the orientation line bundle of S. Then the projectification $N=\mathbb{R}P(\xi_S\oplus\underset{\sim}{\mathbb{R}})$ is an orientable closed 3-manifold smoothly fibered over S by a map π. The subbundle ξ_S of $\xi_S\oplus\underset{\sim}{\mathbb{R}}$ determines a section of π whose image we identify with S. Moreover, there is a canonical line bundle $\gamma\subset\pi^*(\xi_S)\oplus\underset{\sim}{\mathbb{R}}$ over N, and the projection $\pi^*(\xi_S)\oplus\underset{\sim}{\mathbb{R}}\longrightarrow\underset{\sim}{\mathbb{R}}$ restricts to a nondegenerate zero-morphism $h_1:\gamma\longrightarrow\underset{\sim}{\mathbb{R}}$ with zeroset $S\subset N$. Next note that $w_1(S)^2=0$ since S is zero-bordant. Hence by 3.10 there is a monomorphism $h_2:\gamma|S=\xi_S\hookrightarrow\underset{\sim}{\mathbb{R}}^2$. Now we can fit h_1 and an extension of h_2 together to represent γ as a subbundle of $\underset{\sim}{\mathbb{R}}^3$, i.e. as the pullback of λ under a map $g_1:N\longrightarrow P^2$. Then the bordism class $x=[g=(g_1,g_2'\circ\pi):N\longrightarrow P^2\times Y]$ has the desired property since

$$g^*(w_1(\lambda)w)[N] = w_1(\gamma)\cdot\pi^*(g_2'^*(w))[N] = g_2'^*(w)[S] = w(v)$$

(apply fact 9.11 to h_1).

If $w_1(\eta)^2;\ w_1(\eta)y_1+\binom{n-3}{2}y_1^2,\dots,w_1(\eta)y_r+\binom{n-3}{2}y_r^2$ are linearly independent in $H^2(Y;\mathbb{Z}_2)$, we may pick a dual system of elements in $H_2(Y;\mathbb{Z}_2)$, and, by the procedure above, corresponding bordism classes $\bar{x};x_1,\dots,x_r\in\Omega_3(P^2\times Y)$; in view of (14.9), (14.10) and (14.11) we see that the elements

$$f_1'\circ\sigma\circ j_3(\bar{x}) = \left(1;\ 1+\binom{n-3}{2};\ 0,\dots,\ 0,\dots,\ 0\right) \text{ and}$$
$$f_1'\circ\sigma\circ j_3(x_i) = (0;\ \underbrace{0}_{0^{th}};\ \underbrace{0}_{1^{th}},\dots,\underbrace{1}_{i^{th}},\dots,\ 0)\ ,\ i=1,\dots,r\ ,$$

of $H_1(P^2\times Y;\mathbb{Z}_2)$ span the image of $f_1'\circ\sigma\circ j_3$; hence it follows from diagram (14.7) that $\ker f_2 \cong \mathbb{Z}_2$ is generated by $\delta_2 \circ f_1^{\prime -1}(z)$ where z is any element of $H_1(P^2 \times Y; \mathbb{Z}_2)$ such that $w_1(\eta)(z) \neq (1+\binom{n-3}{2})w_1(\lambda)(z)$.

Clearly $Y=BO(n)$ satisfies the independence condition above. Thus $(id \times \tau)_*$ is injective on $\ker f_2$ provided M also satisfies this condition.

If, in particular, $n \equiv 1(4)$, and if we just assume that $w_1(M)^2$ is not of the form $w_1(M)y+y^2$, $y \in H^1(M;\mathbb{Z}_2)$, then $(1;0;0,..0)$ lies still in the image of $f_1^{\prime} \circ \sigma \circ j_3$, and therefore $\Theta_M(\Omega_2(P^2;(n-3)\lambda)) = \mathbb{Z}_2(\delta_2 \circ f_1^{\prime -1}(1;0;0,..0)) = \{0\}$ in $\Omega_2(P^2 \times M;\phi_M)$. Thus we can combine lemma 14.8, proposition 13.8 and theorem 13.3 with the preceding discussion and we obtain the following result.

<u>Theorem 14.14.</u> <u>Let M be a connected closed smooth manifold of dimension $n \geq 9$, $n \equiv 1(4)$. Assume that $w_1(M)^2$ is not in the image of the homomorphism $e: H^1(M;\mathbb{Z}_2) \longrightarrow H^2(M;\mathbb{Z}_2)$ defined by $e(y) = w_1(M)\cdot y + y^2$.</u>

<u>If (i) $w_{n-1}(M) = 0$, $W_{n-2}(M) = 0$ and $\ker e = \mathbb{Z}_2(w_1(M))$,</u>

<u>or if (ii) M has a 3-field with finite singularities,</u>

<u>then:</u> <u>$\operatorname{span}(M) \geq 3$</u>.

<u>Remark 14.15.</u> Poincaré duality and the Wu relations 9.6 show that the condition $\ker e = \mathbb{Z}_2(w_1(M))$ above holds if and only if the sequence

$$H^{n-2}(M;\mathbb{Z}_2) \xrightarrow{Sq^1} H^{n-1}(M;\mathbb{Z}_2) \xrightarrow{Sq^1} H^n(M;\mathbb{Z}_2)$$

is exact. ∎

We return to the previous discussion and we consider the case $n \equiv 3(4)$. As before we exploit the naturality of the exact sequences in the diagram 14.7 with respect to the maps involved in 13.7. If M has a 3-field u with finite singularities, then clearly $\sigma(\mathrm{Index}(u,or)) \in \Omega_2(P^2;(n-3)\lambda)$ must be of the form $\delta_2(\delta_1^{\prime}(a)+bt)$, where $a \in \mathbb{Z}_2$, $b=0$ or 1, and $t \in \Omega_1(P^2 \times BO(2);(n-3)\lambda+\Gamma)$ satisfies $f_1^{\prime}(t) \neq 0$ in $H_1(P^2;\mathbb{Z}_2) \cong \mathbb{Z}_2$. The discussion preceding theorem 14.14 implies that

$\Theta(\delta_2(t))\in\Omega_2(P^2\times BO(n);\phi)$ is nontrivial. On the other hand, we know from proposition 13.8 that $\omega_3(M)=\Theta_M(\sigma(\text{Index}(u,or)))$. Hence $f_2(\omega_3(M))=0$, and by lemma 14.8

$$\begin{aligned}\omega_3'(M) &= \Theta(\sigma(\text{Index}(u,or))) = \\ &= \Theta(\delta_2\circ\delta_1'(a) + b\delta_2(t)) \\ &= b\cdot\Theta(\delta_2(t)) \\ &= 0 \quad .\end{aligned}$$

Therefore $b=0$, and $\omega_3(M)=\Theta_M\circ\delta_2(\delta_1'(a))$ vanishes if $w_1(M)\neq 0$.

We obtain the following result along the same lines as the previous theorems.

Theorem 14.16. Let M be a nonorientable connected closed smooth manifold of dimension $n\geq 7$, $n\equiv 3(4)$.

If (i) $W_{n-2}(M)=0$ and the homomorphism

$w_1(M)\cdot : H^1(M;\mathbb{Z}_2)\longrightarrow H^2(M;\mathbb{Z}_2)$ is injective,

or if (ii) M has a 3-field with finite singularities,

then: $\text{span}(M)\geq 3$.

Note here that for $n\equiv 3(4)$ $w_{n-1}(M)$ vanishes if $w_{n-2}(M)$ does; indeed, then the Wu relations imply that we have for every $x\in H^1(M;\mathbb{Z}_2)$

$$\begin{aligned}x\cdot w_{n-1}(M) &= (x^2w_1(M) + xw_1(M)^2)w_{n-3}(M) \\ &= Sq^1(xw_1(M)w_{n-3}(M)) + xw_1(M)Sq^1(w_{n-3}(M)) \\ &= xw_1(M)^2w_{n-3}(M) + xw_1(M)^2w_{n-3}(M) + xw_1(M)w_{n-2}(M) \\ &= 0\end{aligned}$$

(use facts 9.5 and 9.6 and/or apply [32], lemma 2.12.(3) to $k=n-3$). ∎

Exercise 14.17. Prove that for $n > 6$

$$\pi_{n-1}(V_{n,3}) \cong \Omega_2(P^2;(n-3)\lambda) \cong \begin{cases} \mathbb{Z}_2 & \text{if } n \equiv 1(4) \\ \mathbb{Z}_2 \oplus \mathbb{Z} & \text{if } n \equiv 2(4) \\ \mathbb{Z}_2 \oplus \mathbb{Z}_2 & \text{if } n \equiv 3(4) \\ \mathbb{Z}_4 \oplus \mathbb{Z} & \text{if } n \equiv 4(4) \end{cases}$$

(use diagram 14.7; in order to decide whether there are elements of order 4, use a bordism involving the - possibly punctured - cylinder $S^1 \times I$).

Describe generators of $\Omega_2(P^2;(n-3)\lambda)$ by explicitly exhibiting surfaces S with line bundles λ and with trivializations of $TS \oplus (n-3)\lambda$.

Remark 14.18. For the sake of completeness and for later use, we reproduce tables of Atiyah and Dupont (see [6], p.25, and [18], p.68) concerning 3-fields on a closed connected oriented n-manifold M, $n>6$.

If M has a 3-field u with finite singularities, then Index(u,or) corresponds to

$\mathfrak{k}(M)$, $(0;\frac{1}{2}\chi(M))$, 0 or $(\frac{1}{2}\chi(M) - (-1)^q$ signature $(M))$, mod 4; $\chi(M))$ resp., according to whether $n \equiv 1, 2, 3$ or $4(4)$ (in the last case put $q = \frac{n}{4}$).

Atiyah and Dupont deduce the following necessary and sufficient conditions for M to have span≥ 3 (provided $n>6$):

$n \equiv 1(4)$	$W_{n-2}(M) = 0$, $L_q(p_1,..,p_q) \equiv 0(4)$, $\mathfrak{k}(M) = 0$
$n \equiv 2(4)$	$w_{n-2}(M) = 0$, $\chi(M) = 0$
$n \equiv 3(4)$	no condition
$n \equiv 4(4)$	$w_{n-2}(M) = 0$, sign$(M) \equiv 0$ mod 8, $\chi(M) = 0$

If $n=4q+1$, it is assumed here that $H_1(M;\mathbb{Z})$ has no 2-torsion. ■

We can weaken this very last assumption (as well as the conditions in theorems 14.14 and 14.16) somewhat and still get precise criteria. The idea, suggested by work of Olk [50], is to use knowledge about

$\omega_2(M)$ in the calculation of $\omega_3(M)$.

Theorem 14.19. Let M be a connected closed smooth manifold of dimension $n\geq 9$, $n\equiv 1(4)$. Assume that $w_1(M)^2=0$ and that

$$\mathbb{Z}_2 w_1(M) = \ker(e := w_1(M)\cdot + Sq^1 : H^1(M;\mathbb{Z}_2) \longrightarrow H^2(M;\mathbb{Z}_2)).$$

Then $\operatorname{span}(M)\geq 3$ if and only if $w_{n-1}(M)$, $W_{n-2}(M)$ and the "twisted" semicharacteristic $R_L(M)\in\mathbb{Z}_2$ (see [6], theorem 7.6) vanish.

Proof. Recall from [6] or [19] that $\operatorname{span}(M)\geq 2$ if and only if $w_{n-1}(M)=0$ and $R_L(M)=0$. If even $\operatorname{span}(M)\geq 3$, then in addition the (twisted) integer Stiefel-Whitney class $W_{n-2}(M)$ vanishes.

Conversely, assume that these three invariants of M are zero. Then, in particular, $f_2(\omega_3(M))=0$ and hence $\omega_3'(M)=0$ (see lemma 14.8), and there is an element $v\in H_1(P^2\times M;\mathbb{Z}_2)$ such that $\delta_2\circ f_1'^{-1}(v)=\omega_3(M)$ (see diagram 14.7). We may actually pick v to lie already in the subgroup $H_1(P^2;\mathbb{Z}_2)$; this follows from the condition on $\ker(e)$ by a suitable modification of the proof of theorem 14.14.

Now consider the commuting diagram (compare 13.15)

$$\begin{array}{ccccc}
H_1(P^2\times M;\mathbb{Z}_2) & \xleftarrow[\cong]{f_1'} & \Omega_1(P^2\times M\times BO(2);\phi_{M,3}+\Gamma) & \xrightarrow{\delta_2} & \Omega_2(P^2\times M;\phi_{M,3}) \\
\downarrow{\scriptstyle w_1(\lambda)} & & \downarrow{\scriptstyle\Delta} & & \downarrow{\scriptstyle\Delta} \\
\mathbb{Z}_2 & \cong & \Omega_0(P^1\times M\times BO(2);\phi_{M,2}+\Gamma) & \xrightarrow{\delta_1} & \Omega_1(P^1\times M;\phi_{M,2})
\end{array} .$$

Since $w_1(M)^2=0$, δ_1 is injective (see the proof of theorem 14.4). The vanishing of $w_{n-1}(M)$ and $R_L(M)$, and theorem 13.11, now imply that

$$0 = \omega_2(M) = \Delta(\omega_3(M)) = \Delta\circ\delta_2\circ f_1'^{-1}(v) = \delta_1(w_1(\lambda)(v)) .$$

Hence v and $\omega_3(M)$ vanish, and $\operatorname{span}(M)\geq 3$.

Corollary 14.20. Let M be an orientable connected closed manifold of dimension $n \geq 9$, $n \equiv 1(4)$. Assume that the order of every element in $H_1(M;\mathbb{Z})$ is finite and not divisible by 4.

Then $\operatorname{span}(M) \geq 3$ if and only if $w_{n-1}(M)$, $W_{n-2}(M)$ and $k(M)$ vanish.

Proof. The condition on $H_1(M;\mathbb{Z})$ is equivalent to the injectivity of Sq^1 on $H^1(M;\mathbb{Z}_2)$. This follows easily from the fact that Sq^1 is a Bockstein homomorphism (cf. [60]) and from the resulting horizontal exact sequence

$$\begin{array}{ccccccc} H^1(M;\mathbb{Z}_2) & \xrightarrow{j_*} & H^1(M;\mathbb{Z}_4) & \longrightarrow & H^1(M;\mathbb{Z}_2) & \xrightarrow{Sq^1} & H^2(M;\mathbb{Z}_2) \\ \downarrow \cong & & \downarrow \cong & & & & \\ \operatorname{Hom}(H_1(M;\mathbb{Z}),\mathbb{Z}_2) & \xrightarrow{j_*} & \operatorname{Hom}(H_1(M;\mathbb{Z}),\mathbb{Z}_4) & & & & \end{array}$$

where $j:\mathbb{Z}_2 \longrightarrow \mathbb{Z}_4$ is the obvious monomorphism.

Now apply theorem 14.19. Since M is orientable here, the condition on $\ker e$ just means that Sq^1 is injective on $H^1(M;\mathbb{Z}_2)$, and $R_L(M)$ is the usual (untwisted) Kervaire semicharacteristic $k(M)$.

Example 14.21. Consider $M=P^{4q+1}$ or $M=P^{4q-1} \times S^2$, $q>1$. Since $H_1(M;\mathbb{Z}) \cong \mathbb{Z}_2$ in both cases, our corollary applies (in contrast, see the assumptions for [18], table 1). However, the conclusion that $\operatorname{span} P^{4q+1}=1$ and $\operatorname{span}(P^{4q-1} \times S^2) \geq 3$ is obvious anyway.

Exercise 14.22. Let M be a nonorientable connected closed n-manifold, $n \equiv 3(4)$, $n \geq 7$, such that

$$\mathbb{Z}_2 w_1(M) = \ker(w_1(M)\cdot \; : H^1(M\,;\mathbb{Z}_2) \longrightarrow H^2(M;\mathbb{Z}_2))$$

(and in particular $w_1(M)^2=0$).

Show: $\operatorname{span}(M) \geq 3$ iff $\operatorname{span}(M) \geq 2$ and $W_{n-2}(M)=0$.

(Use the approach of the proof of theorem 14.19).

§15. Four linearly independent vector fields.

In this section we are mainly concerned with existence criteria for four linearly independent vector fields on a closed connected n-manifold M. If M is orientable, there is some overlap with the work of D. Randall (see e.g. [53]).

As in the last section, we will consider the cases Y = point, M or $B(S)O(n)$ simultaneously, and the n-plane bundle μ over Y is $\mathbb{R}^n$, TM or γ respectively. Then the virtual vector bundle

$$\phi_Y = \lambda \otimes \eta - 4\lambda - \eta$$

over $P^3 \times Y$ is the relevant coefficient bundle in our obstruction theory for 4-fields, i.e. ϕ_Y is just $(n-4)\lambda$, ϕ_M or ϕ respectively. Since $w(4\lambda) = 1+w_1(\lambda)^4 = 1$, we know from lemma 9.10 that

$$
\begin{aligned}
& w_1(\phi_Y) = n\cdot w_1(\lambda) \\
& w_2(\phi_Y) = \tbinom{n}{2}\cdot w_1(\lambda)^2 + w_1(\lambda)w_1(\eta) \quad , \\
(15.1)\qquad & w_3(\phi_Y) = \tbinom{n}{3}\cdot w_1(\lambda)^3 + (n-1)w_1(\lambda)^2 w_1(\eta) + w_1(\lambda)w_1(\eta)^2 \\
& w_4(\phi_Y) = \tbinom{n-1}{2}\, w_1(\lambda)^3 w_1(\eta) + w_1(\lambda)^2(w_2(\eta) + (n-1)w_1(\eta)^2)) + \\
& \qquad\qquad w_1(\lambda)(w_3(\eta) + w_1(\eta)w_2(\eta) + w_1(\eta)^3).
\end{aligned}
$$

According to theorem 9.3 the relevant obstruction group fits into the following diagram of exact sequences

(15.2)

$$\begin{array}{ccccccc}
 & & 0 & & & & \\
 & & \downarrow & & & & \\
 & & \ker f_2' & & & & \\
 & & \downarrow & & & & \\
\overline{\Omega}_4(P^3\times Y;\phi_Y) & \xrightarrow{\sigma\circ j_4} & \Omega_2(P^3\times Y\times BO(2);\phi_Y+\Gamma) & \xrightarrow{\delta_3} & \Omega_3(P^3\times Y;\phi_Y) & \xrightarrow{f_3} & \overline{\Omega}_3(P^3\times Y;\phi_Y) \longrightarrow \cdots \\
 & \searrow^{\overline{\sigma}=(\overline{\sigma}_1,\overline{\sigma}_2,\overline{\sigma}_3)} & \downarrow f_2' & & & & \\
 & & (\ker : H_2(P^3\times Y;\mathbb{Z}_2) \xrightarrow{\binom{n}{2}w_1(\lambda)^2+w_1(\lambda)w_1(\eta)} \mathbb{Z}_2)\oplus\mathbb{Z}_2\oplus\begin{cases}\mathbb{Z} & n \text{ even}\\ \mathbb{Z}_2 & n \text{ odd}\end{cases} & & & & \\
 & & \downarrow & & & & \\
 & & 0 & & & &
\end{array}$$

$$j : \overline{\Omega}_3(P^3\times Y;\phi_Y) \cong \begin{cases} \mathbb{Z}\oplus H_2(Y;\mathbb{Z}_2)\oplus H_3(Y;\mathbb{Z}) & n\equiv 0(2) \\ H_1(Y;\mathbb{Z}_2)\oplus H_3(Y;\mathbb{Z}_2) & n\not\equiv 0(2) \end{cases}$$

The isomorphism j to the right is defined as follows.

If n is even, represent an element $x\in\overline{\Omega}_3(P^3\times Y;\phi_Y) = \Omega_3(P^3\times Y)$ by an oriented 3-manifold N together with a map $(g_1,g_2):N\longrightarrow P^3\times Y$; let $Z^2\subset N$ be the zero set of a non-degenerate section of $g_1^*(\lambda)$, and put

$$j(x) = (\text{mapping degree of } g_1;\ g_{2*}[Z^2];\ g_{2*}[N]) .$$

E.g. given a singular surface $h : S\longrightarrow Y$, let ξ_S denote the orientation bundle $\Lambda^2 TS$ and consider the projective space bundle $\pi : RP(\xi_S\oplus \underset{\sim}{\mathbb{R}}) \longrightarrow S$; then $RP(\xi_S\oplus \underset{\sim}{\mathbb{R}})$, together with a classifying map g_1 of the canonical line bundle over this projectification, and with $g_2 = h\circ\pi$, represents an element in $\Omega_3(P^3\times Y)$; its image under j has the form $(?\ ;\ h_*[S];\ 0)$ (see also the construction following theorem 14.13). Since $\mu : \Omega_3(Y)\xrightarrow{\cong} H_3(Y;\mathbb{Z})$ (see fact 9.4), examples of this type, as well as the examples $x = [P^3,(\text{id},\text{constant})]$, show that j is onto. On the other hand, according to standard results of algebraic topology, there is an isomorphism j' from $\Omega_3(P^3\times Y) \cong H_3(P^3\times Y;\mathbb{Z})$, to

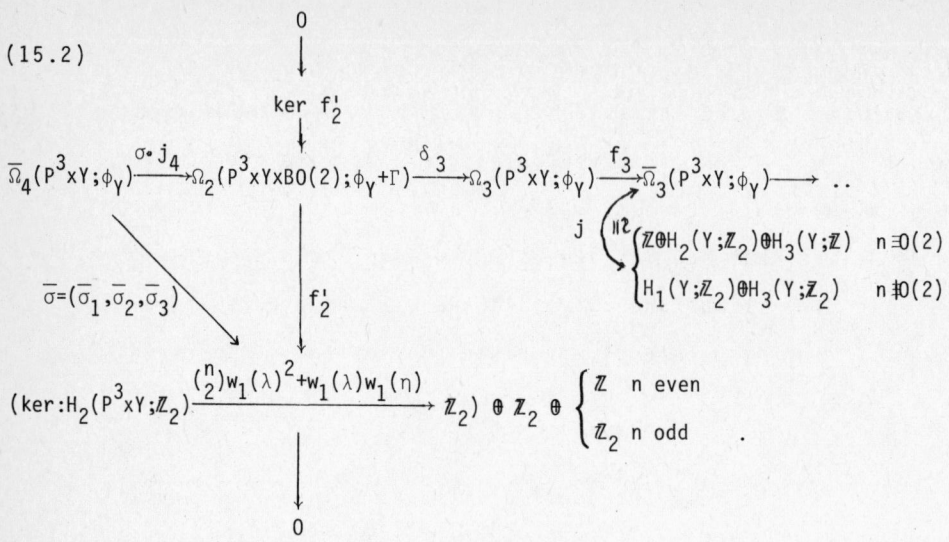

$$\underbrace{H_3(P^3;\mathbb{Z})}_{\cong\,\mathbb{Z}} \oplus \underbrace{\mathbb{Z}_2 \otimes H_2(Y;\mathbb{Z}) \oplus \mathrm{Tor}(\mathbb{Z}_2, H_1(Y;\mathbb{Z}))}_{\cong\, H_2(Y;\mathbb{Z}_2)} \oplus H_3(Y;\mathbb{Z})$$

The epimorphism $j \circ j'^{-1}$ agrees with the identity on $\mathbb{Z}$ and on $H_3(Y;\mathbb{Z})$; hence it is bijective, and so is j for even n.

If n is odd, then by proposition 9.18 and fact 9.4 we have the isomorphism

$$j : \bar{\Omega}_3(P^3 \times Y;\phi_Y) \xrightarrow[\cong]{\pi_{2*}} \mathfrak{N}_3(Y) \cong H_1(Y;\mathbb{Z}_2) \oplus H_3(Y,\mathbb{Z}_2) .$$

Given $x = [N^3,(g_1,g_2),or]$ in the domain of j, let $Z^1 \subset N^3$ be the zero set of a non-degenerate section of $g_1^*(\lambda) \oplus g_1^*(\lambda)$; then it is not hard to see that

$$j(x) = (g_{2*}[Z^1],\ g_{2*}[N^3]) .$$

Lemma 15.3. Assume $n \geq 4$ and $Y = M$. Then the homomorphism $j \circ f_3$ (cf. 15.2.) maps our obstruction $\omega_4(M)$ to the tripel

$$(\chi(M);\ PD(w_{n-2}(M));\ PD(W_{n-3}(M)))$$

(when n is even), respectively to the pair

$$(PD(w_{n-1}(M));\ PD(w_{n-3}(M)))$$

(when n is odd), where PD denotes appropriate Poincaré duality isomorphisms.

Moreover, if $f_3(\omega_4(M))$ vanishes, then so does $\omega_4'(M) \in \Omega_3(P^3 \times BO(n);\phi)$ (here we take $BO(n)$ even if M is orientable).

Proof. Basically by definition, $j \circ f_3$ is made up from iterates of Δ (see 13.9) and from Hurewicz and forgetful maps. The formula $\Delta^i(\omega_4(M)) = \omega_{4-i}(M)$ (see proposition 13.11), and propositions 5.3 and 13.5 give the indicated values for $j \circ f_3(\omega_4(M))$.

If $f_3(\omega_4(M))=0$, then all Stiefel-Whitney numbers of M involving $w_i(M)$, $i>n-4$, vanish (use the relations 12.4 and $nw_{n-2} = w_1w_{n-3}+Sq^1(w_{n-3})$). Hence a result of Stong (see theorem 12.13) implies that there exists a manifold M' with $\text{span}(M') \geq 4$ which is bordant to M' (even in $\mathfrak{N}_n(0,0)$ since $\chi(M) = \chi(M') = 0$, see theorem 12.1). According to theorem 13.4, $\omega_4'(M) = \omega_4'(M') = 0.$ ∎

Now we turn our attention to the lefthand part of diagram 15.2. We define $\overline{\sigma}_1$, $\overline{\sigma}_2$, $\overline{\sigma}_3$ to be the component homomorphisms of $\overline{\sigma}:=f_2\circ\sigma\circ j_4$.

First assume that n is odd.

The obvious homomorphism

$$\pi_{2*} : \overline{\Omega}_4(P^3xY;\phi_Y) \longrightarrow \ker(w_1(N)^4 : \mathfrak{N}_4(Y) \longrightarrow \mathbb{Z}_2)$$

is onto (where $w_1(N)^4$ assigns to every singular manifold in Y the indicated Stiefel-Whitney number of the underlying 4-manifold). Indeed, given a bordism class $t = [N^4 \longrightarrow Y]$ in the group to the right hand side, we may assume that N^4 is connected and nonorientable. Then the bundle $\underset{\sim}{\text{Hom}}(\xi_N, \underset{\sim}{\mathbb{R}}^4)$ allows a nowhere vanishing section since its Euler class $w_1(N)^4$ is zero (see proposition 3.10). Therefore the orientation bundle ξ_N occurs as a subbundle of $\underset{\sim}{\mathbb{R}}^4$ and hence as a pullback from P^3. This implies that t lies in the image of π_{2*}.

The following calculation shows that $\overline{\sigma}$ factorizes through the epimorphism π_{2*}. Given

$$x = [(g_1,g_2) : N^4 \longrightarrow P^3xY, \text{ or}:\xi_N \cong g_1^*(\lambda)] \in \overline{\Omega}_4(P^3xY;\phi) ,$$

recall that Sq^1 acts by multiplication with $w_1(N)=w_1(\lambda)$ on $H^3(N;\mathbb{Z}_2)$ (cf. fact 9.6; we omit indicating pullbacks) and that

$$0 = w_1(N)^4 = w_4(N) + w_2(N)^2$$

(check the second equality on the generators $[P^2xP^2]$ and $[P^4]$ of $\mathfrak{N}_4$).

The formulas in theorem 9.3 show that for

$$z \in H^2(P^3 \times Y;\mathbb{Z}_2) \cong \mathbb{Z}_2 w_1(\lambda)^2 \oplus w_1(\lambda)H^1(Y;\mathbb{Z}_2) \oplus H^2(Y;\mathbb{Z}_2)$$

we have

$$z(\bar{\sigma}_1(x)) = (z^2 + \tbinom{n}{2}w_1(N)^2 z + w_1(\eta)^2 z + w_1(\eta)Sq^1(z))[N] .$$

We obtain

$$w_1(\lambda)^2(\bar{\sigma}_1(x)) = w_1(N)^2 w_1(\eta)^2[N] ;$$
$$w_1(\lambda)y(\bar{\sigma}_1(x)) = w_1(N)^2((1+\tbinom{n}{2}))y^2 + w_1(\eta)y)[N],\ y\in H^1(Y;\mathbb{Z}_2)$$

(15.4) $$z(\bar{\sigma}_1(x)) = (\tbinom{n}{2}w_1(N)^2 z+Sq^2(z)+w_1(\eta)Sq^1(z)+w_1(\eta)^2 z)[N], z\in H^2(Y;\mathbb{Z}_2)$$

and

$$\bar{\sigma}_2(x) = (w_4(N) + w_1(N)^2(w_2(\eta) + \tbinom{n-1}{2}w_1(\eta)^2) + w_1(\eta)^4)[N]$$

and

$$\bar{\sigma}_3(x) = (w_4(N) + w_1(N)^2 w_1(\eta)^2)[N] .$$

It is important to calculate also $\ker f_2'$. Thus consider the condition

(*) $$w_1(\lambda)c^2 + w_1(\lambda)w_1(\eta)c + \tbinom{n}{2}w_1(\lambda)^2 c = 0$$

defined for elements $c\in H^1(P^3 \times Y;\mathbb{Z}_2)$. If we write $c = aw_1(\lambda)+y$, $y\in H^1(Y;\mathbb{Z}_2)$, we obtain: c satisfies (*) if and only if

$$w_1(\lambda)^3(a^2+\tbinom{n}{2}a) + w_1(\lambda)^2(aw_1(\eta)+\tbinom{n}{2}y) + w_1(\lambda)(y^2+yw_1(\eta))$$

vanishes, i.e. if and only if $a=0$ and $y^2+yw_1(\eta)=0$ (when $n\equiv 1(4)$), respectively iff $c\in\mathbb{Z}_2(w_1(\lambda)+w_1(\eta))$ (when $n\equiv 3(4)$). According to theorem 9.3 we have now

(15.5) $$\ker f_2' \cong H_1(P^3 \times Y;\mathbb{Z}_2)/\{s\in H_1(P^3 \times Y;\mathbb{Z}_2)\,|\,c(s)=0 \text{ for all } c \text{ satisfying } (*)\}.$$

Now we study the cases $n\equiv 1(4)$ and $n\equiv 3(4)$ separately in more detail. If $n\equiv 3(4)$, then $\ker f_2' \cong \mathbb{Z}_2$ is generated by $[(\mathrm{id},\ \mathrm{constant})_*(g)]$, where g is the nontrivial element in $H_1(P^3;\mathbb{Z}_2)$ (cf. 15.5); clearly maps such as $\mathrm{id}\times\tau$, relating the cases Y=point, M and $B(S)O(n)$ with one another (compare 13.7), induce isomorphisms on $\ker f_2'$.

<u>Proposition 15.6. Assume $n\equiv 3(4)$, $n\geq 11$. Then $\delta_3(\ker f_2')$ (cf. diagram 15.2) is isomorphic to $\mathbb{Z}_2$. Moreover, for Y=point we have</u>

$$\pi_{n-1}(V_{n,4}) \cong \Omega_3(P^3;(n-4)\lambda) = \delta_3(\ker f_2') \oplus \ker \Theta_{BSO(n)} ,$$

<u>where the kernel of</u>

$$\Theta_{BSO(n)} : \Omega_3(P^3;(n-4)\lambda) \longrightarrow \Omega_3(P^3\times BSO(n);\phi)$$

<u>consist of 0 and the element ε given by the Liegroup S^3, together with a constant map and an obvious stabilization of the left (or right) invariant parallelization.</u>

<u>Proof.</u> If Y=point, the diagram 15.2 takes the following form (we write ϕ_{pt} for $(n-4)\lambda$) :

(15.7)

$$\begin{array}{ccccccc}
 & & 0 & & & & \\
 & & \downarrow & & & & \\
 & & \ker f_2' \cong \mathbb{Z}_2 & & & & \\
 & & \downarrow & & & & \\
\overline{\Omega}_4(P^3;\phi_{pt}) & \xrightarrow{\sigma\circ j_4} & \Omega_2(P^3\times BO(2);\phi_{pt}+\Gamma) & \xrightarrow{\delta_{3,pt}} & \Omega_3(P^3;\phi_{pt}) & \longrightarrow & 0 \\
 & & \downarrow f_2' & & \cong & & \\
 & & & & \pi_{n-1}(V_{n,4}) & & \\
 & & \{0\}\oplus\mathbb{Z}_2\oplus\mathbb{Z}_2 & & & & \\
 & & \downarrow & & & & \\
 & & 0 & & & &
\end{array}$$

(diagonal arrow $\overline{\sigma}_{pt}=(0,w_4,w_4)$: $\overline{\Omega}_4(P^3;\phi_{pt}) \to \{0\}\oplus\mathbb{Z}_2\oplus\mathbb{Z}_2$)

According to Paechter [51], $\pi_{n-1}(V_{n,4}) = \mathbb{Z}_2\oplus\mathbb{Z}_2$. Since $\overline{\sigma}_{pt}[\mathbb{C}P(2),\mathrm{trivial},\mathrm{or}] = (0,1,1)$, the claim concerning $\delta_3(\ker f_2')$ holds for Y=point.

Next consider the corresponding diagram for $Y = BSO(n)$. The manifold $\mathbb{C}P(1) \times P^2 \amalg \mathbb{C}P(2)$, together with the Whitney sum of a trivial bundle with the canonical complex line bundle, gives rise to an element $x \in \overline{\Omega}_4(P^3 x BSO(n);\phi)$ such that $\overline{\sigma}(x)=f_2'(\sigma\circ j_4(x))=(0,0,1)$. Thus on one hand $\sigma\circ j_4(x)$ lies in the kernel of $\delta_{3,BSO(n)}$, on the other hand $\sigma\circ j_4(x)$ comes from an element z in $\Omega_2(P^3 x BO(2);\phi_{pt}+\Gamma)$ such that $f_2'(z) = (0,0,1) \notin \mathrm{Im}\overline{\sigma}_{pt}$ and hence $\delta_{3,pt}(z) \neq 0$.

We have the following inclusions

$$\delta_{3,p}(z) \in \ker \Theta_{BSO(n)} \subset \ker \Theta_{BO(n)} \subset \Delta(\Omega_4(P^4;(n-5)\lambda));$$

the last one follows readily from the diagram in theorem 13.11, applied to $k=4$ and extended to the right. But according to Paechter [51], $\Omega_4(P^4;(n-5)\lambda) \cong \pi_{n-1}(V_{n,5})$ is isomorphic to $\mathbb{Z}_2$, and therefore so are the three groups above.

If $\delta_3(\mathrm{Ker} f_2')$ were trivial for any choice of Y, then the same would hold for $Y = BO(n)$, and $\ker \Theta_{BO(n)}$ would contain all four elements of $\Omega_3(P^3;\phi_{pt})$. This would contradict the conclusion above. The first statement of our proposition follows in full generality.

Now consider the following part of the commuting diagram in 13.11 (applied to $k=3$)

$$\begin{array}{ccccc}
\Omega_3(\text{point};\text{trivial}) & \xrightarrow{\text{incl}_*} & \Omega_3(P^3;(n-4)\lambda) & \xrightarrow{\Delta} & \Omega_2(P^2;(n-3)\lambda) \\
 & & \Big\downarrow {\scriptstyle \Theta_{BSO(n)}} & & \Big\downarrow \\
 & & \Omega_3(P^3 x BSO(n);\phi_4) & \xrightarrow{\Delta} & \Omega_2(P^2 x BSO(n);\phi_3)
\end{array} .$$

The vertical arrow to the right is injective (this follows readily from an analysis of the two groups to the right via (14.7); note that δ_2 is injective in both cases, by the footnote to theorem 9.3). A similar comparison of diagrams 15.7, and 14.7 for Y = point, shows that the upper homomorphism Δ above is injective on $\delta_3(\mathrm{Ker} f_2')$; Hence its kernel has at most two elements. On the other hand, it contains $\ker \Theta_{BSO(n)} \cong \mathbb{Z}_2$.

We conclude that

(15.8) $$\ker \Theta_{BSO(n)} = \text{incl}_*(\Omega_3(\text{point, trivial})) .$$

Finally recall the wellknown fact (see [7]) that the Liegroup S^3, with a left (or right) invariant parallelization, represents a generator of the framed bordism group Ω_3(point; trivial). This implies the desired description of $\ker \Theta_{BSO(n)}$. ∎

Proposition 15.9. Let M be a closed connected n-manifold, $n\equiv 3(4)$, $n \geq 11$, which admits a 4-field with finite singularities.

Then $\omega_4(M) \doteq e\cdot \Theta_M(\varepsilon)$ for some $e \in \mathbb{Z}_2$. Furthermore, if

(15.10) $$w_2(M)^2 + w_1(M)w_3(M) + w_1(M)^4 = 0,$$

then e is a welldefined invariant of M.

Proof. According to propositions 13.8. and 15.6, $\omega_4(M)$ is of the form $\Theta_M(d+e\varepsilon)$, $d\in\delta_3(\ker f_2')$. In particular, $\omega_4(M)$ lies in the image of δ_3, so $f_3(\omega_4(M)) = 0$, and therefore

$$\omega_4'(M) = (\text{id}\times\tau)_*(\omega_4(M)) = \Theta_{BO(n)}(d) = 0$$

(see 15.3 and 13.7). Since $\Theta_{BO(n)}$ is injective on $\delta_3(\ker f_2')$ by the previous proposition, $d=0$ and $\omega_4(M)=e\cdot\Theta_M(\varepsilon)$.

Now note that, in the situation of formula 15.4, we have in general

(15.11) $$w_2(\eta)\bar{\sigma}_1(x)+\bar{\sigma}_2(x)+\bar{\sigma}_3(x)=(w_2(\eta)^2+w_1(\eta)w_3(\eta)+w_1(\eta)^4)[N],$$

provided $n\equiv 3(4)$. In particular, if the vanishing condition (15.10) for the Stiefel-Whitney classes of M holds, then $\Theta_M(\varepsilon)\neq 0$. Indeed, in the terminology of the previous proof, $\varepsilon = \delta_{3,pt}(z)$, where $f_2'(z) = (0,0,1)$; thus $\Theta_M(\varepsilon)$ is of the form $\delta_3(\tilde{z})$, where

$f_2'(\tilde{z}) = (0,0,1) \notin \overline{\sigma}(\overline{\Omega}_4(P^3 \times M;\phi_M))$, by relation (15.11), and hence $\tilde{z} \notin \mathrm{Im}\, \sigma \circ j_4 = \ker \delta_3$. Therefore $\Theta_M(\varepsilon)$ has order 2.■

Theorem 15.12. Let M be a closed connected manifold of dimension $n \geq 11$, $n \equiv 3(4)$. Assume that

(i) $H^1(M;\mathbb{Z}_2) \cong \mathbb{Z}_2$

or (ii) M has a 3-field with finite singularities, and $w_1(M)\cdot y = 0$ for all $y \in H^1(M;\mathbb{Z}_2)$.

Then M allows a 4-field with finite singularities if and only if $\underline{w}_{n-3}(M) = 0$.

If in addition $w_2(M)^2+w_1(M)w_3(M)+w_1(M)^4 \neq 0$, then:

$\mathrm{span}(M) \geq 4$ if and only if $w_{n-3}(M) = 0$.

Note that the assumption (ii) is fulfilled when $w_1(M)=0$, since then $\mathrm{span}(M)\geq 3$ by the work of Atiyah and Dupont (see e.g. the table in remark 14.18).

Recall a few facts (basically due to E. Thomas, see also section 14) concerning 2-fields on an arbitrary closed connected manifold M of odd dimension $n\geq 5$. If $w_1(M)^2 \neq 0$, then: $\mathrm{span}(M)\geq 2$ if $w_{n-1}(M)=0$. If $w_1(M)^2=0$ and $w_{n-1}(M)=0$, then

$$\omega_2(M) = \tilde{e}(M)\cdot[S^1, \text{constant, invariant parallelization}]$$

for some welldefined invariant $\tilde{e}(M) \in \mathbb{Z}_2$ (e.g. $\tilde{e}(S^n)=\frac{1}{2}(n+1)$, mod 2). The following strikingly analogous result is an immediate consequence of theorem 15.12 and proposition 15.9 (and of the work of Adams [1]).

Corollary 15.13. Let M be a closed connected orientable n-manifold, $n \geq 11$, $n \equiv 3(4)$.

If $w_2(M)^2 \neq 0$, then:

$\mathrm{span}(M) \geq 4$ if and only if $w_{n-3}(M) = 0$.

<u>If $w_2(M)^2=0$, and $w_{n-3}(M)=0$ (i.e., M has a 4-field with finite singularities), then:</u>

$$\omega_4(M) = e(M)\cdot[S^3, \text{constant, invariant parallelization}]$$

<u>where $e(M) \in \mathbb{Z}_2$ is the invariant defined in proposition 15.9. (E.g. $e(S^n) = \frac{1}{4}(n+1)$, taken modulo 2).</u>

This result was also obtained by very different methods by Duane Randall (cf. [53] and [54], who actually calculated $e(M)$ for spin manifolds M (i.e. $w_1(M) = 0$ and $w_2(M) = 0$) as follows: if $n \equiv 3(8)$ and $w_{n-3}(M) = 0$ (and we write $n = 2s+1$), then $e(M)$ equals the mod 2 Kervaire semicharacteristic

(15.14) $$\chi_2(M) = \sum_{i=0}^{s} \dim H_i(M;\mathbb{Z}_2) \mod 2,$$

if $n \equiv 7(8)$, then $w_{n-3}(M) = 0$ and $e(M) = 0$.

<u>Proof of theorem 15.12.</u> If M has a 4-field with finite (or without any) singularities, then clearly $w_{n-3}(M)=0$ (see also exercise 13.14).

Conversely, assume that $w_{n-3}(M)$ vanishes. Then also

$$0 = Sq^2(w_{n-3}(M)) + w_2(M)\cdot w_{n-3}(M) = w_{n-1}(M)$$

(cf. fact 9.5). Therefore it follows from lemma 15.3 that $\omega_4(M) = \delta_3(v)$ for some $v \in \Omega_2(P^3 x M x BO(2);\phi_M+\Gamma)$ and that

$$\omega_4'(M) = (id x \tau)_*(\omega_4(M)) = 0.$$

Now we have two approaches. The first one is to examine under what conditions $(id x \tau)_*$ is injective on the image of δ_3 (in this case $\omega_4(M)=0$, and $\text{span}(M)\geq 4$ by theorem 13.3).

Consider the diagram 15.2 and the formulas 15.4 for general Y. Given an element $a \in H_2(Y;\mathbb{Z}_2)$, there is a singular surface $g : T \longrightarrow Y$ such that $g_*[T] = a$ (cf. fact 9.4); then the map $g\circ\pi_2 : P^2 x T \longrightarrow Y$ gives rise to an element x in $\overline{\Omega}_4(P^3 x Y;\phi_Y)$ such that

$$z(\overline{\sigma}_1(x)) = w_1(P^2xT)^2\cdot(g\ \pi_2)^*(z)[P^2xT]$$
$$= g^*(z)[T]$$
$$= z(a)$$

for all $z\in H^2(Y;\mathbb{Z}_2)$, and hence $\pi_{2*}\circ\overline{\sigma}_1(x) = a$. This, together with other simple examples of elements in $\overline{\Omega}_4(P^3xY;\phi_Y)$, implies that $H_2(Y;\mathbb{Z}_2)\oplus\mathbb{Z}_2\oplus\mathbb{Z}_2$ is the direct sum of the image of $(\pi_{2*}\circ\overline{\sigma}_1,\overline{\sigma}_2,\overline{\sigma}_3)$ and of $\{0\}\oplus\{0\}\oplus\mathbb{Z}_2$ if $w_2(\eta)^2+w_1(\eta)w_3(\eta)+w_1(\eta)^4=0$, and that $(\pi_{2*}\circ\overline{\sigma}_1,\overline{\sigma}_2,\overline{\sigma}_3)$ is surjective otherwise (use the relation 15.11).

Furthermore, note that for all $x\in\overline{\Omega}_4(P^3xY;\phi_Y)$ and for all $y\in H^1(Y;\mathbb{Z}_2)$ we have

$$w_1(\lambda)y(\overline{\sigma}_1(x)) = w_1(\eta)\cdot y(\overline{\sigma}_1(x)) .$$

In particular

$$w_1(\lambda)^2(\overline{\sigma}_1(x)) = w_1(\lambda)w_1(\eta)(\overline{\sigma}_1(x)) = w_1(\eta)^2(\overline{\sigma}_1(x)) .$$

When $Y=BO(n)$, this is the only relation valid in $\mathrm{im}\overline{\sigma}$. When $Y = M$, any cohomology class $y\in H^1(M;\mathbb{Z}_2) - \{0,w_1(M)\}$ leads to an additional relation and hence to a new element in $\mathrm{coker}\sigma\circ j_4 \cong \mathrm{im}\delta_3$. It follows from proposition 15.6 that $(id\times\tau)_*$ is injective on the image of δ_3 if and only if $w_2(M)^2+w_1(M)w_3(M)+w_1(M)^4 \neq 0$ and if $H^1(M;\mathbb{Z}_2) = \{0,w_1(M)\}$. If only the last condition holds, then $\omega_4(M) \in \mathrm{im}\delta_3\cap\ker(id\times\tau)_*$ is of the form $\delta_3(v)$, where $f_2'(v) \in \{0\}\oplus\{0\}\oplus\mathbb{Z}_2$; therefore $\omega_4(M)$ lies already in $\Theta_M(\Omega_3(P^3;\phi_{pt}))$ (compare 15.7), and it can be shown easily (see 13.16) that M carries a 4-field with finite singularities. This implies the claims of our theorem when $H^1(M;\mathbb{Z}_2) \cong \mathbb{Z}_2$ and $w_1(M) \neq 0$. (The case when $w_1(M)=0$ is included in the following discussion).

Our second approach, suggested by Christof Olk, is to use the relation

$$\Delta(\omega_4(M) = \omega_3(M)$$

(see theorem 13.11).

Assume that condition (ii) holds. Then it follows from the work of Atiyah and Dupont (see e.g. our remark 14.18) and from theorem 14.16 that $\omega_3(M)=0$. Hence it is worthwile to compare the obstruction groups $\Omega_{k-1}(P^{k-1}\times M;\phi_{M,k})$, $k=3,4$, and the diagrams (14.7) and (15.2) into which they fit. In particular, we obtain the following commuting diagram (cf. 13.15 and 9.11)

(15.15)

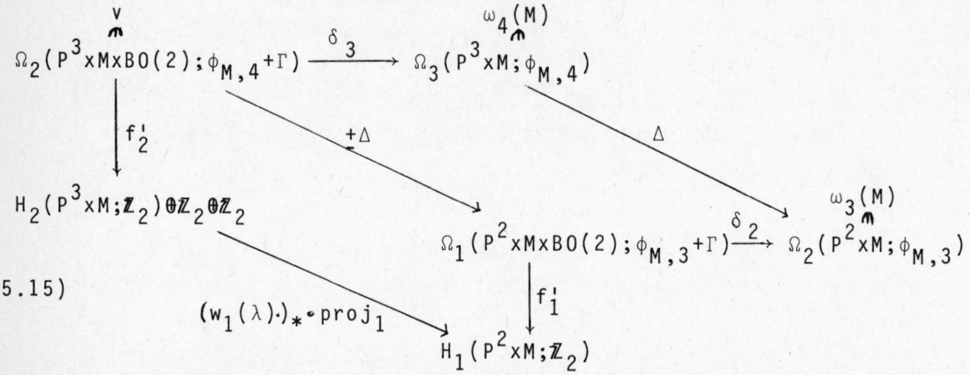

Here f_1' vanishes on $\ker\delta_2 = \operatorname{im}\sigma\circ j_3$; indeed, an inspection of formulas (14.10) and (14.11) shows that condition (ii) implies $f_1'\circ\sigma\circ j_3=0$.

Since $\delta_2(\pm\Delta(v)) = \Delta\circ\delta_3(v) = \Delta(\omega_4(M)) = \omega_3(M) = 0$, we conclude that $f_1'(\pm\Delta(v))$ vanishes and hence $f_2'(v)\in\ker((w_1(\lambda)\cdot)_*\circ proj_1) = H_2(M;\mathbb{Z}_2)\oplus\mathbb{Z}_2\oplus\mathbb{Z}_2$. On the other hand, this group contains also the image of $\bar\sigma$, again due to condition (ii) (and formula 15.4). Thus we can apply our analysis of $(\pi_{2*}\circ\bar\sigma_1,\bar\sigma_2,\bar\sigma_3)$ above to complete the proof. ∎

<u>Exercise 15.16.</u> Show that the manifold

$$M = (\mathbb{C}P(2q)\ \#\ \mathbb{C}P(2q)) \times S^1 \times S^1 \times S^1$$

has $\operatorname{span}\geq 4$ for $q\geq 1$. ∎

Next assume $n\equiv 1(4)$. We restrict our attention to the case when the homomorphism f_2' in diagram 15.2 is bijective, or, equivalently, when $w_1(\eta)=0$ and $Sq^1:H^1(Y;\mathbb{Z}_2)\longrightarrow H^2(Y;\mathbb{Z}_2)$ is injective (see the discussion of $\ker f_2'$ in 15.5).

This condition is satisfied e.g. if Y = point or $Y = BSO(n)$. Formulas 15.4, together with the example

$$\overline{\sigma}[\mathbb{C}P(2), \text{constant}, \text{or}] = (0,\ 1,\ 1)\ ,$$

show that $\Omega_3(P^3;\phi_{pt}) \cong \mathbb{Z}_2 \oplus \mathbb{Z}_2$ is generated by $\delta_3\circ(f_2')^{-1}\ (1,0,0)$ and $\delta_3\circ(f_2')^{-1}\ (0,1,0)$ (see diagram 15.2). On the other hand, the obvious (classifying) map $P^2 x \mathbb{C}P(1) \longrightarrow P^3 x BSO(n)$ gives rise to an element $x \in \overline{\Omega}_4(P^3 x BSO(n);\phi)$ such that $\overline{\sigma}(x) = (0,1,0)$. We may conclude (and we note for later use) that

$$(15.17) \qquad \mathbb{Z}_2 \cong \ker\ (\Theta\ :\ \Omega_3(P^3;\phi_{pt}) \longrightarrow \Omega_3(P^3 x BSO(n);\phi)).$$

Theorem 15.18. Let M be a closed connected orientable manifold of dimension $n \geq 9$, $n \equiv 1(4)$, such that the homomorphism $Sq^1 : H^1(M;\mathbb{Z}_2) \longrightarrow H^2(M;\mathbb{Z}_2)$ is injective and its image does not contain $w_2(M)$. Assume

(i) $w_{n-1}(M) = 0$, $w_{n-3}(M) = 0$ and $Sq^2 : H^2(M;\mathbb{Z}_2) \longrightarrow H^4(M;\mathbb{Z}_2)$ is injective,

or (ii) M has a 4-field with finite singularities.

Then $\text{span}(M) \geq 4$ if and only if the real Kervaire semicharacteristic $k(M)$ vanishes.

Proof. According to lemma 15.3 or the calculation of $\Omega_3(P^3;\phi_{pt})$, the assumptions above imply that $\omega_4(M) = \delta_3(v)$ for some $v \in \Omega_2(P^3 x M x BO(2);\phi_M + \Gamma)$.

Now pick a suitable singular surface $g: T \longrightarrow M$ such that the homology class $g_*[T]$ is annihilated by $Sq^1(H^1(M;\mathbb{Z}_2))$, but not by $w_2(M)$. Then the map $P^2 x T \xrightarrow{\pi_2} T \xrightarrow{g} M$ gives rise to an element $x \in \overline{\Omega}_4(P^3 x M;\phi_M)$ such that $\overline{\sigma}(x) = (0,1,0)$ (for a similar computation, see the proof of theorem 15.12). If (i) holds, similar and other easy examples (use fact 9.4!) show that the image of $\overline{\sigma}$ is a complement of

$H_2(P^3;\mathbb{Z}_2)$.

It is clear now that we always may assume v to lie in $(f_2')^{-1}(H_2(P^3;\mathbb{Z}_2))$. Compare the diagrams 15.2 and 14.2 via the homomorphism $\Delta\circ\Delta$ which connects them (see 13.15), and note that $\Delta\circ\Delta$ maps $\delta_3(f_2')^{-1}(H_2(P^3;\mathbb{Z}_2))$ isomorphically onto $\delta_1(\mathbb{Z}_2)$. Thus $\omega_4(M)=0$ if and only if $\omega_2(M)=0$, i.e. if and only if $k(M)=0$ (see remark 14.5).∎

Exercise 15.19. Consider the manifold

$$M = (\mathbb{C}P(2q) \# \mathbb{C}P(2q)) \times S^1 ,$$

where $q \geq 2$.

(i) Use the results quoted in the remarks 14.5 and 14.18 to show that $\text{span}(M)=2$.

(ii) Show that, for even q, the assumption (i) in theorem 15.18 holds, and $k(M)=0$ and $w_2(M) \notin Sq^1(H^1(M;\mathbb{Z}_2))$.

Conclude that the injectivity requirement for Sq^1 in theorem 15.18 cannot be dropped.∎

Now assume that n is even. Given an element $x = [N,g]$ of $\Omega_4(P^3 \times Y) = \overline{\Omega}_4(P^3 \times Y;\phi_Y)$ observe that (when pulled back to N)

$$w_1(\lambda)^2 z = w_1(\lambda)Sq^1(z)$$

for all $z \in H^2(P^3 \times Y;\mathbb{Z}_2)$, and, in particular, $w_1(\lambda)^2 y^2$ vanishes for all $y \in H^1(P^3 \times Y;\mathbb{Z}_2)$; indeed,

$$w_1(\lambda)^2 z + w_1(\lambda)Sq^1(z) = Sq^1(w_1(\lambda)z) = w_1(N)w_1(\lambda)z = 0$$

(cf. 9.6). Thus we can simplify the formulas for $\overline{\sigma}(x)$ in theorem 9.3, e.g.

$$z\overline{\sigma}_1(x) = (z^2+w_1(\lambda)w_1(\eta)z + \tbinom{n}{2}w_1(\lambda)^2 z)[N] ,$$

and we obtain

$$w_1(\lambda)^2\bar\sigma_1(x) = w_1(\lambda)^3 w_1(\eta)[N] \quad ;$$

$$w_1(\lambda)y\bar\sigma_1(x) = (\tbinom{n}{2}w_1(\lambda)^3y+w_1(\lambda)(w_1(\eta)^2y+w_1(\eta)y^2))[N],\ y\in H^1(Y;\mathbb{Z}_2);$$

$$z\bar\sigma_1(x) = (z^2+w_1(\lambda)(w_1(\eta)z+\tbinom{n}{2}Sq^1(z)))[N],\ z\in H^2(Y;\mathbb{Z}_2) \tag{15.20}$$

and

$$\bar\sigma_2(x) = (w_4(N)+w_1(\lambda)w_1(\eta)^3+\tbinom{n-1}{2}w_1(\lambda)^3w_1(\eta))[N]$$

and

$$\bar\sigma_3(x) = \pm\ 3 \text{ signature } (N) \quad .$$

Here we have used the fact that in $H^4(N;\mathbb{Z})\cong\mathbb{Z}^r$ the Pontrjagin class

$$\begin{aligned} p_1(\phi_Y) &= p_1(\lambda\otimes\eta) - p_1(\eta) \\ &= -(c_2(\lambda_{\mathbb{C}}\otimes\eta_{\mathbb{C}}) - c_2(\eta_{\mathbb{C}})) \end{aligned}$$

is a multiple of $c_1(\lambda_{\mathbb{C}})$ (see the formula 4.4.3 III in [27]; since $2c_1(\lambda_{\mathbb{C}})$ vanishes, so does $p_1(\phi_Y)$.

Next we study $\ker f_2'$. If $n\equiv0(4)$ and $w_1(\eta)=0$, then by theorem 9.3

$$\ker f_2' \cong H_1(P^3\times Y;\ \mathbb{Z}_2) \oplus \mathbb{Z}_4. \tag{15.21}$$

If $n\equiv2(4)$ or $w_1(\eta)\neq0$, consider the condition

$$(*) \quad ((\tbinom{n}{2}w_1(\lambda)^2+w_1(\lambda)w_1(\eta))(bw_1(\lambda)+y)+(w_1(\lambda)^2w_1(\eta)+w_1(\lambda)w_1(\eta)^2)c_o = 0$$

defined for any element $c = bw_1(\lambda) + y + c_o w_1(\gamma)$ of $H^1(P^3\times Y\times BO(2);\mathbb{Z}_2)$. We have : c satisfies (*) if and only if

$$w_1(\lambda)^3 b\tbinom{n}{2} + w_1(\lambda)^2((b+c_o)w_1(\eta) + \tbinom{n}{2}y) + w_1(\lambda)(w_1(\eta)y + c_o w_1(\eta)^2)$$

vanishes, i.e. if and only if $c\in\mathbb{Z}_2(w_1(\eta)+w_1(\gamma))$ (when $n\equiv2(4)$), resp. iff $b=c_o$ and $w_1(\eta)(y+c_o w_1(\eta))=0$ (when $n\equiv0(4)$ and $w_1(\eta)\neq0$). According to theorem 9.3 we have in both cases that

$$\ker f_2' \cong H_1(P^3\times Y\times BO(2);\mathbb{Z}_2)\big/\{\ s \mid c(s) = 0 \ \text{ for all } \ c \ \text{ satisfying } (*)\}.$$

Therefore, if $n\equiv2(4)$, or if $n\equiv0(4)$ and $w_1(\eta)\neq0$ and multiplication with $w_1(\eta)$ is injective on $H^1(Y;\mathbb{Z}_2)$, then

(15.22) $$\ker f_2' \cong \mathbb{Z}_2$$

is generated by $[(0,0,\text{ generator of } H_1(BO(2);\mathbb{Z}_2))]$ (in the second case, $[(\text{generator of } H_1(P^3;\mathbb{Z}_2),0,0)]$ will also do).

Exercise 15.23. Show that for $n\equiv2(4)$, $n\geq10$, $\pi_{n-1}(V_{n,4})$ is the direct sum of an infinite cyclic group and of a group of order 12.

Theorem 15.24. *Let M be a closed connected manifold of dimension $n\equiv2(4)$, $n\geq10$, such that the homomorphism*

$$H^2(M;\mathbb{Z}_2)\xrightarrow{((w_1(M)\cdot+Sq^1),\ Sq^2)} H^3(M;\mathbb{Z}_2)\oplus \mathrm{Hom}(H_4(M;\mathbb{Z}),\mathbb{Z}_2)$$

is injective.

Then:

(i) *M has a 4-field with finite singularities if and only if the Stiefel-Whitney classes $w_{n-2}(M)\in H^{n-2}(M;\mathbb{Z}_2)$ and $W_{n-3}(M)\in H^{n-3}(M;\tilde{\mathbb{Z}})$ vanish.*

(ii) *$\mathrm{span}(M)\geq4$ if and only if $w_{n-2}(M)$, $W_{n-3}(M)$ and the Euler number $\chi(M)$ vanish.*

Example 15.25. For $q\geq2$ complex projective space $\mathbb{C}P(2q+1)$ has a 4-field with finite singularities if and only if q is odd (i.e. $\dim \mathbb{C}P(2q+1)\equiv6(8)$). On the other hand, $\mathrm{span}\ \mathbb{C}P(2q+1)=0$.

Proof of theorem 15.24. Note that the diagram (15.2) is natural with respect to such homomorphisms as Θ_M and Θ (see 13.7), and that the groups $\ker f_2'\cong\mathbb{Z}_2$ (see 15.22), formed for the cases Y=point, M and $B(S)O(n)$, correspond to one another bijectively.

Also, recall the (Wu) relations

(15.26) $$w_n(M) = w_1(M)w_{n-1}(M) = w_1(M)^2w_{n-2}(M)$$

(apply lemma 12.4 to $i = n-2$).

If $w_{n-2}(M)$ and $W_{n-3}(M)$ vanish, then so does the mod 2 Euler number $w_n(M)[M]$, and according to lemma 15.3 $j \cdot f_3(\omega_4(M))$ has the form $(2q,0,0)$ for some $q \in \mathbb{Z}$. But this element comes already from Y=point (in fact, it equals $j \circ f_3 \circ \Theta_M(qz)$, where $z \in \Omega_3(P^3;(n-4)\lambda)$ is characterized by the equation $\Theta_{S^n}(z) = \omega_4(S^n)$). Thus $\omega_4(M)$ lies in $\operatorname{im} \Theta_M + \ker f_{3,M}$. If we can show that $\ker f_{3,M} = \operatorname{im} \delta_{3,M} \subset \operatorname{im} \Theta_M$, then clearly M has a 4-field with finite singularities by 13.16.

Now observe that the injectivity assumption in our theorem just implies this. Indeed, we may choose a basis $z_1',\ldots,z_r',z_1'',\ldots,z_s''$ of $H^2(M;\mathbb{Z}_2)$ with the following properties:

a) the elements $w_1(M)z_1' + Sq^1(z_1'),\ldots,w_1(M)z_r' + Sq^1(z_r')$ are linearly independent in $H^3(M;\mathbb{Z}_2)$ while the elements $w_1(M)z_i'' + Sq^1(z_i'')$, $i = 1,..s$, vanish; and

b) $z_1''^2,\ldots, z_s''^2$ define linearly independent elements in $\operatorname{Hom}(H_4(M;\mathbb{Z}),\mathbb{Z}_2) \cong \operatorname{Hom}(\tilde{\Omega}_4(M),\mathbb{Z}_2)$ (cf. fact 9.4). Thus on one hand we may pick bordism classes $[L_i,h_i] \in \mathfrak{N}_3(M)$, $i = 1,\ldots,r$, such that the corresponding homology classes $h_{i*}[L_i]$ are dual to the cohomology classes $w_1(M)z_j' + Sq^1(z_j')$, $j = 1,\ldots r$. Now apply the construction following theorem 14.13. According to 3.10, the orientation bundle ξ_i of L_i can be mapped monomorphically into $\underset{\sim}{\mathbb{R}}^3$ (since $\underset{\sim}{\operatorname{Hom}}(\xi_i,\underset{\sim}{\mathbb{R}}^3)$ has a section). Hence the canonical line bundle γ over the projectification $N_i' = RP(\xi_i \oplus \underset{\sim}{\mathbb{R}})$ allows a monomorphism into $\underset{\sim}{\mathbb{R}}^4$, thus determining a map $g_{i1} : N_i' \longrightarrow P^3$. The oriented 4-manifold N_i', together with the map $g_i = (g_{i1}, h_i \circ \pi)$, represents a bordism class x_i' in $\Omega_4(P^3 \times M)$. We have

$$\begin{aligned} z_j'\bar{\sigma}_1(x_i') &= g_i^*(w_1(\lambda)(w_1(M)z_j' + Sq^1(z_j'))[N_i'] \\ &= w_1(\gamma)\cdot\pi^*(h_i^*(w_1(M)z_j' + Sq^1(z_j'))[N_i'] \\ &= h_i^*(w_1(M)z_j' + Sq^1(z_j'))[L_i] \\ &= \delta_{ij} \quad , \end{aligned}$$

and similarly $z_j''\bar{\sigma}(x_i') = 0$ (see 15.20).

On the other hand, we may pick bordism classes $[N_i'', g_{i2}] \in \tilde{\Omega}_4(M)$, $i=1,\dots,s$, such that the corresponding mod 2 homology classes $g_{i2*}[N_i'']$ are dual to $z_1''^2, \dots z_s''^2$. Then we have for the elements $x_i'' = [N_i'', (\text{constant}, g_{i2})]$ of $\Omega_4(P^3 \times M)$ that

$$z_j''\bar{\sigma}_1(x_i'') = g_{i2}^*(z_j''^2)[N_i''] = \delta_{ij} .$$

The examples $x_1', \dots, x_r', x_1'', \dots, x_s''$, as well as bordism classes of the form $[P^3 \times S^1 \xrightarrow{id \times \ell} P^3 \times M]$ imply that

$$\bar{\sigma}_1(\Omega_4(P^3 \times M)) = \ker(w_1(\lambda)^2 + w_1(\lambda)w_1(M)).$$

Thus the image of $\bar{\sigma}$, and elements coming from Y=point, generate already the target group of f_2' in diagram 15.2 (written down for $Y=M$). Therefore, $\Omega_2(P^3 \times M \times BO(2); \phi_M + \Gamma)$ is spanned by the images of $\sigma \circ j_4$ and of Θ_M; and the image of δ_3 must lie in $\Theta_M(\Omega_3(P^3;(n-4)\lambda))$. Thus we have completed the proof of the first part of our theorem.

If $w_{n-2}(M)$, $W_{n-3}(M)$ and $\chi(M)$ vanish, then so do all the Stiefel-Whitney numbers of M which contain $w_n(M)$, $w_{n-1}(M)$, $w_{n-2}(M)$ or $w_{n-3}(M)$ as a factor (see 15.26). It follows from theorem 12.13 that there is a manifold M' with $\text{span}M' \geq 4$ such that M' is bordant to M (first in $\mathfrak{N}_n$, and then even in $\mathfrak{N}_n(0,0)$ since $\chi(M')=\chi(M)=0$, see theorem 12.1). Therefore $\omega_4'(M)=0$, (see theorem 13.4). On the other hand, $\omega_4'(M)=(id \times \tau)_*(\omega_4(M))$, and $\omega_4(M)$ lies in the image of Θ_M. So actually $\omega_4(M)=\Theta_M(v)$ for some $v \in \ker \Theta$ (see 13.7). The second statement of our theorem follows now from theorem 13.3 and from the following observation.

Proposition 15.27. Assume $n \equiv 2(4)$, $n \geq 10$. Then $\Theta: \Omega_3(P^3;(n-4)\lambda) \longrightarrow \Omega_3(P^3 \times BO(n);\phi)$ is injective.

Proof. Consider the diagram 13.11, applied to $k=4$ and extended slightly to the right. Given an element v in the kernel of $\Theta=(id \times \tau)* \circ \Theta_M$, we have

$$d(v) = (\text{const}_* \circ \Theta)(d(v)) = (\text{const}_* \circ d)(\Theta(v)) = 0,$$

and hence

$$v \in \operatorname{im} \Delta = \operatorname{im} \Delta \circ \sigma = \sigma(\text{proj}_*(\pi_{n-1}(V_{n,5}))) .$$

It follows from the tables of Paechter [51] that this last group is infinite cyclic (see also the discussion following 16.16).

Now let u be some 5-field with finite singularity on the sphere S^n, and let $z \in \text{proj}_*(\pi_{n-1}(V_{n,5}))$ be the index of the corresponding 4-field. Lemma 15.3 implies that

$$\begin{aligned} j \circ f_3 \circ \Theta(\sigma(z)) &= j \circ f_3 \circ (\text{id} \times \tau_S)_*(\omega_4(S^n)) \\ &= (2, 0, 0) \end{aligned}$$

in $\mathbb{Z} \oplus H_2(BO(n);\mathbb{Z}_2) \oplus H_3(BO(n);\mathbb{Z})$. Thus the first component of $j \circ f_3 \circ \Theta$ is injective on $\sigma(\text{proj}_*(\pi_{n-1}(V_{n,5})))$, and so is Θ. We conclude that

$$\ker \Theta = \ker \Theta \cap \sigma(\text{proj}_*(\pi_{n-1}(V_{n,5}))) = 0 .$$

Finally we turn to the case $n \equiv 0(4)$. If $w_1(\eta)=0$, then also $w_2(\phi_Y) = 0$, and diagram (15.2) takes the following form

(15.28)

$$\begin{array}{ccccccccc}
 & & & & 0 & & & & \\
 & & & & \downarrow & & & & \\
 & & & & \mathbb{Z}_2 \oplus H_1(Y;\mathbb{Z}_2) \oplus \mathbb{Z}_4 & & & & \\
 & & & & \downarrow & & & & \\
\cdots \longrightarrow \Omega_4(P^3 \times Y) & \xrightarrow{\sigma \circ j_4} & & & \Omega_2(P^3 \times Y \times BO(2); \phi_Y + \Gamma) & \xrightarrow{\delta_3} & \Omega_3(P^3 \times Y; \phi_Y) & \xrightarrow{f_3} & \Omega_3(P^3 \times Y) \longrightarrow 0 \\
\wr\Vert & & & & & & & & \wr\Vert \\
H_1(Y;\mathbb{Z}) \oplus H_3(Y;\mathbb{Z}_2) \oplus H_4(Y;\mathbb{Z}) \oplus \Omega_4 & & \overline{\sigma} \searrow & & \downarrow f_2' & & & & \mathbb{Z} \oplus H_2(Y;\mathbb{Z}_2) \oplus H_3(Y;\mathbb{Z}_2) \\
 & & & & (\mathbb{Z}_2 \oplus H_1(Y;\mathbb{Z}_2) \oplus H_2(Y;\mathbb{Z}_2)) \oplus \mathbb{Z}_2 \oplus \mathbb{Z} & & & & \\
 & & & & \downarrow & & & & \\
 & & & & 0 & & & &
\end{array}$$

(see also 9.4, 15.21, and the footnote of theorem 9.3).

Theorem 15.29. Let M be a closed connected manifold of dimension $n>8$, $n\equiv 0(4)$, and assume that $H^1(M;\mathbb{Z}_2)=0$ and that the homomorphism

$$Sq^2 : H^2(M;\mathbb{Z}_2) \longrightarrow \mathrm{Hom}(H_4(M;\mathbb{Z});\ \mathbb{Z}_2)$$

is injective.

Then M has a 4-field with finite singularities, if and only if the Stiefel-Whitney classes $w_{n-2}(M) \in H^{n-2}(M;\mathbb{Z}_2)$ and $W_{n-3}(M) \in H^{n-3}(M;\mathbb{Z})$ vanish.

If in addition $n\equiv 0(8)$, we have also: $\mathrm{span}(M)\geq 4$ if and only if $w_{n-2}(M)$, $W_{n-3}(M)$ and the Euler number $\chi(M)$ vanish and $\mathrm{signature}(M)\equiv 0(16)$.

Example 15.30. For $q>2$ complex projective space $\mathbb{C}P(2q)$ has a 4-field with finite singularities if and only if q is even (i.e. $\dim\mathbb{C}P(2q)\equiv 0(8)$). Moreover, quaternionic projective space $\mathbb{H}P(q)$ has always a 4-field with finite singularities.

On the other hand, $\mathrm{span}\,\mathbb{C}P(2q)=\mathrm{span}\,\mathbb{H}P(q)=0$.

Proof of theorem 15.29. Our injectivity condition and formula 15.20 imply that $\overline{\sigma}(\Omega_4(P^3 x M))$ projects onto the factor $H_2(M;\mathbb{Z}_2)$ (see 15.28). Hence for $Y=M$ $\ker f_3=\mathrm{im}\,\delta_3$ lies already in $\Theta_M(\Omega_3(P^3;\phi_{pt}))$.

If $w_{n-2}(M)$ and $W_{n-3}(M)$ vanish, then $f_3(\omega_4(M))$ has the form $(s,0,0)$ and hence comes from $\Omega_3(P^3 x\,\mathrm{point})$ and even from $\Omega_3(P^3;\phi_{pt})$. Thus $\omega_4(M)$ lies in the image of Θ_M. The first claim in our theorem follows now from 13.16.

The second claim can be deduced from the work of Atiyah and Dupont [6] and from the results in the next section (see 16.28 for more details).

Proposition 15.31. Assume $n>0$, $n\equiv 0(4)$. Then

$$\Theta : \Omega_3(P^3;(n-4)\lambda) \longrightarrow \Omega_3(P^3 x BSO(n);\phi)$$

(cf. 13.7) is injective, while the kernel of the unoriented analogue

$$\Theta : \Omega_3(P^3;(n-4)\lambda) \longrightarrow \Omega_3(P^3 \times BO(n);\phi)$$

contains at least eight elements.

Proof. The first (oriented) Θ fits into the following commuting diagram of exact sequences

$$\begin{array}{ccccc} \Omega_3(\text{point};\text{trivial}) & \longrightarrow & \Omega_3(P^3;(n-4)\lambda) & \longrightarrow & \Omega_2(P^2;(n-3)\lambda) \\ \downarrow & & \downarrow\Theta & & \downarrow \\ \Omega_3(BSO(n);\text{trivial}) & \xrightarrow{\text{incl}_*} & \Omega_3(P^3 \times BSO(n);\phi_{(4)}) & \xrightarrow{\Delta} & \Omega_2(P^2 \times BSO(n);\phi_{(3)}) \end{array}$$

extracted from 13.11 (put k=3). Note that the right hand vertical arrow is injective; this follows easily from the naturality of diagram 14.7 and from the footnote in theorem 9.3. Clearly, the vertical arrow to the left is (split) injective. We will show presently that incl_* is also a monomorphism. The injectivity of Θ itself follows then from some standard diagram chasing.

In order to study the homomorphism incl_* above, we fit it into the following commuting diagram (where we denote $BSO(n)$ by Y, and $\phi_{(4)}$ by ϕ, for brevity)

(15.32)
$$\begin{array}{ccccccccc} & & H_4(Y;\mathbb{Z})\oplus\Omega_4 & \longrightarrow & \Omega_2(Y \times BO(2);\Gamma) & \longrightarrow & \Omega_3(Y;\text{trivial}) & \longrightarrow & H_3(Y;\mathbb{Z}) \\ & & \cap & & \downarrow \text{incl}_* & & \downarrow \text{incl}_* & & \cap \\ \Omega_4(P^3 \times Y;\phi) & \xrightarrow{f_4} & \mathbb{Z}_2\oplus H_4(Y;\mathbb{Z})\oplus\Omega_4 & \xrightarrow{\sigma j_4} & \Omega_2(P^3 \times Y \times BO(2);\phi+\Gamma) & \xrightarrow{\delta_3} & \Omega_3(P^3 \times Y;\phi) & \longrightarrow & \mathbb{Z}\oplus\mathbb{Z}_2\oplus H_3(Y;\mathbb{Z}). \end{array}$$

Both lines are special cases of the exact sequence in theorem 9.3 (for the lower line, see also 15.28). This theorem also implies that the left hand arrow labelled incl_* is injective.

Again standard diagram chasing (as in the proof of the five lemma) shows that the right hand arrow incl_* (which is the relevant one for us) has trivial kernel. To see this, we merely need to establish that $\sigma j_4(H_4(Y;\mathbb{Z})\oplus\Omega_4)$ is already the whole image of σj_4, i.e. that $f_4(\Omega_4(P^3 \times Y;\phi))$ is not entirely contained in $H_4(Y;\mathbb{Z})\oplus\Omega_4$.

Consider the four-dimensional manifold $P^3 x S^1$, and over it the pullbacks λ and κ of the canonical (nontrivial) line bundles over P^3 and S^1. Note that

$$\lambda \otimes \underset{\sim}{\mathbb{R}}^4 \cong T(P^3 x S^1) \cong \underset{\sim}{\mathbb{R}}^4 ,$$

and that the oriented 4-plane bundle

$$\zeta := \underset{\sim}{\mathbb{R}} \oplus \lambda \oplus \kappa \oplus (\lambda \otimes \kappa)$$

also enjoys the property $\lambda \otimes \zeta \cong \zeta$. Thus, if

$$g : P^3 \times S^1 \longrightarrow P^3 \times BSO(n) = P^3 \times Y$$

is defined by classifying maps of λ and of $\zeta \oplus \underset{\sim}{\mathbb{R}}^{n-4}$, we have an isomorphism $\overline{g}$ from

$$T(P^3 x S^1) \oplus g^*(\phi^+) \cong \lambda \otimes \underset{\sim}{\mathbb{R}}^4 \oplus \lambda \otimes (\zeta \oplus \underset{\sim}{\mathbb{R}}^{n-4})$$

to

$$g^*(\phi^-) \cong \lambda \otimes \underset{\sim}{\mathbb{R}}^4 \oplus (\zeta \oplus \underset{\sim}{\mathbb{R}}^{n-4}).$$

The image of the resulting bordism class $\omega = [P^3 x S^1, g, \overline{g}] \in \Omega_4(P^3 x Y; \phi)$ under f_4 (see 15.32) does not lie in the subgroup $\{o\} \oplus H_4(Y;\mathbb{Z}) \oplus \Omega_4$ of $\mathbb{Z}_2 \oplus H_4(Y;\mathbb{Z}) \oplus \Omega_4 \cong \Omega_4(P^3 x Y)$. Indeed, this subgroup is annihilated by the characteristic number defined by $w_1(\lambda) w_3(\gamma) \in H^4(P^3 x Y; \mathbb{Z}_2)$, while we have

$$\begin{aligned} w_1(\lambda) w_3(\gamma) \mu(f_4(\omega)) &= w_1(\lambda) w_3(\zeta \oplus \underset{\sim}{\mathbb{R}}^{n-4})[P^3 x S^1] \\ &= w_1(\lambda) w_1(\lambda) w_1(\kappa)(w_1(\lambda) + w_1(\kappa))[P^3 x S^1] \\ &= w_1(\lambda)^3 w_1(\kappa)[P^3 x S^1] \\ &= 1 . \end{aligned}$$

This completes the proof of the first claim in our proposition.

Next, interrelate diagram 15.2, written down for Y=point on one hand (see also 15.28) and for Y=BO(n) on the other hand, via the unoriented version of Θ. If Y=BO(n), then $\ker f_2^! \cong \mathbb{Z}_2$, (see 15.22) and the map

$$\mathrm{id} \times \ell : P^3 \times S^1 \longrightarrow P^3 \times BO(n) ,$$

ℓ not nulhomotopic, represents a bordism class $x \in \overline{\Omega}_4(P^3 \times BO(n);\phi)$ such that $\overline{\sigma}(x)=(a,1,0)$ where a denotes the generator of $H_2(P^3;\mathbb{Z}_2)$ (see 15.20). If Y=point, then $\ker f_2' \cong \mathbb{Z}_2 \oplus \mathbb{Z}_4$ (see 15.21), and $\overline{\sigma}$ maps the generator $[\mathbb{C}P(2), \text{constant}]$ of $\overline{\Omega}_4(P^3 \times \text{point};(n-4)\lambda)=\Omega_4(P^3)$ to $\pm(0,1,3)$; thus $\ker f_2'$ contributes four elements to $\ker \Theta$, and $\delta_3(f_2'^{-1}\{(a,1,0)\}$ contains further elements of this kernel. The second claim of our proposition follows. ∎

<u>Theorem 15.33. Let M be a closed connected non-orientable manifold of dimension $n>8$, $n \equiv 0(4)$, such that the homomorphism $w_1(M)\cdot : H^1(M;\mathbb{Z}_2) \longrightarrow H^2(M;\mathbb{Z}_2)$ is injective. Assume that M has a 4-field with finite singularities. Then:</u>

$$\underline{\operatorname{span}(M) \geq 4 \text{ if and only if } \chi(M) = 0 .}$$

<u>Proof.</u> The last argument in the preceding proof applies also to the homomorphism

$$\pi_{n-1}(V_{n,4}) \xrightarrow[\cong]{\sigma} \Omega_3(P^3;(n-4)\lambda) \xrightarrow{\Theta_M} \Omega_3(P^3 \times M;\phi_M)$$

(cf. 13.7). Thus $\ker \Theta_M$ (which lies in $\ker \Theta$), still contains at least eight elements.

On the other hand, we will see in the next section (cf. the tables in 16.15) that $\ker \Theta$ has precisely eight elements and contains $\sigma(\mathrm{Index}(u,or))$ for any 4-field u with finite singularities on any closed n-manifold having zero Euler number.

Therefore, if $\chi(M)=0$, then the index of the fourfield on M lies in $\ker \Theta\circ\sigma = \ker \Theta_M \circ \sigma$. It follows from proposition 13.8 that $\omega_4(M)=0$, and hence $\operatorname{span}(M) \geq 4$. ∎

<u>Theorem 15.34. Let M be a closed connected n-manifold, $n \equiv 0(4)$, $n>8$. Assume that</u> $w_1(M)^2 \neq 0$, $H^1(M;\mathbb{Z}_2) \cong \mathbb{Z}_2$ <u>and that the homomorphism</u>

$$H^2(M;\mathbb{Z}_2) \xrightarrow{w_1(M)\circ \oplus Sq^2} H^3(M;\mathbb{Z}_2) \oplus \mathrm{Hom}(H_4(M;\mathbb{Z}),\mathbb{Z}_2)$$

is injective. Then we have:

(i) M has a 4-field with finite singularities if and only if the Stiefel-Whitney classes $w_{n-2}(M) \in H^{n-2}(M;\mathbb{Z}_2)$ and $W_{n-3}(M) \in H^{n-3}(M;\tilde{\mathbb{Z}})$ vanish.

(ii) $\mathrm{span}(M)\geq 4$ if and only if $w_{n-2}(M)$, $W_{n-3}(M)$ and $\chi(M)$ vanish.

Proof. Consider diagram 15.2 for $Y=M$. The assumptions on M guarantee that $\ker f_2' \cong \mathbb{Z}_2$ (cf. 15.22) and that the image of $\bar{\sigma}_1$ contains at least a complement of $H_2(P^3;\mathbb{Z}_2)$ (see the proof of theorem 15.24 for a very similar argument). It follows that the image of δ_3 lies already in $\Theta_M(\Omega_3(P^3;(n-4)\lambda))$, and so does $\omega_4(M)$ if $w_{n-2}(M) = W_{n-3}(M) = 0$. Then 13.16 implies claim (i). Claim (ii) follows immediately from theorem 15.33.

Example 15.35. For $q>2$ real projective space P^{4q} has a 4-field with finite singularities if and only if q is even, i.e. $\dim P^{4q} \equiv 0(8)$.

Indeed, $M=P^{4q}$ satisfies the assumptions of the last theorem. Note also that $w_{4q-2}(P^{4q}) = \binom{4q+1}{3} w_1(\lambda)^{4q-2} = 0$, and that the reduction homomorphism $H_3(P^{4q};\mathbb{Z}) \longrightarrow H_3(P^{4q};\mathbb{Z}_2)$ is bijective. Hence P^{4q} admits a 4-field with finite singularities if and only if

$$w_{4q-3}(P^{4q}) = \binom{4q+1}{4} w_1(\lambda)^{4q-3}$$

vanishes.∎

Exercise 15.36. Show that real projective space P^n, $n\geq 4$, has a 4-field with finite singularities if and only if $n \equiv 7$ or $8(8)$. (Hint: use 13.14, 15.12, the preceding example and the parallelizability of P^7).

§16. Indices of framefields with finite singularities.

Throughout this section we assume that $n>2k>0$. Thus we may identify $\pi_{n-1}(V_{n,k})$ with $\Omega_{k-1}(P^{k-1};(n-k)\lambda)$ via σ (see e.g. the diagram 13.7). We will study various subsets of this group, e.g. the set of all elements in $\pi_{n-1}(V_{n,k})$ which can occur as indices of k-fields with finite singularities on closed n-dimensional manifolds (see §13 for definitions), or the kernel of the homomorphism $\Theta\circ\sigma$ in 13.7 (which we will denote briefly by $\ker\Theta^{(or)}$). It turns out that our results are closely related to a question raised by Atiyah and Dupont in [6].

Most of the time we will treat two theories (oriented and unoriented) simultaneously. In the oriented theory, we consider only k-fields with finite singularities on oriented manifolds, and the local orientations (used to identify the local obstructions with elements in $\pi_{n-1}(V_{n,k})$) are induced from the global orientation of the manifold. In the unoriented theory, global orientation questions are entirely ignored, and the local orientations may be picked arbitrarily even if the manifolds happen to be orientable.

Proposition 16.1. Assume $n>2k>0$. Let u be a k-field with finite singularities on an arbitrary, resp. oriented, closed n-manifold M, and let the system or of local orientations at the singularities of u be arbitrary, resp. induced from the global orientation of M.

Then there exists a k-field $\bar{u}$ with just one singular point on an arbitrary, resp. oriented, connected closed n-manifold $\bar{M}$ and a suitable local orientation $\overline{or}$ so that $\mathrm{Index}(\bar{u},\overline{or})=\mathrm{Index}(u,or)$ (and hence $\chi(\bar{M})=\chi(M)$).

Proof. We use a "connected sum operation along embedded tori". If M_1, M_2 are connected components of M, choose embedded k-dimensional tori $T_i \subset M_i$, $i=1,2$, lying in small balls which avoid the singularity of u.

Because of the dimension conditions we may deform $u|T_i$ into the standard k-framefield along the tori T_i, and simultaneously we may deform a trivialization of the complement of $im(u)|T_i$ into a trivialization of the normal bundle of T_i in M_i. Hence we may assume that, in a whole tubular neighborhood $T_i \times D$ of T_i, u is given by the standard k-field in the direction of $T_i \times \{x\}$, $x \in D$. Thus, if we glue $M_1 - T_1 \times \mathring{D}$ and $M_2 - T_2 \times \mathring{D}$ together along their common boundary, we get a new manifold with one less connected component and carrying a k-field with the same index as before.

We can iterate this procedure to obtain the desired connected manifold $\overline{M}$; in the unoriented theory we can make sure that $\overline{M}$ is nonorientable by first adding (Kleinbottle x $(S^1)^{n-2}$, equipped with a suitable k-framefield), to (M,u). Now our claim follows from proposition 13.6. Note also that the Euler number $\chi(M)$ is the image of Index(u,or) under the homomorphism

$$(16.2) \qquad pr_* : \pi_{n-1}(V_{n,k}) \longrightarrow \pi_{n-1}(S^{n-1}) \xrightarrow[\cong]{\deg} \mathbb{Z}$$

induced by the projection which associates to every k-frame in $\mathbb{R}^n$ its first vector.

<u>Definition 16.3.</u> Define subsets $\pi^{or}(n,k)$, $\tilde{\pi}^{or}(n,k)$, $\pi(n,k)$ and $\tilde{\pi}(n,k)$ of $\pi_{n-1}(V_{n,k})$ as follows. $z \in \pi_{n-1}(V_{n,k})$ lies in $\pi^{or}(n,k)$ (resp. in $\tilde{\pi}^{or}(n,k)$) if there exists a closed oriented n-manifold M with arbitrary Euler number (resp. with $\chi(M)=0$) and carrying a k-field u with finite singularities such that Index(u,or)=z, where or is induced by the global orientation of M. $\pi(n,k)$ and $\tilde{\pi}(n,k)$ are defined in the same fashion, but without any orientability assumptions on M and without any assumptions whatever on the system or of local orientations.

Note that clearly $\tilde{\pi}(n,k)=\pi(n,k) \cap \ker pr_*$ and $\tilde{\pi}^{or}(n,k)=\pi^{or}(n,k) \cap \ker pr_*$ (see 16.2).

Theorem 16.4. Assume $n>2k>0$.

Then the four subsets

$$\begin{array}{ccc} \tilde{\pi}^{or}(n,k) & \subset & \pi^{or}(n,k) \\ \cap & & \cap \\ \tilde{\pi}(n,k) & \subset & \pi(n,k) \end{array}$$

are actually subgroups of $\pi_{n-1}(V_{n,k})$.

If n is odd, then $\pi(n,k)=\tilde{\pi}(n,k)$ and $\pi^{or}(n,k)=\tilde{\pi}^{or}(n,k)$.

If n is even, we have

$$\pi(n,k) = \tilde{\pi}(n,k) \oplus \begin{cases} \mathbb{Z}\cdot z_1 & \text{if } k\le \operatorname{span}(S^{n-1}), \\ \mathbb{Z}\cdot z_2 & \text{if } k>\operatorname{span}(S^{n-1}). \end{cases}$$

Here we may choose z_1 to be given by $S^{n-1}\times\mathbb{R}^k \xrightarrow{v} TS^{n-1} \subset S^{n-1}\times\mathbb{R}^n$, where v is the lifting of an arbitrary k-framefield on P^{n-1}; we may choose z_2 to be the index of a k-framefield with finite singularities on S^n; then $z_i\in \operatorname{proj}_*(\pi_{n-1}(V_{n,k+1}))$ and $\operatorname{pr}_*(z_i)=i$, $i=1,2$.

For even n there is also a welldefined integer $h^{or}(n-1)$, $0\le h^{or}(n-1)\le \operatorname{span}(S^{n-1})$, such that

$$\pi^{or}(n,k) = \tilde{\pi}^{or}(n,k) \oplus \begin{cases} \mathbb{Z}\cdot z_1^{or} & \text{if } k\le h^{or}(n-1), \\ \mathbb{Z}\cdot z_2^{or} & \text{if } k>h^{or}(n-1), \end{cases}$$

where $z_i^{or}\in\pi_{n-1}(V_{n,k})$ satisfies $\operatorname{pr}_*(z_i^{or})=i$ for $i=1,2$, and we may choose $z_2^{or}=z_2$ as above. E.g. $h^{or}(n-1)=0$ for $n\equiv 2(4)$, $h^{or}(n-1)=2$ for $n\equiv 4(8)$, and $h^{or}(n-1)\ge 4$ for $n\equiv 0(8)$.

Remark 16.5. Given an even integer $n>0$, write it in the form

$$n = (\text{odd integer}) \cdot 2^{c+4d}$$

where $0\le c\le 3$ and $0\le d$, and define

$$h(n-1) = 2^c + 8d - 1 ,$$

so that e.g. $h(n-1)=1$ for $n\equiv 2(4)$, $h(n-1)=3$ for $n\equiv 4(8)$, and $h(n-1)\ge 7$ for $n\equiv 0(8)$.

Then, according to the theorem of Adams and Hurwitz-Radon (see [1], [5 , prop. 15.14]) we have

$$\operatorname{span}(P^{n-1}) = \operatorname{span}(S^{n-1}) = h(n-1).$$

We will need only the first of these two equations in the following proof.

Proof of theorem 16.4. Clearly the four subsets contain 0 and are closed under addition, so we need to worry only about inverses.

Given an element z of the reduced group $\tilde{\pi}(n,k)$ or $\tilde{\pi}^{or}(n,k)$, by proposition 16.1 we may represent it as the index of a k-field $u=(u_1,..,u_k)$ with just one singular point on a closed connected manifold M such that $\chi(M)=0$. Hence we may assume that the first component vectorfield u_1 has no zero. Then the index of $\bar{u}=(-u_1,u_2,..u_k)$ with respect to reversed orientations is $-z$. Indeed, by definition we can represent the indices of u and $\bar{u}$ by maps $h, \bar{h}: S^{n-1} \longrightarrow W^k(\mathbb{R}^k,\mathbb{R}^n)$ which are related as follows. For all $x \in S^{n-1}$ $\bar{h}(x)$ is the monomorphism

$$\mathbb{R}^k \xrightarrow{\ b\ } \mathbb{R}^k \xrightarrow{h(a(x))} \mathbb{R}^n \xrightarrow{\ a\ } \mathbb{R}^n \quad ,$$

where a and b are the reflections with respect to the first factors $\mathbb{R}^1$ in $\mathbb{R}^n$ and $\mathbb{R}^k$. We may assume that $h(a(x))$ maps these first factors identically into one another so that $\bar{h}(x)=h(a(x))$. Hence

$$\operatorname{Index}(\bar{u},-or) = [\bar{h}] = [h \circ a] = -[h] = -z.$$

This implies that $\tilde{\pi}(n,k)$ and $\tilde{\pi}^{or}(n,k)$ are subgroups of $\pi_{n-1}(V_{n,k})$. If n is odd, we have nothing else to show.

So let n be even for the rest of the proof. Consider the "self-connected sum" of the torus $(S^1)^n$: remove two disjoint small open balls from one copy of this torus and identify the two boundary spheres suitably with one another. The resulting closed n-manifold has Euler number -2 and admits a k-field with finite singularities. Clearly S^n

has Euler number +2 and also admits such a k-field whose index we call z_2. It follows that $2\mathbb{Z}$ is contained in the image of $\pi(n,k)$ and of $\pi^{or}(n,k)$ under the homomorphism $pr_*:\pi_{n-1}(V_{n,k}) \longrightarrow \mathbb{Z}$ (cf. 16.2). Moreover, given any element $z \in \pi^{(or)}(n,k)$, $2z$ has an inverse in $\pi^{(or)}(n,k)$, and hence so does z itself. Thus $\pi(n,k)$ and $\pi^{or}(n,k)$ are also subgroups of $\pi_{n-1}(V_{n,k})$.

Next, we want to consider the following conditions in both theories (oriented or not):

(i) $pr_*(\pi^{(or)}(n,k)) = \mathbb{Z}$;

(ii) there is a closed arbitrary (resp. orientable) n-manifold M with Euler number 1 and carrying a k-field u with finite singularities;

(iii) there is a (k-1)-framefield v on S^{n-1} such that the bordism class $[S^{n-1},v]$ in $\mathfrak{N}_{n-1}(k-1,k-1)$ (resp. in $\Omega_{n-1}(k-1,k-1)$) vanishes (cf. §12);

(iv) $\operatorname{span}(S^{n-1}) \geq k$.

We claim that in both theories the conditions (i), (ii) and (iii) are equivalent, and that they imply (iv). The equivalence of (i) and (ii) follows from the context of (16.2). If (ii) holds, we may assume, by proposition 16.1, that u has just one singular point which is the center of an embedded unit ball $B \subset M$, and that on $\partial B = S^{n-1}$ u is given by the outward pointing vector field together with a (k-1)-framefield v tangential to S^{n-1}; clearly $u|M-\mathring{B}$ provides the required zero-bordism. Conversely, if (iii) holds and (N,u) is a zero-bordism of (S^{n-1},v), then $M=N \cup_{S^{n-1}} B$ carries a k-field with one singularity (extend u!) and has Euler number 1 since $\chi(N) = 0$. On the other hand, if (iii) holds, then also $\chi''(S^{n-1},v)$ vanishes (see 13.10 and 13.11), and so does $\chi(S^{n-1},v)$ since $\theta_{S^{n-1}}$ is an isomorphism; hence

$$\omega_k(S^{n-1}) = \operatorname{incl}_*(\chi(S^{n-1},v)) = 0 ,$$

and, according to theorem 13.3, $\mathrm{span}(S^{n-1}) \geq k$.

In the unoriented theory, condition (iv) is actually equivalent to conditions (i), (ii) and (iii). Indeed, if $\mathrm{span}(S^{n-1}) \geq k$, then also $\mathrm{span}(P^{n-1}) \geq k$ (see remark 16.5), and a suitable k-framefield on P^{n-1} extends to a k-field with Index z_1 on $P^n = P^{n-1} \cup \mathrm{cell}$; in particular, (ii) holds.

The description of $\pi(n,k)$ follows now from the exact sequence

(16.6) $$0 \longrightarrow \tilde{\pi}(n,k) \subset \pi(n,k) \xrightarrow{pr_* |} \mathbb{Z} .$$

The monomorphisms which represent z_1 and z_2 can clearly be extended to monomorphisms $S^{n-1} \times \mathbb{R}^{k+1} \hookrightarrow S^{n-1} \times \mathbb{R}^n$; hence $z_1 \in \mathrm{proj}_*(\pi_{n-1}(V_{n,k+1}))$, while z_2 lies even in $\mathrm{proj}_*(\pi_{n-1}(V_{n,n}))$.

Finally, define $h^{or}(n-1)$ to be the largest integer k', $0 < k' \leq \mathrm{span}(S^{n-1})$, for which condition (ii) above is satisfied in the oriented theory; if no such k' exists, put $h^{or}(n-1) = 0$. E.g. for $n \equiv 2(4)$ there is no orientable closed n-manifold with odd Euler number, and (ii) cannot hold. If $0 < k \leq h^{or}(n-1)$, let z_1^{or} be the Index of a k-field with finite singularities on a closed oriented n-manifold with Euler number 1. The description of $\pi^{or}(n,k)$ now follows from the oriented analogue of the exact sequence (16.6).

Here are a few more comments on $h^{or}(n-1)$. If $\mathbb{K}$ is the field of complex numbers (resp. of quaternions) of real dimension $d=2$ (resp. $d=4$), note that the projective space $\mathbb{K}P(2q)$, $q>0$, is an orientable closed 2qd-manifold of odd Euler number and carrying a d-field with finite singularities; indeed, $\mathbb{K}P(2q) = \mathbb{K}P(2q-1) \cup \mathrm{cell}$, and any d-framefield of $T(\mathbb{K}P(2q)) \mid \mathbb{K}P(2q-1)$ extends to a d-field on $\mathbb{K}P(2q)$ with a possible singular point at the center of the top-dimensional cell. This example shows that $h^{or}(n-1) \geq 2$ for $n \equiv 0(4)$ and $h^{or}(n-1) \geq 4$ for $n \equiv 0(8)$.

Actually, $h^{or}(n-1) = 2$ for $n \equiv 4(8)$. This follows from the relation

(16.7) $$\binom{n+3}{4} w_n(M) = w_2(M) w_{n-2}(M)$$

valid for every closed orientable manifold M of even positive dimension n. Indeed, if $n \equiv 4(8)$ and if M carries a 3-field with finite singularities, then $w_{n-2}(M)$ vanishes, and so does the mod 2 Euler number $(\chi(M))_2 = w_n(M)[M]$; hence the oriented version of condition (ii) does not hold for $k=3$.

It remains to deduce (16.7) from the Wu relations 9.5 and 9.6. So let M be a closed orientable manifold of even dimension $n>0$, and denote its Stiefel-Whitney classes briefly by w_j. The first few Wu classes are given by

$$U_1 = 0, \quad U_2 = w_2, \quad U_3 = 0, \quad U_4 = w_4 + w_2^2 .$$

We have

$$0 = Sq^3(w_{n-3}) = w_3 w_{n-3},$$

and hence

$$\begin{aligned} w_2^2 w_{n-4} &= Sq^2(w_2 w_{n-4}) \\ &= w_2^2 w_{n-4} + w_3 w_{n-3} + w_2(w_2 w_{n-4} + \tbinom{n-1}{2} w_{n-2}) \\ &= \tbinom{n-1}{2} w_2 w_{n-2} \end{aligned}$$

and

$$\begin{aligned} w_2^2 w_{n-4} &= Sq^4(w_{n-4}) - w_4 w_{n-4} \\ &= \tbinom{n+1}{2} w_2 w_{n-2} + \tbinom{n+3}{4} w_n \end{aligned}$$

The last two identities imply (16.7).∎

<u>Problem 16.8.</u> Determine $h^{or}(n-1)$ for $n \equiv 0(8)$. Essentially this amounts to finding out for which k some closed orientable n-manifold with odd Euler number can carry a k-field with finite singularities.∎

Given a homotopy class $[v] \in \pi_{n-1}(V_{n,k})$, we may assign to it the bordism class of the monomorphism

$$v : S^{n-1} \times \mathbb{R}^k \hookrightarrow S^{n-1} \times \mathbb{R}^n = TS^{n-1} \oplus \underset{\sim}{\mathbb{R}}$$

or of the corresponding "dual" epimorphism. This procedure defines a homomorphism

$$\varepsilon^{(or)} : \pi_{n-1}(V_{n,k}) \longrightarrow \Omega_{n-1}(B(S)O(1),(\underset{\sim}{\mathbb{R}},\underset{\sim}{\mathbb{R}}^k),\gamma_1,k)$$

where γ_1 denotes the canonical line bundle. The target group is isomorphic, via destabilization, to the framefield bordism group $\mathfrak{N}_{n-1}(k-1,k-1)$ (resp. $\Omega_{n-1}(k-1,k-1)$), see proposition 7.17.

Theorem 16.9. We have

$$\pi^{(or)}(n,k) = \ker \varepsilon^{(or)}.$$

Moreover

$$\pi^{(or)}(n,k) \subset \text{proj}_*(\pi_{n-1}(V_{n,k+1}))$$

for $n>2k>0$ (except possibly when $n=2k+1$, and k is odd and ≥ 9).

On the other hand, we have

$$2\cdot\text{proj}_*(\pi_{n-1}(V_{n,k+1})) \subset \pi(n,k)$$

and

$$(\text{Id}+i)(\text{proj}_*(\pi_{n-1}(V_{n,k+1}))) \subset \pi^{or}(n,k)$$

(where i is the involution on $\pi_{n-1}(V_{n,k})$ which applies the automorphism $(-\text{id})\times\text{id}$ of $\mathbb{R}^n=\mathbb{R}\times\mathbb{R}^{n-1}$ both to the domain S^{n-1} and to the k-frames in $\mathbb{R}^n$).

Furthermore, if the homomorphism

$$(f,\chi') : \mathfrak{N}_{n-1}(k-1,k-1) \longrightarrow \mathfrak{N}_{n-1}\oplus\Omega_{k-1}(BO(n-k),\text{trivial})$$

is injective (this holds e.g. for $k\leq 3$, see theorems 12.1 and 12.8), then

$$\text{proj}_*(\pi_{n-1}(V_{n,k+1})) \subset \pi(n,k) ;$$

the analogous conclusion holds in the oriented theory.

Remark 16.10. As the proof will show, all but the second statement in the theorem above still holds if we merely assume that $1 \leq k < n$. In particular, the characterization of $\pi^{(or)}(n,k)$ as kernel of $\varepsilon^{(or)}$ gives an alternate, more general (though maybe less elementary) proof of the first statement of theorem 16.4: $\pi^{(or)}(n,k)$ and $\tilde{\pi}^{(or)}(n,k)$ are actually subgroups of $\pi_{n-1}(V_{n,k})$ whenever $1 \leq k < n$.

Proof of theorem 16.9. Consider the diagram

$$\begin{array}{ccccccc}
\cdots \to \pi_{n-1}(V_{n,k+1}) & \xrightarrow{proj_*} & \pi_{n-1}(V_{n,k}) & \xrightarrow{\partial} & \pi_{n-2}(S^{n-k-1}) & \longrightarrow & \cdots \\
 & & \downarrow \varepsilon^{(or)} & & \downarrow \sigma & & \\
 & & \Omega_{n-1}(B(S)O(1),(\tilde{\mathbb{R}},\tilde{\mathbb{R}}^k),\gamma_1,k) & \xrightarrow{\chi''} & \Omega_{k-1}(\text{point};\text{trivial}) & & \\
 & & \cong \uparrow St & \nearrow \chi'' & & & \\
 & & \left.\begin{matrix}\mathfrak{N} \\ (\Omega)\end{matrix}\right\} {}_{n-1}(k-1,k-1) & & & &
\end{array}$$

where the top line is the exact sequence of the fibration proj. Both homomorphisms χ'' are defined as in (13.10) (for the upper χ'' we have to replace TM by $TM \oplus \tilde{\mathbb{R}}$ etc.); so the bottom triangle commutes. The square commutes also up to sign. Indeed, given an element in $\pi_{n-1}(V_{n,k})$, we may represent it by a smooth monomorphism

$$v : S^{n-1} \times \mathbb{R}^k \hookrightarrow S^{n-1} \times (\mathbb{R}^k \times \mathbb{R}^{n-k})$$

which restricts to the obvious inclusion over the upper halfsphere $S_+^{n-1} = \{(x_1,\ldots x_n) \in S^{n-1} \mid x_n \geq 0\}$. Let ζ denote the orthogonal complement of the image of v so that e.g. $\zeta | S_+^{n-1} = S_+^{n-1} \times (\{0\} \times \mathbb{R}^{n-k})$.

If s is a suitable nondegenerate section of ζ whose singularity S lies in the interior of S_+^{n-1}, then the restricition of s to the equator $S^{n-2} = \partial S_+^{n-1}$ represents $\partial[v]$, and the stably framed singularity S represents both $\sigma \circ \partial[v]$ and $\pm \chi'' \circ \varepsilon^{(or)}[v]$; so these two elements of $\Omega_{k-1}(\text{point,trivial})$ coincide.

Next, given $z\in\pi^{(or)}(n,k)$, let u be a k-field with finite singularity on a closed manifold M such that $\text{Index}(u,or)=z$. Remove an open ball around each singular point to obtain a compact manifold $\dot{M}$. If we modify the monomorphism over $\dot{M}$

$$(u|\dot{M}) \oplus \text{id} : \mathbb{R}^k \oplus \mathbb{R} \hookrightarrow T\dot{M} \oplus \mathbb{R}$$

suitably in a collar neighborhood of $\partial\dot{M}$ (rotate the inward pointing vectorfield into the extra factor $\mathbb{R}$!), we get a zero bordism which shows that $\varepsilon^{(or)}(z)=0$. Hence

$$\pi^{(or)}(n,k) \subset \ker \varepsilon^{(or)} \subset \ker \sigma\circ\partial .$$

If $n>2k+1$, then σ is an isomorphism (cf. proposition 5.4), and

$$\pi^{(or)}(n,k) \subset \ker \sigma\circ\partial = \ker \partial = \text{proj}_*(\pi_{n-1}(V_{n,k+1}))$$

follows trivially. We will show below that $\ker \sigma\circ\partial=\ker \partial$ in the remaining dimension cases when $n=2k+1$.

On the other hand, for any $[v]\in\pi_{n-1}(V_{n,k})$ there is a (k-1)-framefield w on a closed (n-1)-manifold N such that the two monomorphisms

$$v : S^{n-1} \times \mathbb{R}^k \hookrightarrow S^{n-1} \times \mathbb{R}^n = TS^{n-1} \oplus \mathbb{R}$$

and

$$w \oplus \text{id}: N \times (\mathbb{R}^{k-1}\oplus\mathbb{R}) \hookrightarrow TN \oplus \mathbb{R}$$

are bordant in $\Omega_{n-1}(B(S)O(1),(\mathbb{R},\mathbb{R}^k),\gamma_1,k)$. If $n>2$, we may apply the method of the proof of the stability theorem 6.6 to a fine bordism between v and $w\oplus\text{id}$; after attaching suitable handles in the interior, we obtain a bordism B between S^{n-1} and N together with a k-framefield on B which restricts to v and $w\oplus\text{id}$ on the boundary; here we identify $TS^{n-1}\oplus\mathbb{R}=TB|S^{n-1}$ (and $TN\oplus\mathbb{R}=TB|N$) via the inward (resp. outward) pointing vectorfield. Gluing B and the ball D^n together along S^{n-1}, we get a compact manifold C carrying a k-field u with index $[v]$ such that $u|\partial C$ is given by a (k-1)-framefield tangential to ∂C, together with the outward pointing vectorfield.

If $[v]$ lies already in $\ker \varepsilon^{(or)}$, then $St[N,w]=\varepsilon^{(or)}[v]$ vanishes, and so does $[N,w]$. Gluing a fine zero bordism of (N,w) together with (C,u) along $\partial C=N$, we obtain a closed n-manifold carrying a k-field with index $[v]$. This shows that also

$$\ker \varepsilon^{(or)} \subset \pi^{(or)}(n,k) .$$

The first statement of our theorem follows.

Similarly, for any $z\in\pi_{n-1}(V_{n,k+1})$ there is a (k+1)-field with index z on a compact manifold C such that, on the boundary ∂C, the first k component vectorfields are tangential to ∂C. Thus we can define a k-field with finite singularities on the double $C \cup_{\partial C}(-C)$; its index with respect to suitable local orientations is $2\,proj_*(z)$ (resp. $proj_*(z)+i\,proj_*(z)$ in the oriented theory, where we have to reverse orientations on the second copy of C in order to orient the double).

Finally assume that

$$(f,\chi') : \left.\begin{matrix}\mathfrak{N}\\(\Omega)\end{matrix}\right\}_{n-1}(k-1,k-1) \longrightarrow \left.\begin{matrix}\mathfrak{N}_{n-1}\\(\Omega_{n-1})\end{matrix}\right\} \oplus \Omega_{k-1}(B(S)O(n-k);\text{trivial})$$

is injective; then the corresponding homomorphism on $\Omega_{n-1}(B(S)O(1),(\mathbb{R},\mathbb{R}^k),\gamma_1,k)$ is also injective. Given $[v]\in proj_*(\pi_{n-1}(V_{n,k+1}))$, we have $\chi''(\varepsilon^{(or)}[v]=\sigma(0)=0$ (see the diagram at the beginning of the proof). Since the complement of the image of $v:S^{n-1}\times\mathbb{R}^k \hookrightarrow TS^{n-1}\oplus\mathbb{R}^n$ is trivial over any strict subset of S^{n-1}, we see actually that $\chi'(\varepsilon^{(or)}[v])$ vanishes, and so does the bordism class $f(\varepsilon^{(or)}[v]=[S^{n-1}]$. Hence $\varepsilon^{(or)}[v]=0$, and we deduce that

$$proj_*(\pi_{n-1}(V_{n,k+1})) \subset \ker \varepsilon^{(or)} = \pi^{(or)}(n,k) .$$

Some of the arguments above fail to work when $n=2$ and $k=1$; however, in this case it is easy to check the statements of the theorem directly.

In order to complete the proof of the second statement of 16.9, it remains to show that $\ker \sigma\circ\partial \subset \ker \partial$ when $n=2k+1$, and k is even or $k<9$. For this purpose, consider the diagram

$$\begin{array}{ccccccccc}
 & & & & \pi_{n-1}(V_{n,k})=\pi_{2k}(V_{2k+1,k}) & & & & \\
 & & & & \downarrow \partial & & & & \\
\cdots\to\pi_{2k+1}(S^{k+1}) & \xrightarrow{H} & \pi_k(S^k) & \xrightarrow{P} & \pi_{2k-1}(S^k) & \xrightarrow{E} & \pi_{2k}(S^{k+1}) & \longrightarrow & 0 \\
 & & \cong & & & \searrow^{\sigma} & \pm\sigma\downarrow\cong & & \\
 & & \mathbb{Z} & & & & \Omega_{k-1}(\text{point};\text{trivial}) & &
\end{array}$$

where the horizontal line is the exact EHP-sequence of G.W. Whitehead [69]; the triangle to the right commutes (see 5.4 and 5.5).

If $k=1$, 3 or 7, then the Hopf invariant map H is onto (cf. e.g. [26]) and σ is bijective; hence $\ker \sigma\circ\partial = \ker \partial$.

If k is even, then $H \equiv 0$ and $\ker \sigma \cong \mathbb{Z}$; again, $\ker \sigma\circ\partial \subset \ker \partial$, or, equivalently,

$$\partial| : \ker \sigma\circ\partial = \partial^{-1}(\ker \sigma) \longrightarrow \ker \sigma \subset \pi_{2k-1}(S^k)$$

vanishes, since $\pi_{2k}(V_{2k+1,k})$ is a torsion group for all $k\geq 2$. This fact can be deduced from the exact sequence

$$\pi_{2k}(S^{2k-j}) \longrightarrow \pi_{2k}(V_{2k+1,j+1}) \longrightarrow \pi_{2k}(V_{2k+1,j}) \xrightarrow{\partial_j} \pi_{2k-1}(S^{2k-j})$$

by induction over $j+1$, $1\leq j<k$; note that all groups are finite here (cf. [58]) except the groups involved in the homomorphism $\partial_1 : \mathbb{Z} \longrightarrow \mathbb{Z}$ which is injective since

$$\partial_1[\mathrm{Id}_{S^{2k}}] = \chi(S^{2k}) = 2$$

(see also the description of $\sigma\circ\partial$ at the beginning of the proof).

If $k=5$, ∂ fits into the following exact sequence

$$\begin{array}{ccccccccccc}
\pi_{10}(V_{11,5}) & \xrightarrow{\partial} & \pi_9(S^5) & \longrightarrow & \pi_9(V_{11,6}) & \longrightarrow & \pi_9(V_{11,5}) & \longrightarrow & \pi_8(S^5) & \longrightarrow & \pi_8(V_{11,6}) \\
\wr\| & & \wr\| & & \wr\| & & \wr\| & & \wr\| & & \wr\| \\
\mathbb{Z}_2 & & \mathbb{Z}_2 & & \mathbb{Z}_2 & & \mathbb{Z}_{12} & & \mathbb{Z}_{24} & & \mathbb{Z}_2
\end{array}$$

(where we got the explicit values from the tables of Toda [66], p.186, and of Paechter [51], p.250). In the order from right to left, the arrows have to be onto, injective, zero, bijective resp., and finally, $\partial\equiv 0$. Thus $\ker \sigma\circ\partial = \ker\partial = \pi_{2k}(V_{2k+1,k})$ for $k=5$. This completes the proof of theorem 16.9. ∎

It may happen that different k-fields with finite singularities on a given closed manifold have different indices; this is possible because in general the index is a "higher order obstruction". We will try to measure this phenomenon and to relate it to the homomorphism

$$\Theta^{(or)} : \pi_{n-1}(V_{n,k}) \underset{\cong}{\xrightarrow{\sigma}} \Omega_{k-1}(P^{k-1};(n-k)\lambda) \xrightarrow{\Theta} \Omega_{k-1}(P^{k-1}\times B(S)O(n);\phi)$$

defined in (13.7) (we use σ to identify the two groups to the left, and we write Θ or Θ^{or} according to whether we are working in the unoriented theory or not).

Theorem 16.11. Assume $n>2k>0$.

Then the kernel of $\Theta^{(or)}$ equals the set of all $z\in\pi_{n-1}(V_{n,k})$ satisfying the following condition: there exists an arbitrary, resp. oriented, closed n-manifold with $\text{span}\geq k$ which carries also a k-field u with finite singularities such that $\text{Index}(u,or)=z$ (where the system or of local orientations may be arbitrary, resp. must be induced from the global orientation).

Therefore, $\Theta^{(or)}$ is universal among homomorphisms θ defined on $\pi_{n-1}(V_{n,k})$ (or on $\pi^{(or)}(n,k)$) and such that $\theta(\text{Index}(u,or))$ is always independent of the k-field u with finite singularities and depends only on the underlying arbitrary, resp. oriented, closed n-manifold M:

$\Theta^{(or)}$ is such a homomorphism, and any other such homomorphism θ factorizes uniquely through $\Theta^{(or)}(\pi_{n-1}(V_{n,k}))$ (or through $\Theta^{(or)}(\pi^{(or)}(n,k)))$. In particular, $\theta(\text{Index}(u,or))$ depends actually only on the following invariants of M: the Stiefel-Whitney numbers, the Euler number and, in the oriented theory, in addition the Pontrjagin numbers and, for $n\equiv 1(4)$, the real Kervaire semicharacteristic.

Clearly, $\ker \Theta^{(or)} \subset \tilde{\pi}^{(or)}(n,k)$. Moreover, if (n,k) is $\mathfrak{N}$-pleasant (e.g. if $k\leq 4$, see definition 12.12 and theorem 12.13), then

$$\ker \Theta = \tilde{\pi}(n,k).$$

Proof. Given $z\in\ker \Theta^{(or)}$, we want to construct a k-field with index z on a closed n-manifold which also carries a k-field without any singularity. We will use arguments from the proofs of theorems 3.1 and 13.4.

Let $u: \underline{\mathbb{R}}^k \longrightarrow \underline{\mathbb{R}}^n = TD^n$ be a nondegenerate (k-1)-morphism over the unit ball D^n such that the restriction of u to $\partial D^n = S^{n-1}$ represents z. Since $\Theta^{(or)}(z)=0$, the singularity data $(S, (\underline{\text{Ker}}, \underline{\mathbb{R}}^n|S), \bar{g})$ of u allows a zero bordism $(\mathcal{S}, (\widetilde{\underline{\text{Ker}}}, \eta), \bar{G})$. As in the proof of theorem 13.4 we can extend the canonical splittings at $S=\partial\mathcal{S}$ to obtain splittings $\mathcal{S}\times\mathbb{R}^k = \widetilde{\underline{\text{Ker}}} \oplus \widetilde{\underline{\text{Coim}}}$ and $\eta = \widetilde{\underline{\text{Coker}}} \oplus \widetilde{\underline{\text{Im}}}$ over all of $\mathcal{S}$. Also, the dimension assumptions allow us to choose a decomposition

$$\underline{\text{Hom}}(\widetilde{\underline{\text{Ker}}}, \widetilde{\underline{\text{Coker}}}) = \nu \oplus (\mathcal{S}\times\mathbb{R}).$$

Moreover, given a collar neighborhood $C=S\times[0,1]$ of $S=S\times\{0\}$ in $\mathcal{S}$, we can extend the inclusion $S\subset \mathring{D}^n$ to an embedding $j:C \hookrightarrow D^n$ with normal bundle $\nu|C$, such that $j(C)$ meets ∂D^n transversely at $j(S\times\{1\})$. Therefore, if $D(\nu)$ is the total space of the unit disk bundle of ν, we may identify $D(\nu)|C$ with a subset of D^n via the tubular neighborhood construction. We form the union of D^n with all of $D(\nu)$, smooth corners and denote the resulting manifold with boundary by N.

Next we proceed somewhat along the lines of the proof of theorem 3.1.

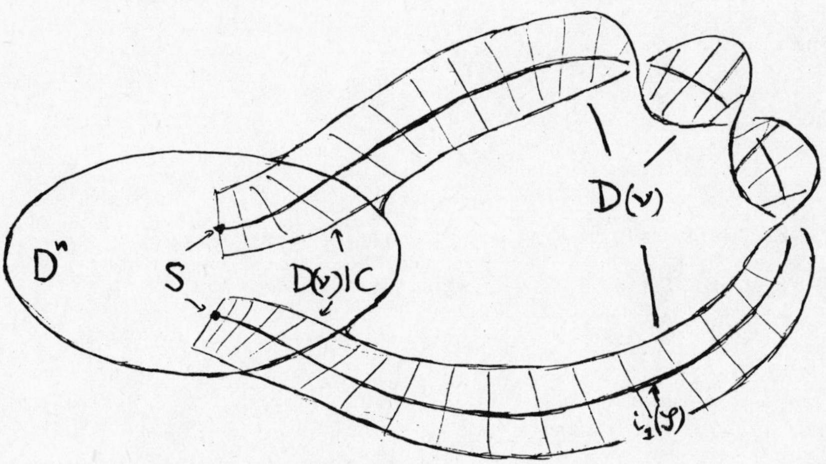

Figure 16.12. The construction of the compact manifold N.

We can find an embedding

$$i = (i_1, i_2) : \mathcal{S} \longrightarrow N \times I$$

which realizes $\mathcal{S}$ as a closed submanifold of $N \times I$ such that $\mathcal{S} \cap \partial(N \times I) = \partial\mathcal{S} = S \subset N \times \{0\}$; indeed, let i_1 be the zero section of $D(\nu)$, and let i_2 be a smooth function which, on the (half) collar $S \times [0, \frac{1}{2}]$ is essentially given by the second projection, and which takes the constant value $\frac{1}{2}$ outside of this half collar. The normal bundle of i is $\nu \oplus \underset{\sim}{\mathbb{R}} = \underset{\sim}{\mathrm{Hom}}(\underset{\sim}{\widetilde{\mathrm{Ker}}}, \underset{\sim}{\widetilde{\mathrm{Coker}}})$. Thus we may apply the model construction (1.9), and we obtain a (k-1)-morphism

$$u_I | : \underset{\sim}{\mathbb{R}}^k \longrightarrow \eta \cong pr_1^*(TN)$$

over a tubular neighborhood of $\mathcal{S}$ in $N \times I$; here we use a destabilization of the stable isomorphism

$$\eta \oplus \underset{\sim}{\mathbb{R}} \xleftarrow[\cong]{\overline{G}} T\mathcal{S} \oplus \underset{\sim}{\mathrm{Hom}}(\underset{\sim}{\widetilde{\mathrm{Ker}}}, \underset{\sim}{\widetilde{\mathrm{Coker}}}) \cong T(N \times I)|\mathcal{S}$$

over $\mathcal{S}$. If we made our choices judiciously, $u_I|$ agrees with the (k-1)-morphism u originally given over $D^n \subset N\times\{0\}$.

We can extend u and $u_I|$ to a (k-1)-morphism

$$u_I : \underset{\sim}{\mathbb{R}}^k \longrightarrow pr_1{}^*(TN)$$

over all of $N\times I$ whose singularity lies at $\mathcal{S}$. Indeed, suitable retractions allow us to define u_I between $N\times\{0\}$ and the graph of a function $h:N \longrightarrow [0,1]$ such that h vanishes near the boundary of N and $h\circ i_1 > i_2$ on $\mathcal{S}$; then u_I extends over the rest of $N\times I$ via the obvious ("vertical") retraction to the graph of h. u_I restricts to k-fields u^ℓ on $N=N\times\{\ell\}$, $\ell=0,1$, which agree near ∂N. While u^1 has no singularity at all, the singularity of u^o lies at $S\subset D^n\subset N$; clearly u^o may be modified within $\mathring{D}^n$ to become a k-field with finite singularities having index z.

Now we write u^ℓ in terms of its component vectorfields $u^\ell=(u_1^\ell,u_2^\ell,\dots,u_k^\ell)$, $\ell=0,1$. If $k>1$ and hence $n>4$, we may assume that $u_1^o|\partial N=u_1^1|\partial N$ is the outward pointing vectorfield and that $u_j^o|\partial N=u_j^1|\partial N$ is tangential to ∂N for $j=2,3,..,k$; this can be guaranteed at least after we attach suitable handles to N according to the procedure in the proof of the stability theorem 6.6. Then the double

$$N \cup_{\partial N} (-N)$$

(possibly oriented!) of N is a closed n-manifold carrying the k-fields

$$(u_1^\ell, u_2^\ell,.., u_k^\ell) \cup_{\partial N} (-u_1^1, u_2^1,.., u_k^1), \quad \ell = 0,1,$$

with finite singularities and index z, resp. with no singularity at all. If $k=1$, then $\Theta^{(or)}$ is an isomorphism, and $z=0$ obviously occurs also as the index of a k-field on a manifold having span $\geq k$.

Conversely, we see (cf. also prop. 13.8) that $\Theta^{(or)}(\mathrm{Index}(u,or)) = \omega'(M)$ is always independent of the k-field u with finite singularities, and depends only on the underlying closed manifold M. This completes the proof of the first statement in theorem 16.11: $\Theta^{(or)}$ annihilates the set of elements z described there, and so does every other homomorphism Θ which always makes $\mathrm{Index}(u,or)$ depend only on M. The second claim in our theorem now follows from theorem 13.4.

Clearly, the description of $\ker \Theta^{(or)}$, given above, implies that this kernel lies in $\tilde{\pi}^{(or)}(n,k)$.

On the other hand, given an element $z \in \tilde{\pi}(n,k)$, there is a closed n-manifold M satisfying $\chi(M)=0$ which carries a k-field u with finite singularities such that $\mathrm{Index}(u,or)=z$. The existence of u implies that (all Stiefel-Whitney numbers involving) $w_{n-k+1}(M),\ldots, w_{n-1}(M)$ vanish, and so does the mod 2 Euler number $w_n(M)[M]$. Therefore, if (n,k) is $\mathfrak{N}$-pleasant, M is bordant to a manifold M' with span$\geq k$; this holds first in $\mathfrak{N}_n$, and then even in the vectorfield bordism group $\mathfrak{N}_n(0,0)$ since $\chi(M)=\chi(M')=0$ (cf. theorem 12.1); thus

$$\Theta(z) = \omega_k'(M) = \omega_k'(M') = 0 ;$$

we conclude that $\tilde{\pi}(n,k) \subset \ker \Theta$ in this case.∎

Exercise 16.13. Let M be an arbitrary, resp. oriented, compact n-manifold with boundary ∂M, and let u be a k-field on M with finite singularities which avoid ∂M. Show that $\Theta^{(or)}(\mathrm{Index}(u,or))$ depends only on M and on the regular homotopy class of the monomorphism $u|\,\partial M \times \mathbb{R}^k$. ∎

Exercise 16.14. Determine the subgoups $\pi(n,1)=\mathrm{proj}_*(\pi_{n-1}(V_{n,2}))$ and $\pi^{or}(n,1)$ of $\pi_{n-1}(V_{n,1}) \cong \mathbb{Z}$, and show that $\tilde{\pi}(n,1) = \ker \Theta = \tilde{\pi}^{or}(n,1) = \ker \Theta^{or} = 0$.

Example 16.15. For low k we tabulate the groups studied in theorems 16.4, 16.9 and 16.11.

$n > 4$	$\pi_{n-1}(V_{n,2})$	$\pi(n,2)$ $=\mathrm{proj}_*(\pi_{n-1}(V_{n,3}))$	$\tilde{\pi}(n,2)$ $=\ker\Theta$	$\pi^{or}(n,2)$	$\tilde{\pi}^{or}(n,2)$	$\ker\Theta^{or}$
$n\equiv 1(4)$	$\mathbb{Z}_2$	$\mathbb{Z}_2$	$\mathbb{Z}_2$	$\mathbb{Z}_2$	$\mathbb{Z}_2$	0
$n\equiv 2(4)$	$\mathbb{Z}_2\oplus\mathbb{Z}\bar{z}_1$	$\mathbb{Z}_2\oplus\mathbb{Z}(2\bar{z}_1)$	$\mathbb{Z}_2$	$\mathbb{Z}z_2$	0	0
$n\equiv 3(4)$	$\mathbb{Z}_2$	$\mathbb{Z}_2$	$\mathbb{Z}_2$	0	0	0
$n\equiv 4(4)$	$\mathbb{Z}_2\oplus\mathbb{Z}\bar{z}_1$	$\mathbb{Z}_2\oplus\mathbb{Z}\bar{z}_1$	$\mathbb{Z}_2$	$\mathbb{Z}_2\oplus\mathbb{Z}\bar{z}_1$	$\mathbb{Z}_2$	0

Here $\mathbb{Z}_2$ is generated by $\sigma^{-1}(\delta_1(1))$ (see 14.2).

$n > 6$	$\pi_{n-1}(V_{n,3})$	$\pi(n,3)$ $=\mathrm{proj}_*(\pi_{n-1}(V_{n,4}))$	$\tilde{\pi}(n,3)$ $=\ker\Theta$	$\pi^{or}(n,3)$	$\tilde{\pi}^{or}(n,3)$	$\ker\Theta^{or}$
$n\equiv 1(4)$	$\mathbb{Z}_2$	$\mathbb{Z}_2$	$\mathbb{Z}_2$	$\mathbb{Z}_2$	$\mathbb{Z}_2$	0
$n\equiv 2(4)$	$\mathbb{Z}_2\oplus\mathbb{Z}z_2$	$\mathbb{Z}z_2$	0	$\mathbb{Z}z_2$	0	0
$n\equiv 3(4)$	$\mathbb{Z}_2\oplus\mathbb{Z}_2$	$\mathbb{Z}_2$	$\mathbb{Z}_2$	0	0	0
$n\equiv 0(8)$ / $n\equiv 4(8)$	$\mathbb{Z}_4\oplus\mathbb{Z}z_1$	$\mathbb{Z}_4\oplus\mathbb{Z}z_1$	$\mathbb{Z}_4$	$\mathbb{Z}_4\oplus\mathbb{Z}z_1$ / $\mathbb{Z}_4\oplus\mathbb{Z}z_2$	$\mathbb{Z}_4$	0

$n > 8$	$\pi_{n-1}(V_{n,4})$	$\pi(n,4)$ $=\mathrm{proj}_*(\pi_{n-1}(V_{n,5}))$	$\tilde{\pi}(n,4)$ $=\ker\Theta$	$\pi^{or}(n,4)$	$\tilde{\pi}^{or}(n,4)$	$\ker\Theta^{or}$
$n\equiv 1(4)$	$\mathbb{Z}_2\oplus\mathbb{Z}_2$	$\mathbb{Z}_2\oplus\mathbb{Z}_2$	$\mathbb{Z}_2\oplus\mathbb{Z}_2$	$\mathbb{Z}_2\oplus\mathbb{Z}_2$	$\mathbb{Z}_2\oplus\mathbb{Z}_2$	$\mathbb{Z}_2$
$n\equiv 2(4)$	$\mathbb{Z}_{12}\oplus\mathbb{Z}z_2$	$\mathbb{Z}z_2$	0	$\mathbb{Z}z_2$	0	0
$n\equiv 3(4)$	$\mathbb{Z}_2\oplus\mathbb{Z}_2$	$\mathbb{Z}_2$	$\mathbb{Z}_2$	$\mathbb{Z}_2$	$\mathbb{Z}_2$	$\mathbb{Z}_2$
$n\equiv 0(8)$	$\mathbb{Z}_{24}\oplus\mathbb{Z}_4\oplus\mathbb{Z}z_1$	$\mathbb{Z}_8\oplus\mathbb{Z}z_1$	$\mathbb{Z}_8$	$\mathbb{Z}_8\oplus\mathbb{Z}z_1$	$\mathbb{Z}_8$	0
$n\equiv 4(8)$	$\mathbb{Z}_{24}\oplus\mathbb{Z}_4\oplus\mathbb{Z}\bar{z}_1$	$\mathbb{Z}_8\oplus\mathbb{Z}z_2$		$\mathbb{Z}_8\oplus\mathbb{Z}z_2$		

In all three tables we choose z_1 and z_2 as in theorem 16.4. When n is divisible by $d=2$ (resp. by $d=4$), the complex (resp. quaternionic) structure on $\mathbb{R}^n$ allows us to define the element $\bar{z}_1=[v:S^{n-1}\longrightarrow V_{n,d}]$ by $v(x)=(x,ix)$ (resp. $v(x)=(x,ix,jx,kx)$). Clearly, the subindices of the elements z_1, $\bar{z}_1$, z_2 also indicate their values under the obvious homomorphism (cf. 16.2)

$$pr_* : \pi_{n-1}(V_{n,k}) \longrightarrow \pi_{n-1}(S^{n-1}) \xrightarrow{\cong} \mathbb{Z} .$$

We know a priori that $\tilde{\pi}(n,k) = \ker \Theta$ for $k \leq 4$ and $\pi(n,k) = proj_*(\pi_{n-1}(V_{n,k+1}))$ for $k \leq 3$ (see theorems 16.11 and 16.9); however, if $k = 4$, we start from the (very powerful) relation $\pi(n,4) \subset proj_*(\pi_{n-1}(V_{n,5}))$, and we obtain the full equality of these two groups only in the course of the calculations given below.

In the first two tables, the values for the full group $\pi_{n-1}(V_{n,k})$ and for the kernels of Θ and Θ^{or} can be deduced from the calculations of §14 (applied to Y=point and $Y=B(S)O(n)$). In order to determine $\tilde{\pi}^{or}(n,k)$, use the tables in Thomas [65], p.647, and Atiyah-Dupont [6], p.25, (see also remarks 14.5 and 14.18), together with the existence of k-fields with finite singularities on S^n, $\mathbb{C}P(2q)$ and $\mathbb{H}P(2q)$ (see the section preceding 16.7; note also that the signature of these projective spaces is ± 1). Finally, $\pi(n,k)$ and $\pi^{or}(n,k)$ are easily deduced from theorem 16.4. *)

Now turn to the case $k=4$. The values for $\pi_{n-1}(V_{n,4})$ are taken from the tables of Paechter [51], p.250. For odd n, the calculation of $\ker \Theta^{or}$ has been done in 15.6 and 15.17.

If $n \equiv 1(4)$, the index z of a 4-field on S^n does not lie in $\ker \Theta^{or}$; otherwise, we know from the commutative diagrams 13.7 and 13.11 and from the injectivity of Θ^{or} for $k=2$ that

$$\Theta^{or}(proj_* \circ proj_*(z)) = \Delta \circ \Delta \circ \Theta^{or}(z) = 0,$$

*) For the calculation of $\tilde{\pi}^{or}(8q+4,3)$ see remark 19.50.

and hence the index $proj_* \circ proj_*(z)$ of the corresponding 2-field vanishes; this contradicts the fact that $\operatorname{span}(S^n)=1$. We conclude that $\tilde{\pi}^{or}(n,4)$ must be the full group $\pi_{n-1}(V_{n,4})=\mathbb{Z}_2 \oplus \mathbb{Z}_2$, and so are all the intermediate groups $\pi^{(or)}(n,4)$ etc.

A similar "pinching" argument disposes of the case $n \equiv 3(4)$: we just have to know that $\pi_{n-1}(V_{n,5})=\mathbb{Z}_2$ (see Paechter [51]).

Next consider the commuting diagram

$$\cdots \to \pi_{n-1}(S^{n-5}) \to \pi_{n-1}(V_{n,5}) \xrightarrow{proj_*} \pi_{n-1}(V_{n,4}) \to \pi_{n-2}(S^{n-5}) \to \cdots$$

(16.16)
$$pr_*: \pi_{n-1}(V_{n,5}) \to \mathbb{Z}, \qquad pr_*: \pi_{n-1}(V_{n,4}) \to \mathbb{Z} \qquad .$$

If $n \equiv 2(4)$, neither pr_* is onto since an element $z \in pr_*^{-1}(1)$ would lead to at least three linearly independent vectorfields on S^{n-1}. Thus, both in

$$\pi_{n-1}(V_{n,5}) \cong \begin{cases} \mathbb{Z} & \text{if } n>10,\ n\equiv 2(4) \\ \mathbb{Z}_2 \oplus \mathbb{Z} & \text{if } n=10 \end{cases}$$

(see [51]) and in $\pi_{n-1}(V_{n,4})$, the factor $\mathbb{Z}$ is generated by z_2. We conclude that $proj_*(\pi_{n-1}(V_{n,5}))=\mathbb{Z}z_2$, which settles the case $n \equiv 2(4)$ (see also the statement preceding theorem 16.4). This conclusion still holds when $n=10$ since in this case the relevant part of the exact homotopy sequence in (16.16) takes the form

$$\to \mathbb{Z}_2 \oplus \mathbb{Z}z_2 \xrightarrow{proj_*} \mathbb{Z}_{12} \oplus \mathbb{Z}z_2 \to \mathbb{Z}_{24} \to \mathbb{Z}_2 \to 0$$

(see [51]).

Finally assume that $n \equiv 0(4)$. Then $\pi_{n-1}(V_{n,5}) \cong \mathbb{Z}_8 \oplus \mathbb{Z}$ (see [51]). Again, we determine the generator of the $\mathbb{Z}$-factor by considering the homomorphism pr_* and the span of S^{n-1} (use remark 16.5!). Also we know from proposition 15.31 that there are at least 8 elements in $\ker \Theta$; on the other hand

$$\ker \Theta = \pi(n,4) \cap \ker pr_* \subset \text{torsion of } proj_*(\pi_{n-1}(V_{n,5})) \subset proj_*(\mathbb{Z}_8),$$

and therefore all these groups are equal and cyclic of order 8. It remains to study $\pi^{or}(n,4)$, its torsion subgroup $\tilde{\pi}^{or}(n,4)$, and $\ker\ \Theta^{or}$. For this purpose consider the commuting diagram

(16.17)
$$\begin{array}{ccc} \pi_{n-1}(V_{n,6}) & & \\ \downarrow \text{proj}_* & & \\ \pi_{n-1}(V_{n,5}) & \xrightarrow{\theta^o} & KR(P_4) \cong \mathbb{Z} \oplus \mathbb{Z}_8 \\ \downarrow \text{proj}_* & \nearrow \theta' & \\ \text{proj}_*(\pi_{n-1}(V_{n,5})) & & \end{array}$$

where θ^o and θ' are homomorphisms defined by Atiyah and Dupont (see [6], 5.10-5.12). Given a 4-field with finite singularities on an oriented n-manifold M, its index z satisfies the equation

(16.18) $$\theta'(z) = (\chi(M); \tfrac{1}{2}(\chi(M) \pm \text{sign}(M)), \bmod 8)$$

(see [6], theorems 6.2 and 1.1, in conjunction with our 13.6 and 16.9). In particular, for the indices z_2 and z of 4-fields with finite singularities on S^{8q} and $\mathbb{H}P(2q)$ respectively, we have $\theta'(z_2)=(2;1)$ and $\theta'(z)=(2q+1;q)$. Since these elements generate $\mathbb{Z}\oplus\mathbb{Z}_8$, we see that for $n \equiv 0(8)$ $\theta'(\pi^{or}(n,4))=KR(P_4)$ and hence θ' is bijective (see also [6], proposition 5.13), and $\pi^{or}(n,4)=\text{proj}_*(\pi_{n-1}(V_{n,5}))$; moreover, θ' factors through $\Theta^{or}|\pi^{or}(n,4)$ (see theorem 16.11), and hence $\ker\ \Theta^{or}=\ker\ \theta'=0$. Anyway, the claim concerning $\ker\ \Theta^{or}$ has already been established in proposition 15.31. This completes our discussion of the tables in example 16.15.∎ *)

Using the tables of Paechter [51] (i.e. essentially the fortunate fact that $\pi_4^S=\pi_5^S=0$), we deduce from the exact homotopy sequence of the fibration proj that for $n>12$

(16.19) $$\text{proj}_* : \pi_{n-1}(V_{n,6}) \longrightarrow \pi_{n-1}(V_{n,5}) \quad \text{is bijective},$$

*) For the calculation of $\tilde{\pi}^{or}(8q+4,4)$ see remark 19.50.

and

(16.20) $\qquad \mathrm{proj}_* : \pi_{n-1}(V_{n,7}) \longrightarrow \pi_{n-1}(V_{n,6})$ is onto.

This allows us sometimes to get information for $k>4$.

E.g. if $k=5$ or 6 and $n\equiv 2(4)$, $n\geq 14$, then

$$\pi_{n-1}(V_{n,k}) = \mathrm{proj}_*(\pi_{n-1}(V_{n,k+1})) = \pi(n,k) = \pi^{or}(n,k) = \mathbb{Z}_2$$

and

$$\tilde{\pi}(n,k) = \ker \Theta = \tilde{\pi}^{or}(n,k) = \ker \Theta^{or} = 0 \ .$$

This together with the tables above and theorem 14.3, has the following direct consequence.

<u>Theorem 16.21. Assume that $n\equiv 2(4)$, $n>2k$ and $1\leq k\leq 6$. Let M be a closed n-manifold allowing a k-field with finite singularities. Then: $\mathrm{span}(M)\geq k$ if and only if the Euler number of every connected component of M vanishes.</u> ∎

<u>Example 16.22.</u> We can use the theorem of Adams and Hurwitz-Radon (see e.g. remark 16.5) to show that the subgroup $\ker \Theta^{or}$ of $\pi_{n-1}(V_{n,h(n)+1})$ is nontrivial for $n \equiv 3(4)$, $n \neq 3$, 7 or 15.

Indeed, consider the commuting diagram

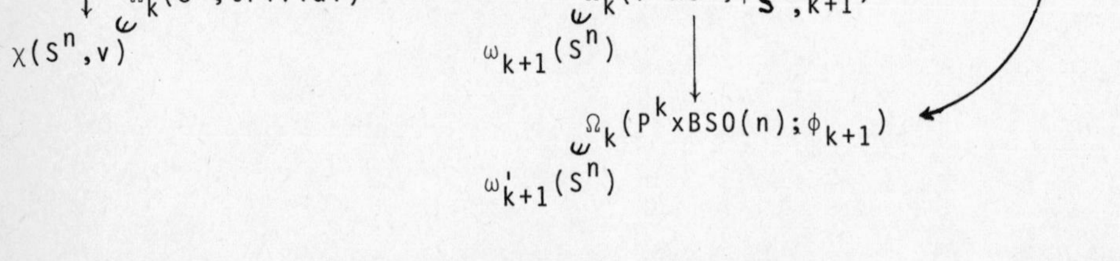

obtained from 13.11. Here k stands for $h(n)=\mathrm{span}(S^n)$; note that for the dimensions n under consideration we have $n > 2k+2$. Thus S^n

carries a k-framefield v, but $\omega_{k+1}(S^n) \neq 0$ by theorem 13.3. On the other hand, S^n is zerobordant in $\Omega_n \cong \Omega_n(0,0)$ (see theorem 12.1), and therefore $\omega'_{k+1}(S^n)=0$ (by theorem 13.4). It follows that

$$\sigma^{-1} \circ \mathrm{incl}_*(\chi''(S^n,v)) \in \ker \Theta^{or} \subset \pi_{n-1}(V_{n,h(n)+1})$$

is nontrivial.∎

The tables in 16.15 illustrate the following fact which we will establish in the next section (in the proof of theorem 17.7).

Theorem 16.23. If $n > 2k > 0$, then the groups

$$\mathrm{proj}_*(\pi_{n-1}(V_{n,k+1})) \cap \ker(\mathrm{pr}_* : \pi_{n-1}(V_{n,k}) \longrightarrow \mathbb{Z}) ,$$

$\tilde{\pi}(n,k)$, $\tilde{\pi}^{or}(n,k)$, $\ker \Theta$ and $\ker \Theta^{or}$ are finite, and all their elements are annihilated by multiplication with 2^{k-1}.∎

Finally we discuss some aspects of the work of Atiyah and Dupont on vectorfields [6]. (One of the very central notions there, the index of elliptic operators, should not be confused with our index of framefields with finite singularities).

First we construct a homomorphism

$$(16.24) \qquad \vartheta^s : \Omega_{k-1}(P^{k-1} \times BSO(n);\phi) \longrightarrow KR^s(P^{k+s-1},P^{s-1})$$

for any $s \geq 0$ so that $n+s \equiv 0(4)$. Given an element x of $\Omega_{k-1}(P^{k-1} \times BSO(n);\phi)$ (cf. 13.2), represent it by a quadruple $(S, \underset{\sim}{Ker}, \eta, \bar{g})$ where $Ker \subset S \times \mathbb{R}^k$ and η are vector bundles of dimensions 1 and n over the closed $(k-1)$-manifold S, and $\bar{g}$ is an isomorphism of the form

$$\bar{g} : TS \oplus (\underset{\sim}{Ker} \otimes \eta) \oplus \underset{\sim}{\mathbb{R}} \cong \eta \oplus \underset{\sim}{Ker} \otimes \underset{\sim}{\mathbb{R}}^k .$$

Let $\underset{\sim}{Im}$ be the complement of $\underset{\sim}{Ker}$ in $S \times \mathbb{R}^k$, and let $\underset{\sim}{Coker}$ be the coker bundle of any monomorphism from $\underset{\sim}{Im}$ into η. Then via destabilization, $\bar{g}$ leads to an isomorphism

$$\eta \cong TS \oplus (\underline{Ker} \otimes \underline{Coker})$$

(see also the discussion at the beginning of §13, with "$TM|S$" replaced by "η"). The model construction (1.9) now gives an oriented compact n-manifold X = the unit disc bundle of $\underline{Hom}(\underline{Ker}, \underline{Coker})$, together with a k-field $\hat{u}$ on X, which is nonsingular over the boundary $Y = \partial X$. Applying the basic construction of Atiyah and Dupont, we define the K-theoretic element

$$\vartheta^s(x) = \text{ind } \alpha_X^s(\hat{u})$$

(see [6], 2.12).

Now let M be an oriented closed smooth n-manifold. Pick a non-degenerate k-field u on M and apply the procedure above to the singularity data of u. Then, according to the structure lemma 1.10, the behavior of u around its singularity is described by $\hat{u}$. Therefore, we get

(16.25) $$\vartheta^s(\omega_k'(M)) = \text{ind } \alpha_{M,k}^s .$$

(cf. theorem 13.4, and [6], 2.13). This sheds some light on the fact that only the following (fine bordism) invariants of M are relevant in the calculation of ind $\alpha_{M,k}^s$: the Euler number, the Kervaire semi-characteristic and the signature (which is a Pontrjagin number).

Similarly, we see that the following diagram commutes

(16.26) $$\begin{array}{ccc} & \xrightarrow{\Theta^{or}} & \Omega_{k-1}(P^{k-1} \times BSO(n); \phi) \\ \pi_{n-1}(V_{n,k}) & & \downarrow \vartheta^s \\ & \xrightarrow{\theta^s} & KR^s(P^{k+s-1}, P^{s-1}) \end{array} ,$$

where θ^s is the homomorphism introduced, and studied in detail, by Atiyah and Dupont. This gives an explicit example of a factorisation as discussed in theorem 16.11 (defined actually not only on $\Theta^{or}(\pi_{n-1}(V_{n,k}))$.

In [6], p.2, the question is raised whether $\Theta^{s}(\pi_{n-1}(V_{n,k}))$ is the largest homomorphic image of $\pi_{n-1}(V_{n,k})$ in which the index of a k-field with finite singularities always becomes dependent only on the underlying closed manifold. The example

$$\Theta^{or}(\pi_{8q-1}(V_{8q,4})) \cong \mathbb{Z}_3 \oplus \mathbb{Z}_4 \oplus \mathbb{Z}_8 \oplus \mathbb{Z} ,$$

$$\Theta^{o}(\pi_{8q-1}(V_{8q,4})) \subset KR(P^3) \cong \mathbb{Z}_4 \oplus \mathbb{Z}$$

for $q>1$ (see 16.15 and [6], p.13) suggests that one should restrict one's attention to the subgroup $\pi^{or}(n,k)$ of $\pi_{n-1}(V_{n,k})$; also if k and n are divisible by 4, one should replace Θ^{o} by the finer homomorphism Θ' (see [6], 5.11) which is defined on $proj_*(\pi_{n-1}(V_{n,k+1}))$ and hence on $\pi^{or}(n,k)$ (see theorem 16.9). In view of 13.6, 16.9 and [6], theorem 6.2, Θ' makes indices become independent of the k-fields.

We can relate this whole question to a bordism problem.

Definition 16.27. The pair (n,k) is called Ω-pleasant if for every bordism class $x \in \Omega_n$ the following two conditions are equivalent:

(i) x can be represented by a manifold M' with $\operatorname{span}(M') \geq k$,

(ii) given any representative M of x, all Stiefel-Whitney numbers of M which contain some $w_i(M)$, $i>n-k$, as a factor, vanish, and so does the signature of M, taken modulo b_k (where b_k is a power of 2, given by the table on p.30 in [6]).

In view of corollary 6.6 in [6], clearly (i) implies (ii). (For the corresponding definition in the unoriented setting, see 12.12).

Proposition 16.28. Assume that (n,k) is Ω-pleasant and $n>2k>0$. Then $\Theta(\pi^{or}(n,k))$ is the largest homomorphic image of $\pi^{or}(n,k)$ in which the

<u>index of a k-field with finite singularities always depends only on the underlying closed oriented manifold. Here θ stands for θ' if $n \equiv k \equiv 0(4)$, and for θ^s (where $n+s \equiv 0(4)$) otherwise.</u>

<u>Proof.</u> We know from theorem 16.11 that $\Theta^{or}|\pi^{or}(n,k)$ has the desired universal property. Hence $\ker \Theta^{or} \subset \ker(\theta|\pi^{or}(n,k))$, and we need to show only that these two kernels are equal.

Assume $n \equiv k \equiv 0(4)$, and let z lie in the kernel of $\theta'|\pi^{or}(n,k)$. Then, by proposition 16.1, z occurs as the index of a k-field with just one singular point on a closed connected oriented n-manifold M. Theorem 6.2 and §3 and 4 of Atiyah and Dupont [6] imply that

$$\begin{aligned} 0 &= \theta'(z) = \operatorname{ind} \alpha^0_{M,k+1} \\ &= (\chi(M);(\tfrac{1}{2}(\chi(M) \pm \operatorname{sign}(M))), \bmod a_{k+1}) . \end{aligned}$$

Hence $\chi(M)=0$, $\operatorname{sign}(M)$ is divisible by $2\,a_{k+1}=b_k$, and clearly condition (ii) in 16.27 holds for $[M]$ (see also 13.14). Since (n,k) is assumed Ω-pleasant, there is a manifold M' with $\operatorname{span}(M') \geq k$ and bordant to M (in Ω_n, and even in $\Omega_n(0,0)$ since $\chi(M)=\chi(M')=0$, see theorem 12.1). Therefore, by theorem 13.4 and proposition 13.8,

$$0 = \omega_k'(M) = \Theta^{or}(z) .$$

We conclude that $\ker(\theta'|\pi^{or}(n,k)) \subset \ker \Theta^{or}$.

The proof for θ^s is very similar. ∎

<u>Exercise 16.29.</u> Deduce the following two results from the work of Atiyah and Dupont [6] and from theorem 16.9.

(i) Let u be a k-field with finite singularities on a closed oriented n-manifold M. Assume $n \equiv k \equiv 0(4)$, $n>2k>0$. Then

$$\operatorname{ind} \alpha^0_{M,k+1} = \theta'(\operatorname{Index}(u,or)).$$

(In other words, one of the assumptions in [6], theorem 6.2, can be dropped in the metastable range).

(ii) Let M be a closed connected oriented manifold of dimension $n \geq 16$, $n \equiv 0(8)$, admitting a k-field with finite singularities. Then for $k=4$, 5 or 6:

$$\mathrm{span}(M) \geq k \longleftrightarrow \chi(M) = 0 \text{ and } \mathrm{sign}(M) \equiv 0(16).$$

<u>Exercise 16.30.</u> Let n be odd and $\mathrm{span}(S^n) < k < n$. Show that the index of a k-field with finite singularities on S^n is a nontrivial element of $\ker \Theta$ (see also example 16.22).

§17. Torsion questions.

The tables in example 16.15 suggest that, while such groups as $\pi_{n-1}(V_{n,k})$ may well contain odd torsion, most of the relevant subgroups do not. In this section we investigate phenomena of this type, and we also try to get some rough estimates on the order of obstructions and other elements in our obstruction groups. For this purpose we adapt a (spectral sequence) argument of H.A. Salomonsen (see [57], § 10).

As before let Y stand for point, M or $B(S)O(n)$, and let η denote the n-plane bundle $\mathbb{R}^n$, TM or γ respectively. For brevity put $\Omega_r(Y;0) = \Omega_r(Y;\text{trivial})$ and

$$(17.1) \qquad \Omega_r(s) = \Omega_r(P^{s-1} \times Y; \lambda \otimes \eta - s\lambda - \eta) .$$

These terms fit into one of the many ("parallel", "crooked") exact sequences defined by 13.9 and indicated or rather suggested in the following diagram (where $t \geq 1$)

$$(17.2) \qquad \begin{array}{ccccccccc}
 & & & & & & \Omega_{r+t}(s+t) & & \\
 & & & & & & \downarrow\Delta & & \\
 & & & & & & \vdots & & \\
 & & \downarrow & & & & \downarrow\Delta & & \\
\cdots\ \Omega_{r+1}(Y;0) & \xrightarrow{incl_*} & \Omega_{r+1}(s) & \longrightarrow & & \longrightarrow & \Omega_{r+1}(s+1) & \longrightarrow & \\
 & & \downarrow\Delta & & & & \downarrow\Delta & & \\
\Omega_r(Y;0) & \xrightarrow{incl_*} & \Omega_r(s-1) & \xrightarrow{d} & \Omega_r(Y;0) & \xrightarrow{incl_*} & \Omega_r(s) & \xrightarrow{d} & \Omega_r(Y;0) \\
 & & \downarrow & & & & \downarrow\Delta & & \\
 & & & & & \longrightarrow & \Omega_{r-1}(s-1) & \xrightarrow{d} & \Omega_{r-1}(Y;0)
\end{array}$$

Next define an involution ι on $\Omega_r(Y;0)$ as follows: given a bordism class x in this group, represent it by a singular manifold $g: N \longrightarrow Y$, together with a stable isomorphism $\bar{g} : TN \oplus g^*(\eta) \cong \mathbb{R}^r \oplus g^*(\eta)$, and put

(17.3) $\iota(x) = [N, g, \bar{g} \circ (\mathrm{Id} \oplus (-\mathrm{Id}))]$.

Then, clearly, for all $s \geq 1$ the homomorphism

(17.4) $$\Omega_r(Y;0) \xrightarrow{\mathrm{incl}_*} \Omega_r(s) \xrightarrow{d} \Omega_r(Y;0)$$

is just $\mathrm{Id}+(-1)^s\iota$. (However, if $s \leq 0$, then $\Omega_r(s)$, and hence $d \circ \mathrm{incl}_*$, vanish).

Now note that

(17.5) $$2 \cdot \ker(\mathrm{Id} + (-1)^s\iota) \subset (\mathrm{Id} + (-1)^{s-1}\iota)\,(\Omega_r(Y;0));$$

indeed, if $x \in \ker(\mathrm{Id}+(-1)^s\iota)$, then $x = (-1)^{s-1}\iota(x)$, and $2x = x+(-1)^{s-1}\iota(x)$ lies in the image to the right. If $s>1$, it follows (cf. 17.2) that

$$2 \cdot \mathrm{incl}_*^{-1}(\Delta(\Omega_{r+1}(s+1))) = 2 \cdot \ker(d \circ \mathrm{incl}_*)$$

is contained in $d(\Omega_r(s-1)) = \ker(\mathrm{incl}_*)$. Hence the subgroup

$$\mathrm{incl}_*(\mathrm{incl}_*^{-1}(\Delta(\Omega_{r+1}(s+1)))) = \Delta(\Omega_{r+1}(s+1)) \cap \mathrm{incl}_*(\Omega_r(Y;0))$$

of $\Omega_r(s)$ is annihilated by 2, and so is

$$\Delta^t(\Omega_{r+t}(s+t)) \cap \mathrm{incl}_*(\Omega_r(Y;0)) \cong \ker \Delta^{t+1} / \ker \Delta^t ,$$

where the iterates of Δ start out from $\Omega_{r+t}(s+t)$. In other words, if $z \in \Omega_{r+t}(s+t)$ satisfies the equation $\Delta^{t+1}(z)=0$, then also $\Delta^t(2z)=0$. Iterating this argument, we obtain the following result which is basically due to Salomonsen.

<u>Proposition 17.6.</u> <u>Let j, k and t be integers such that $k>t>0$. Then for any element z in the kernel of the iterate</u>

$$\Delta^{t+1} : \Omega_j(P^k \times Y; \phi_{k+1}) \longrightarrow \Omega_{j-t-1}(P^{k-t-1} \times Y; \phi_{k-t})$$

<u>(where</u> $\phi_q := \lambda \otimes \eta - q\lambda - \eta$) <u>we have</u>

$$2^t \cdot \Delta(z) = 0 .$$

In particular, $2^{k-1} \cdot \Delta \equiv 0$ on $\ker(\Delta^k : \Omega_j(P^k \times Y; \phi_{k+1}) \longrightarrow \Omega_{j-k}(Y;0))$.

The following application is similar to some work of J.C. Becker (see [8], theorem 5.7), a.o.

Theorem 17.7. Let M be a closed connected n-manifold with vanishing Euler number $\chi(M)$, and assume that $n > 2k > 0$.

Then

$$2^{k-1} \cdot \omega_k(M) = 0 ,$$

and hence $\quad 2^{k-1} \cdot \omega_k'(M) = 0 .$

If in addition M has a k-field u with finite singularity, then

$$2^{k-1} \cdot \mathrm{Index}(u,or) = 0$$

for any system or of local orientations.

Proof. We will establish theorems 17.7 and 16.23 in one wash-up.

Note first that $\Delta^k(\omega_{k+1}(M)) = \omega_1(M) = \chi(M) = 0$, and therefore

$$2^{k-1} \cdot \omega_k(M) = 2^{k-1} \cdot \Delta(\omega_{k+1}(M)) = 0$$

(cf. theorem 13.11 and proposition 17.6). The same argument applies to $\omega_k'(M)$ and to all elements of the group (put Y=point)

$$\Delta(\Omega_k(P^k \times \mathrm{point}; \phi_{k+1})) \cap \ker \Delta^{k-1} .$$

But this group contains the image of

$$\mathrm{proj}_*(\pi_{n-1}(V_{n,k+1})) \cap \ker(\mathrm{pr}_* : \pi_{n-1}(V_{n,k}) \longrightarrow \pi_{n-1}(S^{n-1}))$$

and of $\tilde{\pi}^{(or)}(n,k)$ ($\supset \ker \Theta^{(or)}$, $\{\mathrm{Index}(u, or)\}$)

under the isomorphism

$$\sigma : \pi_{n-1}(V_{n,k}) \xrightarrow{\cong} \Omega_{k-1}(P^{k-1} \times \text{point}; \phi_k)$$

(use the diagram 13.11 and the relation

$$\pi^{(or)}(n,k) \subset \ker \sigma \circ \partial = \sigma^{-1}(\Delta(\Omega_k(P^k \times \text{pt}; \phi_{k+1})))$$

obtained in the proof of theorem 16.9).

It remains only to establish the finiteness claims in theorem 16.23; just note that $\pi_{n-1}(V_{n,k})$ is finitely generated, as can be seen by induction on k.

<u>Remark 17.8.</u> The powers of 2 occurring in the previous theorem can often be made considerably smaller. E.g. $2^{k-2} \cdot \omega_k(M) = 0$ for $n \equiv 2(4)$ and $k \geq 3$ (use 17.6 and the proof of 14.12) and, in the unoriented setting or when n is odd, $2 \cdot \omega_k'(M) = 0$ for all n and k (cf. 13.4 and 12.1). Moreover, an inspection of Paechter's tables [51] shows that $2^{\rho} \cdot \tilde{\pi}(n,k) = 0$ for $k = 5$ or 6, where

$$\rho = \begin{cases} 1 & \text{if } n \text{ is odd} \\ 0 & \text{if } n \equiv 2(4) \\ 3 & \text{if } n \equiv 0(4) \end{cases}$$

and hence $2^{k-(6-\rho)} \cdot \tilde{\pi}(n,k) = 0$ for $k \geq 6$. Actually, this (and example 16.15) implies that 2^{k-1} is the lowest power of 2 annihilating $\tilde{\pi}(n,k)$ if and only if $k = 1$, or $k = 2$, or $n \equiv 0(4)$ and $k = 3$ or 4.

<u>Exercise 17.9.</u> Let M be a connected <u>compact</u> manifold of dimension $n > 2k > 0$, and let $u^{\partial} : \partial M \times \mathbb{R}^k \hookrightarrow TM|\partial M$ be a monomorphism. Assume that there is a nowhere zero vectorfield v over all of M such that $v(x) \notin \operatorname{Im}(u^{\partial})$ for all $x \in \partial M$ (this is satisfied e.g. if $\chi(M)=0$ and $\operatorname{Im}(u^{\partial}) \subset T(\partial M)$).

Show that the obstruction $\omega_k(M \times \mathbb{R}^k, TM, u^{\partial})$ (to extending u^{∂} to a k-field without singularities over all of M, cf. theorem 3.7) is annihilated by multiplication with 2^{k-1}. ∎

Next consider the commuting diagram

(17.10)
$$\begin{array}{ccccccc} & & \ker \Delta^k & \xrightarrow{\Delta|} & \ker \Delta^{k-1} & & \\ & & \cap & & \cap & & \\ \to \Omega_i(Y;0) & \xrightarrow{incl_*} & \Omega_i(P^k xY;\phi_{k+1}) & \xrightarrow{\Delta} & \Omega_{i-1}(P^{k-1}xY;\phi_k) & \xrightarrow{d} & \Omega_{i-1}(Y;0) \\ & & \Delta^k \searrow & & \swarrow \Delta^{k-1} & & \\ & & \Omega_{i-k}(P^o xY;\phi_1) & = & \Omega_{i-k}(Y;\text{trivial}) & & \end{array}$$

where the middle line is the exact sequence 13.9 (see also 17.2). We know from proposition 17.6 that $2^{k-1} \cdot \Delta| \equiv 0$.

<u>Proposition 17.11.</u> <u>Let j, k be integers</u>, $k>0$, <u>and assume that the odd torsion of $\Omega_{j-k+1}(Y;\text{trivial})$ is zero. Consider the homomorphisms</u>

$$\Omega_j(Y;\text{trivial}) \underset{d}{\overset{incl_*}{\rightleftarrows}} \Omega_j(P^{k-1}xY;\phi_k) .$$

<u>Then d is injective on the odd torsion subgroup which in turn lies in the image of $incl_*$.</u>

<u>If in addition η is stably trivial over the j-skeleton of Y, then</u>

$$d \circ incl_* = (1 + (-1)^{n+k}) \text{ Id} ,$$

<u>and hence both d and $incl_*$ restrict to isomorphisms between the odd torsion subgroups provided $n \equiv k(2)$; if $n \not\equiv k(2)$,</u> then $\Omega_j(P^{k-1}xY;\phi_k)$ <u>has no odd torsion.</u>

<u>Proof.</u> The first claim follows from diagram chasing in 17.10 and from the fact that multiplication with powers of 2 gives an isomorphism on odd torsion groups. E.g. given an element z in the odd torsion of $\Omega_{i-1}(P^{k-1}xY;\phi_k)$ (put j=i-1!) such that $d(z)=0$, then $\Delta^{k-1}(z)$ lies in the odd torsion of $\Omega_{i-k}(Y;\text{trivial})$ which vanishes by assumption. Thus z lies in

$$\ker \Delta^{k-1} \cap \Delta(\Omega_i(P^k \times Y;\ \phi_{k+1})) = \Delta(\ker \Delta^k)$$

which is annihilated by 2^{k-1}. Hence $z = 0$.

If η is stably trivial over the j-skeleton of Y with respect to any CW-decomposition, then we note in formula 17.3 that $\iota(x) = (-1)^n x$ since we can deform g into the j-skeleton and therefore $g^*(\eta)$ is trivial. The rest of the theorem follows easily from 17.4.∎

Corollary 17.12. Let ℓ be an integer such that the stable homotopy group π^S_ℓ has no odd torsion (e.g. $\ell = 0$, 1 or 2), and assume that $n > 2k + \ell$.

If $n \not\equiv k(2)$, then the odd torsion of $\pi_{n-1+\ell}(V_{n,k})$ is trivial.

If $n \equiv k(2)$, then each of the homomorphisms

$$\pi^S_{k-1+\ell} \xleftarrow[\cong]{E} \pi_{n-1+\ell}(S^{n-k}) \xrightarrow{incl_*} \pi_{n-1+\ell}(V_{n,k})$$

$$E \circ \partial : \pi_{n-1+\ell}(V_{n,k}) \longrightarrow \pi^S_{k-1+\ell}$$

restricts to an isomorphism between the odd torsion subgroups. (Here E denotes (infinite) suspension, and ∂ comes from the homotopy sequence of the fibration $S^{n-k-1} \subset V_{n,k+1} \longrightarrow V_{n,k}$).

In order to deduce this wellknown result from proposition 17.11, note that the isomorphism

$$\sigma : \pi_{n-1+\ell}(V_{n,k}) \longrightarrow \Omega_{k-1+\ell}(P^{k-1} \times pt;\phi_k)$$

(cf. 5.4) is compatible with the homomorphisms above (see also 13.11).

Corollary 17.13. Let $n > 2k > 0$. Then $\Theta^{(or)}$ maps the odd torsion of $\pi_{n-1}(V_{n,k})$ isomorphically onto a direct summand of the odd torsion subgroup of $\Omega_{k-1}(P^{k-1} \times B(S)O(n);\phi_k)$.

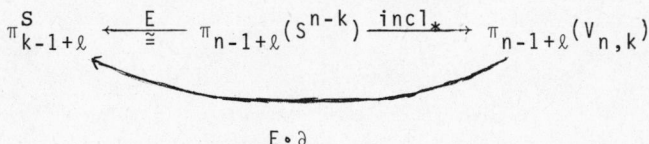

We know already from theorem 16.23 that $\Theta^{(or)}$ is injective on the odd torsion. The full strength of the corollary follows from the commuting diagram

$$\begin{array}{ccccc}
\pi_{n-1}(V_{n,k}) & \xrightarrow[\cong]{\sigma} & \Omega_{k-1}(P^{k-1}\times pt;\phi_k) & \xrightarrow{d} & \Omega_{k-1}(point;0) \\
 & \searrow^{\Theta^{(or)}} & \downarrow & & \downarrow \\
 & & \Omega_{k-1}(P^{k-1}\times B(S)O(n);\phi_k) & \xrightarrow{d} & \Omega_{k-1}(B(S)O(n);0)
\end{array}$$

(cf. 13.11). If $n \equiv k(2)$ (and this is the only nontrivial case), then the upper (resp. lower) d is iso-(resp. mono-)morphic on the odd torsion, so that $\sigma^{-1} \circ d^{-1} \circ const_* \circ d$ defines a left inverse of $\Theta^{(or)}$ there.∎

§18. Counting homotopy classes of framefields.

So far in this chapter we were mainly concerned with existence questions. Now we turn our attention to the homotopy classification of framefields.

Let M be a connected closed smooth manifold of dimension $n>2k+1$, and assume that M carries a k-framefield

$$u_o : M \times \mathbb{R}^k \hookrightarrow TM .$$

Recall from theorem 4.14 that the difference invariant $d_k(u_o,-)$ defines a one-to-one onto correspondence between the homotopy classes of all such k-framefields on M and the elements of the group

$$\Omega_k(P^{k-1} \times M; \phi_{M,k}) , \tag{18.1}$$

where $\phi_{M,k} = \lambda \otimes TM - k\lambda - TM$ (compare also 13.2). We are going to study this group in a few simple cases.

E.g. if $k=1$, the group above is just $\Omega_1(M;\text{trivial})$. According to theorem 9.3 it fits into the following (splitting) short exact sequence

$$0 \longrightarrow \Omega_1(\text{point};\text{trivial}) \hookrightarrow \Omega_1(M;\text{trivial}) \xrightarrow{\mu \cdot f_1} H_1(M;\mathbb{Z}) \longrightarrow 0$$
$$\Omega_1(\text{point};\text{trivial}) \cong \mathbb{Z}_2$$

(see also fact 9.4). We obtain the following (probably wellknown) results.

<u>Proposition 18.2.</u> <u>Let M be a closed connected manifold of dimension $n>3$ such that the Euler number $\chi(M)$ vanishes.</u>

<u>Then the homotopy classes of nowhere vanishing vectorfields on M correspond in a one-to-one way to the elements of</u> $H_1(M;\mathbb{Z})\oplus\mathbb{Z}_2$.

<u>Example 18.3.</u> The sphere S^{2q+1}, $q\neq 1$, $q\geq 0$, has precisely two homotopy classes of vectorfields (but S^3 has as many such homotopy

classes as there are elements in $\pi_3(S^2)\cong\mathbb{Z}$).

Projective space P^{2q+1}, $q>1$, has precisely four homotopy classes of vectorfields. ∎

Before we turn our attention to 2- and 3-framefields, we discuss the auxiliary groups $\overline{\Omega}_i(P^q\times Y;\lambda)$, where i and q are nonnegative integers and Y is pathconnected, paracompact and homotopy equivalent to a CW-complex with compact skeletons in all dimensions. Pick an integer $r>>i,q$, and let P^q, $P'^{r-q-1}\subset P^r$ be the real projective spaces corresponding to the decomposition $\mathbb{R}^{r+1}=\mathbb{R}^{q+1}\oplus\mathbb{R}^{r-q}$. Clearly, P^q is a deformation retract of $P^r-P'^{r-q-1}$. Since the normal bundle of $P'^{r-q-1}\subset P^r$ is $(q+1)\lambda$, the normal bordism Thom-Gysin sequence (see [57], theorem 5.2) of the pair $(P^r,P^r-P'^{r-q-1})\times Y\times BSO(r)$ with coefficients in the virtual vector bundle $\lambda-\gamma_r$ takes the following form

$$
(18.4)\qquad
\begin{array}{ccccccc}
\cdots\to\Omega_i(P^q\times Y\times BSO(r);\lambda-\gamma_r) & \longrightarrow & \Omega_i(P^r\times Y\times BSO(r);\lambda-\gamma_r) & \longrightarrow & \Omega_{i-q-1}(P'^{r-q-1}\times Y\times BSO(r);(q+2)\lambda-\gamma_r) & & \\
\downarrow\cong & & \downarrow\cong & & \downarrow\cong & & \\
\cdots\to\overline{\Omega}_i(P^q\times Y;\lambda) & \xrightarrow{f} & \mathfrak{N}_i(Y) & \xrightarrow{\Delta} & \begin{cases}\mathfrak{N}_{i-q-1}(Y) & \text{if } q \text{ is odd}\\ \Omega_{i-q-1}(P^\infty\times Y) & \text{if } q \text{ is even.}\end{cases} & &
\end{array}
$$

Here all vertical arrows and f are the obvious forgetful maps, and Δ is defined as follows. Given $x=[g:N\longrightarrow Y]\in\mathfrak{N}_i(Y)$, let $Z\subset N$ be the zero set of a nondegenerate section of $(q+1)\xi_N$; then $\Delta(x)$ is given by Z together with $g|Z$ (and with a canonical orientation of Z as well as a classifying map h for $\xi_N|Z$, if q is even).

In particular, for $q=1$ we get the exact sequence

$$
\begin{array}{ccccccccccc}
\overline{\Omega}_3(P^1\times Y;\lambda) & \xrightarrow{f} & \mathfrak{N}_3(Y) & \xrightarrow{\Delta} & \mathfrak{N}_1(Y) & \longrightarrow & \overline{\Omega}_2(P^1\times Y;\lambda) & \xrightarrow{f} & \mathfrak{N}_2(Y) & \xrightarrow{\Delta} & \mathbb{Z}_2\\
 & & \downarrow(\mu,\Delta) & & & & & & \downarrow(\mu,\Delta) & & \\
 & & H_3(Y;\mathbb{Z}_2)\oplus\mathfrak{N}_1(Y) & & & & & & H_2(Y;\mathbb{Z}_2)\oplus\mathbb{Z}_2 & & .
\end{array}
$$

Note that on $N=P^2$ there is a nondegenerate section of $\xi_N\oplus\xi_N$ with just one zero. Hence it follows from fact 9.4 that both vertical arrows (μ,Δ) are isomorphisms. We conclude easily that the homomorphism

$$(18.5)\qquad \mu\circ f : \bar{\Omega}_i(P^1\times Y;\lambda) \longrightarrow H_i(Y;\mathbb{Z}_2)$$

is bijective for $i=2$ and onto for $i=3$.

We are now ready to attack the homotopy classification of 2-framefields. Define homomorphisms

$$\begin{aligned}\bar{e} : H^1(M;\mathbb{Z}_2) &\longrightarrow H^3(M;\mathbb{Z}_2)\\ y &\longrightarrow w_1(M)^2y+w_1(M)y^2\end{aligned}$$

(see also the lines following 14.12), and

$$\bar{\bar{e}} : H^1(M;\mathbb{Z}_2) \xrightarrow{\ red^t\circ(w_1(M)\cdot)\ } \mathrm{Hom}(H_2(M;\mathbb{Z}),\mathbb{Z}_2);$$

given $y\in H^1(M;\mathbb{Z}_2)$, $\bar{\bar{e}}(y)$ is the homomorphism which first reduces mod 2 and then applies $w_1(M)\cdot y$.

<u>Theorem 18.6.</u> <u>Let M be a closed connected manifold of dimension</u> $n>5$ <u>with</u> $\mathrm{span} M\geq 2$.

<u>If n is odd, the number of homotopy classes of 2-framefields on M is</u>

$$|\ker \bar{e}|\cdot|H_2(M;\mathbb{Z}_2)|\cdot\begin{cases}4 & \text{if } w_1(M) = 0\\ 2 & \text{if } w_1(M) \neq 0, \text{ but } w_1(M)^2=0\\ 1 & \text{if } w_1(M)^2 \neq 0\end{cases}.$$

<u>If n is even, the (finite or infinite) number of homotopy classes of 2-framefields on M equals</u>

$$|\ker \bar{\bar{e}}|\cdot|H_1(M;\mathbb{Z})|\cdot|H_2(M;\mathbb{Z})|\cdot\begin{cases}4 & \text{if } w_1(M) = 0\\ 1 & \text{if } w_1(M) \neq 0\end{cases}.$$

Note that $\ker \bar{e} = \ker \bar{\bar{e}} = H^1(M;\mathbb{Z}_2)$ if $w_1(M)=0$.

Example 18.7. The sphere S^{4q+3}, $q>0$, has precisely four homotopy classes of 2-framefields; but the homotopy classes of 2-fields on S^3 are classified by $\pi_3(SO(3))\cong\mathbb{Z}$.

Projective space P^{4q+3}, $q>0$, has 16 homotopy classes of 2-framefields.

The n-dimensional torus $(S^1)^n$, $n>5$, carries infinitely many nonhomotopic 2-framefields if n is even. If n is odd, the number of homotopy classes of 2-framefields on $(S^1)^n$ is

$$4\cdot 2^{\binom{n+1}{2}} .$$

Thus e.g. $(S^1)^7$ carries 2^{30}, $(S^1)^9$ carries 2^{47} and $(S^1)^{11}$ carries 2^{68} 2-framefields.

Proof of theorem 18.6. Homotopy classes of 2-framefields on M correspond to the elements of $\Omega_2(P^1\times M;\phi_{M,2})$. According to theorem 9.3, this group fits into the following diagram

(18.8)

$$\begin{array}{ccccccccc}
 & & \mathbb{Z}_2 & & & & & & \\
 & & \downarrow{\scriptstyle \delta_1'} & & & & & & \\
\overline{\Omega}_3(P^1\times M;n\lambda) & \xrightarrow{\sigma\circ j_3} & \Omega_1(P^1\times M\times BO(2);\phi_{M,2}+\Gamma) & \xrightarrow{\delta_2} & \Omega_2(P^1\times M;\phi_{M,2}) & \xrightarrow{f_2} & \overline{\Omega}_2(P^1\times M;n\lambda) & \xrightarrow{\sigma\circ j_2} & \mathbb{Z}_2 \\
\downarrow{\scriptstyle \mu} & & \downarrow{\scriptstyle f_1'}\ \cong \text{ iff } w_1(M)\neq 0 & & & & \cong\ \begin{cases} H_2(M;\mathbb{Z}_2) & n \text{ odd} \\ H_1(M;\mathbb{Z})\oplus H_2(M;\mathbb{Z}) & n \text{ even}\end{cases} & & \\
H_3(P^1\times M;\mathbb{Z}_2) & \xrightarrow{w_2(\phi_{M,2})\cdot *} & H_1(P^1\times M;\mathbb{Z}_2) & & & & & & \\
 & & \downarrow & & & & & & \\
 & & 0 & & & & & &
\end{array}$$

Recall from 14.1 that $w_1(\phi_{M,2})=nw_1(\lambda)$ and $w_2(\phi_{M,2})=w_1(\lambda)w_1(M)$. It follows from 18.5 and from fact 9.4 that $\overline{\Omega}_2(P^1\times M;\phi_{M,2})$ has the indicated form.

If $w_1(M)$ vanishes, then so does $w_2(\phi_{M,2})$ as well as $\sigma\circ j_2$ and $\sigma\circ j_3$, and the theorem follows in this case.

So we may assume that $w_1(M)\neq 0$ and hence f_1' is bijective. Then $\ker f_2$ is isomorphic to the cokernel of $f_1'\circ\sigma\circ j_3$. Now, given elements

$x=[g:N^3 \longrightarrow P^1 xM, or:\xi_N \cong \lambda^{\oplus n}]$ of $\overline{\Omega}_3(P^1xM;n\lambda)$ and $y \in H^1(M;\mathbb{Z}_2)$, we have

$$(18.9) \qquad y(f_1^! \circ \sigma \circ j_3(x)) = g^*(w_1(\lambda)w_1(M)y)[N]$$

and

$$w_1(\lambda)(f_1^! \circ \sigma \circ j_3(x)) = 0 .$$

If n is odd, then $g^*(w_1(\lambda))=w_1(N)$, and we obtain that

$$y(f_1^! \circ \sigma \circ j_3(x)) = Sq^1(g^*(w_1(M)y))[N]$$
$$= \overline{e}(y)(\mu \circ f(x)) ,$$

where we recall from 18.5 that the forgetful map

$$\mu \circ f : \overline{\Omega}_3(P^1xM;\lambda) \longrightarrow H_3(M;\mathbb{Z}_2)$$

is onto. Let $y_1,..y_r,y'_{r+1},..y'_s$ be a basis of $H^1(M;\mathbb{Z}_2)$ so that the elements $\overline{e}(y_1),..\overline{e}(y_r)$ form a basis of the image of $\overline{e}$ in $H^3(M;\mathbb{Z}_2)$, and that $y'_{r+1},..y'_s \in \ker \overline{e}$. Moreover, pick $x_1,..x_r \in \overline{\Omega}_3(P^1xM;\lambda)$ so that the elements $\mu \circ f(x_1),...,\mu \circ f(x_r)$ of $H_3(M;\mathbb{Z}_2)$ are dual to $\overline{e}(y_1),..,\overline{e}(y_r)$. Then, clearly, the image of $f_1^! \circ \sigma \circ j_3$ is spanned by the linearly independent elements $f_1^! \circ \sigma \circ j_3(x_1),...,f_1^! \circ \sigma \circ j_3(x_r)$ (dual to $y_1,...y_r$). We conclude that

$$\ker f_2 \cong \mathbb{Z}_2 \oplus \ker \overline{e} .$$

If n is even, then clearly $f_1^! \circ \sigma \circ j_3$ vanishes on the first factor of

$$\overline{\Omega}_3(P^1xM;n\lambda) = \Omega_3(P^1xM) = \Omega_3(*xM) \oplus ([Id_{P^1}]x\Omega_2(M))$$

(see fact 9.4, and 18.9). Since the second factor is isomorphic to $H_2(M;\mathbb{Z})$, it is not hard to show that

$$\ker f_2 \cong \mathbb{Z}_2 \oplus \ker \overline{\overline{e}} .$$

The "classifying" group $\Omega_2(P^1 xM;\phi_{M,2})$ is built up from $\ker f_2$ and from $\operatorname{im} f_2 = \ker \sigma\circ j_2$ (see diagram 18.8); so it remains only to calculate $\sigma\circ j_2$.

If n is odd, consider an element $x=[g:N^2 \longrightarrow P^1 xM, \text{ or}:\xi_N \cong \lambda]$ of $\overline{\Omega}_2(P^1 xM;n\lambda)$. We have

$$\begin{aligned}\sigma\circ j_2(x) &= g^*(w_1(\lambda)w_1(M))[N] \\ &= Sq^1(g^*(w_1(M)))[N] \\ &= w_1(M)^2(g_*[N]) .\end{aligned}$$

So $\sigma\circ j_2$ corresponds to evaluating $w_1(M)^2$ on $H_2(M;\mathbb{Z}_2)$.

If n is even, then $\sigma\circ j_2$ corresponds to the homomorphism

$$H_1(M;\mathbb{Z}) \oplus H_2(M;\mathbb{Z}) \xrightarrow{\text{red}\circ\text{proj}} H_1(M;\mathbb{Z}_2) \xrightarrow{w_1(M)} \mathbb{Z}_2 .$$

The theorem follows from all these data.∎

If we apply our methods also to the enumeration of k-fields, k=3,..., the calculations get increasingly complicated. We content ourselves with just one sample of a result in this direction.

Theorem 18.10. Let M be a closed connected manifold of dimension $n\equiv 2(4)$, $n\geq 10$, such that $\operatorname{span} M \geq 3$ and $H_1(M;\mathbb{Z})$ has no elements of order four.

Then the (finite or infinite) number of homotopy classes of 3-framefields on M equals

$$2|\ker \ell|\cdot|H_1(M;\mathbb{Z})|\cdot|H_3(M;\mathbb{Z}_2)|$$

where $\ell:H^1(M;\mathbb{Z}_2)\oplus\mathbb{Z}_2 \longrightarrow H^4(M;\mathbb{Z}_2)$ is defined by

$$\ell(y,a) = y^4+y^2w_1(M)^2+a(w_2(M)^2+w_1(M)w_3(M)+w_1(M)^4).$$

Example 18.11. If $r, s>0$, then the product of spheres $S^{4r+3}xS^{4s+3}$ carries four homotopy classes of 3-framefields while the product $P^{4r+3}xP^{4s+3}$ of projective spaces carries 256 such homotopy classes.

If $n \equiv 2(4)$, $n \geq 10$, then the torus $(S^1)^n$ carries infinitely many nonhomotopic 3-framefields.

<u>Proof of theorem 18.10.</u> As in sections 14 and 15, let Y stand for point, M or $B(S)O(n)$, and define η and ϕ_Y accordingly.

First we have to discuss some auxiliary groups. As a special case of 18.4 we have the exact horizontal sequence

(18.12)

$$\cdots \to \bar{\Omega}_4(P^2 \times Y;\lambda) \xrightarrow{f} \mathfrak{N}_4(Y) \xrightarrow{\Delta} \Omega_1(P^\infty \times Y) \longrightarrow \bar{\Omega}_3(P^2 \times Y;\lambda) \xrightarrow{f} \mathfrak{N}_3(Y) \xrightarrow{0} \mathbb{Z} .$$

$$\Omega_1(P^\infty \times Y) \xrightarrow[\cong]{\mu} \mathbb{Z}_2 \oplus H_1(Y;\mathbb{Z}) \xrightarrow{(\mathrm{id},\mathrm{red})} \mathbb{Z}_2 \oplus H_1(Y;\mathbb{Z}_2) ,$$

$$\Delta' = (\Delta_1', \Delta_2') : \mathfrak{N}_4(Y) \to \mathbb{Z}_2 \oplus H_1(Y;\mathbb{Z}_2) .$$

Here we define $\Delta' = (\mathrm{id},\mathrm{red}) \circ \mu \circ \Delta$. Given $x = [g : N \longrightarrow Y] \in \mathfrak{N}_4(Y)$, we have (in the notation of 18.4)

$$\begin{aligned} \Delta_1'(x) &= w_1(\lambda)(h_*[Z]) \\ &= (w_1(N)|Z)[Z] \\ &= w_1(N)^4[N] \qquad \text{(cf. fact 9.11)}, \end{aligned}$$

and similarly for $y \in H^1(Y;\mathbb{Z}_2)$ (dropping pullbacks from our notation as usual)

$$\begin{aligned} y(\Delta_2'(x)) &= w_1(N)^3 y[N] \\ &= w_1(N)^2 y^2[N] \qquad \text{(cf. fact 9.6)}. \end{aligned}$$

Using elements of $\mathfrak{N}_4(Y)$ of the form $[P^4 \longrightarrow * \subset Y]$ and $[P^2 \times L^2 \longrightarrow L^2 \longrightarrow Y]$, we see that

$$|\Delta'(\mathfrak{N}_4(Y))| = 2 \cdot |Sq^1(H^1(Y;\mathbb{Z}_2))| .$$

Since all elements of $\mathfrak{N}_4(Y)$ and of the 2-primary component of $H_1(Y;\mathbb{Z})$ have order 2, $(\mathrm{id},\mathrm{red})$ is injective on the image of $\mu \circ \Delta$.

Also, $\overline{\Omega}_3(P^2 x Y;\lambda)$ is built up from $\mathfrak{N}_3(Y) \cong H_1(Y;\mathbb{Z}_2) \oplus H_3(Y;\mathbb{Z}_2)$ and from the cokernel of Δ. We conclude that

$$(18.13) \qquad |\overline{\Omega}_3(P^2 x Y;\lambda)| = |H_1(Y;\mathbb{Z}_2)| \cdot |H_3(Y;\mathbb{Z}_2)| \cdot \frac{|H_1(Y;\mathbb{Z})|}{|Sq^1(H^1(Y;\mathbb{Z}_2))|} .$$

Furthermore, elements in

$$(18.14) \qquad f(\overline{\Omega}_4(P^2 x Y;\lambda)) = \ker \Delta = \ker \Delta'$$

are characterized by the equations $w_1(N)^4[N]=0$ and $w_1(N)^2 y^2[N]=0$ for all $y \in H^1(Y,\mathbb{Z}_2)$.

Now consider the diagram

$$(18.15) \qquad \begin{array}{ccccccccc}
 & & 0 & & & & & & \\
 & & \downarrow & & & & & & \\
 & & \ker f_2' \cong \mathbb{Z}_2 & & & & & & \\
 & & \downarrow & & & & & & \\
\overline{\Omega}_4(P^2 x Y;\lambda) & \xrightarrow{\sigma\circ j_4} & \Omega_2(P^2 x Y x BO(2);\phi_{Y,3}+\Gamma) & \xrightarrow{\delta_3} & \underline{\Omega_3(P^2 x Y;\phi_{Y,3})} & \xrightarrow{f_3} & \overline{\Omega}_3(P^2 x Y;\lambda) & \xrightarrow{f_1'\circ\sigma\circ j_3} & H_1(P^2 x Y;\mathbb{Z}_2) \\
\downarrow f & & \downarrow f_2' & & & & & & \\
\ker\Delta \subset \mathfrak{N}_4(Y) & \overset{\tilde{\sigma}=(\tilde{\sigma}_1,\tilde{\sigma}_2,\tilde{\sigma}_3)}{\dashrightarrow} & (\ker : H_2(P^2 x Y;\mathbb{Z}_2) \xrightarrow{w_2(\phi_{Y,3})} \mathbb{Z}_2) \oplus \mathbb{Z}_2 \oplus \mathbb{Z}_2 & & & & & & \\
 & & \downarrow & & & & & & \\
 & & 0 & & & & & &
\end{array}$$

deduced from theorem 9.3 and from the equations (valid for $n \equiv 2(4)$, see 9.10)

$$(18.16) \qquad \begin{aligned}
w_1(\phi_{Y,3}) &= w_1(\lambda) \\
w_2(\phi_{Y,3}) &= w_1(\lambda)^2 + w_1(\lambda) w_1(\eta) \\
w_3(\phi_{Y,3}) &= w_1(\lambda) w_1(\eta)^2 \\
w_4(\phi_{Y,3}) &= w_1(\lambda)^2 w_2(\eta) + w_1(\lambda)(w_3(\eta) + w_1(\eta) w_2(\eta) + w_1(\eta)^3).
\end{aligned}$$

Clearly, the "classifying group" $\Omega_3(P^2 x Y;\phi_{Y,3})$ is built up from

$\operatorname{im} f_3 = \ker f_1^! \circ \sigma \circ j_3$ and from $\ker f_3 \cong \operatorname{coker} \sigma \circ j_4$. We will now compute the cardinalities of these components.

Given $x=[N \longrightarrow P^2 xY; \xi_N \cong \lambda] \in \bar{\Omega}_3(P^2 xY;\lambda)$ and $y \in H^1(Y;\mathbb{Z}_2)$, we know from 9.3 and 9.6 that

$$w_1(\lambda)(f_1^! \circ \sigma \circ j_3(x)) = w_1(N)^2 w_1(\eta)[N]$$

and

$$y(f_1^! \circ \sigma \circ j_3(x)) \qquad = (w_1(N)^2 y + (w_1(\eta)^2 y + w_1(\eta) y^2))[N] .$$

Elements of $\bar{\Omega}_3(P^2 xY;\lambda)$ of the form $[P^2 x S^1 \xrightarrow{\ id x \ell\ } P^2 xY; \xi_{P^2} = \lambda]$ show that the image of $f_1^! \circ \sigma \circ j_3$ is just the kernel of

$$w_1(\lambda) + w_1(\eta) : H_1(P^2 xY;\mathbb{Z}_2) \longrightarrow \mathbb{Z}_2 ,$$

and hence has the same cardinality as $H_1(Y;\mathbb{Z}_2)$. We conclude from (18.13) and (18.15) that

$$\text{(18.17)} \qquad |\operatorname{im} f_3| = \frac{|H_3(Y;\mathbb{Z}_2)| \cdot |H_1(Y;\mathbb{Z})|}{|Sq^1(H^1(Y;\mathbb{Z}_2))|} .$$

In order to compute the cardinality of $\ker f_3 \cong \operatorname{coker} \sigma \circ j_4$, we study the lefthand side of the diagram 18.15. According to theorem 9.3, $\ker f_2^!$ is canonically isomorphic to a quotient of $H_1(P^2 xY;\mathbb{Z}_2)$, and it is not hard to see that this quotient consists of precisely two elements, namely 0 and the class of the generator of $H_1(P^2;\mathbb{Z}_2)$. Also, we can use 9.3, 9.5, 9.6 as well as 18.14 to show that $f_2^! \circ \sigma \circ j_4$ factors through $\ker \Delta$. In fact, diagram 18.15 commutes if we define $\tilde{\sigma}$ as follows. Given $v=[h:N^4 \longrightarrow Y] \in \ker \Delta$, put

$$\begin{aligned}
& w_1(\lambda)^2 \tilde{\sigma}_1(v) = 0 ; \\
& w_1(\lambda) y \tilde{\sigma}_1(v) = w_1(N)^2 y w_1(\eta)[N] , \quad y \in H^1(Y;\mathbb{Z}_2), \\
\text{(18.18)} \qquad & z \tilde{\sigma}_1(v) = (z^2 + w_1(\eta) Sq^1(z) + w_1(\eta)^2 z + w_1(N)^2 z)[N], \quad z \in H^2(Y,\mathbb{Z}_2);
\end{aligned}$$

and

$$\tilde{\sigma}_2(v) = (w_4(N) + w_1(N)^2 w_2(\eta) + w_1(\eta)^4)[N],$$

and

$$\tilde{\sigma}_3(v) = w_4(N)[N]$$

(see also the very similar calculations in 15.4, applied to $n+1 \equiv 3(4)$ and $k+1 = 4$).

E.g. if $Y=B(S)O(n)$ and v is given by $N=\mathbb{C}P(2) \amalg \mathbb{C}P(1)\times P^2$, together with the classifying map of $\lambda_{\mathbb{C}} \oplus \underset{\sim}{\mathbb{R}}^{n-2}$ ($\lambda_{\mathbb{C}}$ = canonical complex line bundle), then $\tilde{\sigma}(v)=(0,0,1)$. On the other hand, if Y=point, then $(0,0,1)$ does not lie in $\mathrm{im}\,(\tilde{\sigma})=\mathrm{im}\,(f_2^! \circ \sigma \circ j_4)$. It follows that there is a nontrivial element w of $\delta_3(f_2^{!\,-1}\{(0,0,1)\}) \subset \Omega_3(P^2 \times \mathrm{point};\phi_{\mathrm{point},3})$ which gets mapped into zero by the homomorphism $\Theta:\Omega_3(P^2\times\mathrm{point};\phi_{\mathrm{point},3}) \longrightarrow \Omega_3(P^2 \times BO(n);\phi_{BO(n),3})$. Now consider the diagram in 13.11 (for $k=3$) which contains Θ as the composite of the lefthand vertical arrows Θ_M and $(\mathrm{id}\times\tau)_*$. Given an element $w' \in \Omega_3(P^2;(n-3)\lambda)$, if $\Theta(w')$ vanishes, then so does $\mathrm{const}_* \circ d \circ \Theta(w') = d(w')$, and $\sigma^{-1}(w')$ must lie in the image of $\mathrm{proj}_*:\pi_n(V_{n,4}) \rightarrowtail \pi_n(V_{n,3})$. But according to Paechter [51], $\pi_n(V_{n,4}) \cong \mathbb{Z}_2$ for $n \equiv 2(4)$, $n \geq 10$. Therefore, $\ker\Theta$ consists precisely of 0 and of the element w constructed above.

Let us return to diagram 18.15. If Y=point, the lefthand vertical homomorphism f is also injective (its kernel being a quotient of $\Omega_2(P^\infty) \cong H_2(P^\infty;\mathbb{Z}) = 0$, by 18.12). Thus $\overline{\Omega}_4(P^2\times\mathrm{point};\lambda) = \mathbb{Z}_2[\mathbb{C}P(2),\mathrm{constant},\mathrm{or}]$, and $f_2^! \circ \sigma \circ j_4$ is also injective. We conclude that δ_3 is injective on $\ker f_2^! \cong \mathbb{Z}_2$; this is clear for Y=point, and it follows for $Y=M$ and $B(S)O(n)$ from the fact, established above, that the subgroups $\ker\Theta$ and $\delta_3(\ker f_2^!)$ of $\Omega_3(P^2\times\mathrm{point};\phi_{\mathrm{point},3})$ have trivial intersection. Therefore we have for all choices of Y that

$$|\ker f_3| = |\mathrm{coker}\,\sigma \circ j_4| = 2|\mathrm{coker}\,\tilde{\sigma}| \,. \tag{18.19}$$

It remains to study the image of $\tilde{\sigma}$. Recall from fact 9.4 and from 18.14 that

$$\ker \Delta \cong T \oplus H_4(Y;\mathbb{Z}_2) \oplus \mathbb{Z}_2 \cdot [\mathbb{C}P(2),\mathrm{const}]$$

where $T=\{t\in H_2(Y;\mathbb{Z}_2) \mid y^2(t)=0$ for all $y\in H^1(Y;\mathbb{Z}_2)\}$. An element $t\in T$ corresponds to any bordism class $[P^2 \times L \xrightarrow{\text{proj}} L \xrightarrow{h} Y] \in \mathfrak{N}_4(Y)$ such that L is a zerobordant surface and $h_*[L]=t$; clearly, by 18.18, the image of T under the $H_2(Y;\mathbb{Z}_2)$-component homomorphism of $\tilde{\sigma}_1$ is again T. An element u of $H_4(Y;\mathbb{Z}_2)$ corresponds to a bordism class $[g:N^4 \longrightarrow Y] \in \tilde{\mathfrak{N}}_4(Y)$ such that $g_*[N]=u$ and $w_1(N)^2 z[N]=0$ for all $z\in H^2(Y;\mathbb{Z}_2)$; $\tilde{\sigma}_1(H_4(Y;\mathbb{Z}_2))$ adds to the total image of $\tilde{\sigma}_1$ a factor which is as large as the image of

$$\ell| : H^1(Y;\mathbb{Z}_2) \longrightarrow H^4(Y;\mathbb{Z}_2)$$
$$y \longrightarrow y^4+y^2 w_1(\eta)^2$$

(use again the formulas 18.18; also extend the definition of ℓ in 18.10 in the obvious way to all choices of Y).

Clearly, $\tilde{\sigma}[\mathbb{C}P(2),\mathrm{constant}]=(0,1,1)$. Furthermore observe that for all $v\in \ker \Delta$

$$\tilde{\sigma}_2(v)+w_2(\eta)\tilde{\sigma}_1(v)+\tilde{\sigma}_3(v) = (w_2(\eta)^2+w_1(\eta)w_3(\eta)+w_1(\eta)^4)[N];$$

so $\tilde{\sigma}_2$ adds a $\mathbb{Z}_2$-factor to the image of $\tilde{\sigma}$ if and only if

$$w_2(\eta)^2+w_1(\eta)w_3(\eta)+w_1(\eta)^4 \notin \ell|(H^1(Y;\mathbb{Z}_2)).$$

Taking dimensions over $\mathbb{Z}_2$, we obtain

$$\dim \operatorname{im} \tilde{\sigma} = \dim H_2(Y;\mathbb{Z}_2) - \dim Sq^1(H^1(Y;\mathbb{Z}_2)) + \dim \operatorname{im} \ell + 1 \quad ,$$

and hence

$$\begin{aligned} \dim \operatorname{coker} \tilde{\sigma} &= \dim H_2(Y;\mathbb{Z}_2) + \dim H_1(Y;\mathbb{Z}_2) + 2 - \dim \operatorname{im} \tilde{\sigma} \\ &= (\dim H^1(Y;\mathbb{Z}_2)\oplus\mathbb{Z}_2 - \dim \operatorname{im} \ell) + \dim Sq^1(H^1(Y;\mathbb{Z}_2)) \\ &= \dim \ker \ell \oplus Sq^1(H^1(Y;\mathbb{Z}_2)) \ . \end{aligned}$$

We conclude

$$|\mathrm{coker}\ \tilde{\sigma}| = |\ker \ell|\cdot|Sq^1(H^1(Y;\mathbb{Z}_2))| \ .$$

Therefore, by formulas 18.17 and 18.19,

$$\begin{aligned}|\Omega_3(P^2 x Y;\phi_{Y,3})| &= |\mathrm{im}\ f_3|\cdot|\ker f_3| \\ &= 2|\ker \ell|\cdot|H_1(Y;\mathbb{Z})|\cdot|H_3(Y;\mathbb{Z}_2)|.\end{aligned}$$

This establishes the claim of the theorem.∎

§19. The invariants χ, χ' and χ'', and a spectral sequence involving bordism groups of framefields.

In this section we study manifolds with framefields, their invariants and their bordisms. We have already mentioned and used the invariants χ, χ' and χ'' occasionally (see e.g. 12.5, 12.8, 13.10 or the proof of 16.4), and now we want to discuss them in more detail. Closely related is a system of exact sequences of framefield bordism groups and a resulting spectral sequence. This gives the set-up for a nearly complete calculation of the bordism groups $\Omega_n(k,k)$ for $k = 1,2$ or 3.

If M is a compact smooth n-manifold and $v : M\times\mathbb{R}^k \hookrightarrow TM$ is a k-framefield such that $v(\partial M\times\mathbb{R}^k) \subset T(\partial M)$, let ζ denote a complement of the image bundle of v in TM, and let $S\subset M$ be the zero set of a generic section of ζ which points outward at ∂M. Then S is a k-dimensional manifold, equipped with the stable parallelization

$$\overline{h} : TS \oplus \zeta|S \cong TM|S \cong \underline{\mathbb{R}}^k \oplus \zeta|S$$

and with the inclusion $S\subset M$, or a classifying map of $\zeta|S$, or the constant map, respectively. We obtain the invariants

$$\begin{aligned} &\chi\,(M,v) \in \Omega_k(M;\text{trivial}) \\ (19.1)\qquad &\chi'(M,v) \in \Omega_k(B(S)O(n-k);\text{trivial}), \quad \text{and} \\ &\chi''(M,v) \in \Omega_k(\text{point};\text{trivial})\,. \end{aligned}$$

Here we treat again an oriented and a nonoriented theory simultaneously. If M is oriented, we orient also the (n-k)-plane bundle ζ via the isomorphism $TM \cong \underline{\mathbb{R}}^k\oplus\zeta$, and $\chi'(M,v)$ lies in the framed bordism of $BSO(n-k)$. If orientations are nonexistent or ignored, $\chi'(M,v)$ is, of course, a framed bordism class of $BO(n-k)$.

Remark 19.2. The invariants $\chi(M,v)$ and $\chi''(M,v)$ were also introduced and studied by E.Y.Miller [45].

The following result explains the geometric meaning of the first two invariants (note the analogy to theorems 13.3 and 13.4).

Theorem 19.3. *Let M be a closed smooth n-dimensional manifold with a k-framefield v.*

(i) *Assume that* $n>2k+2$. *Then:* $\chi(M,v) = 0$ *if and only if the k-framefield v can be extended to a (k+1)-framefield* $\mathbb{R}^{k+1} \hookrightarrow TM$ *over all of M.*

(ii) *For* $n \geq k \geq 0$ *we have:* $\chi'(M,v) = 0$ *if and only if* (M,v) *is bordant (in the "fine" bordism group* $\Omega_n(k,k)$, *resp.* $\mathfrak{N}_n(k,k)$) *to a manifold M' with a k-framefield v' which can be extended to a (k+1)-framefield.*

This result follows directly from theorems 3.7 and 19.5 (given below). It is a worthwhile exercise to formulate and prove the corresponding claim in the case when M is only compact, having a possibly nonempty boundary ∂M.

Example 19.4. Given an odd natural number n, write

$$n + 1 = (2b + 1) \cdot 2^c + 4d$$

(where b, c and d are nonnegative integers and $c \leq 3$), and put

$$k = h(n) = 2^c + 8d - 1 .$$

According to the theorem of Adams and Hurwitz-Radon (see [1], or also remark 16.5), there exists a k-framefield v on the sphere S^n, but no (k+1)-framefield. If b is large enough, it follows from theorem 19.3 that $\chi(S^n,v)$, and hence $\chi'(S^n,v)$ and $\chi''(S^n,v)$, are nonzero (since $\Omega_k(S^n;\text{trivial}) \cong \Omega_k(\text{point};\text{trivial})$); actually, $\chi(S^n,v)$ and $\chi''(S^n,v)$ don't even lie in the kernel of incl_* (see theorem 13.11), while $\chi'(S^n,v)$ lies in $\ker \text{incl} \circ i_*$ iff $n \equiv 3(4)$ (see also example 16.22).

In particular, for any $k > 0$, $k \equiv 0,1,3$ or $7(8)$, the stable stem $\pi_k^S \cong \Omega_k(\text{point};\text{trivial})$ is nonzero. Of course, this is well-known; in fact, these are precisely the dimensions where even the image of the $\mathfrak{J}$-homomorphism

$$\mathfrak{J} : \pi_k(SO) \longrightarrow \pi_k^S$$

is nontrivial (see Adams [2]). ∎

As we already remarked before, the invariant $\chi'(M,v)$ induces a homomorphism between bordism groups which we will now fit into an exact sequence. For this purpose, let us first define one further homomorphism

$$\partial : \Omega_k(B(S)O(n-k);\text{trivial}) \longrightarrow \Omega_{n-1}(k,k) \ (\text{resp.}\ \mathfrak{N}_{n-1}(k,k))$$

as follows. Given an element x in the domain of ∂, represent it by a k-manifold L, an (n-k)-plane bundle ξ over L, and a stable parallelization $\bar{h} : TL \oplus \mathbb{R} \xrightarrow{\cong} \mathbb{R}^{k+1}$; then put

$$\partial(x) = St^{-1}[S(\xi), T(S(\xi)) \oplus \mathbb{R} \xrightarrow{T\pi \oplus Id} TL \oplus \mathbb{R} \xrightarrow{\bar{h}} \mathbb{R}^{k+1}]$$

(cf. proposition 7.17), where π denotes the projection of the spherebundle $S(\xi)$ of ξ onto L.

<u>Theorem 19.5.</u> <u>For</u> $n > k$ <u>there are long exact sequences (unbounded at both ends)</u>

$$..\to \Omega_n(k+1,k+1) \xrightarrow{f} \Omega_n(k,k) \xrightarrow{\chi'} \Omega_k(BSO(n-k);\text{trivial}) \xrightarrow{\partial} \Omega_{n-1}(k,k) \to \cdots$$

and

$$..\to \mathfrak{N}_n(k+1,k+1) \xrightarrow{f} \mathfrak{N}_n(k,k) \xrightarrow{\chi'} \Omega_k(BO(n-k);\text{trivial}) \xrightarrow{\partial} \mathfrak{N}_{n-1}(k,k) \to \cdots,$$

<u>where f forgets about the last component vectorfield of a framefield.</u>

Proof. Consider the diagram

$$\begin{array}{ccccc} \cdots\longrightarrow \Omega_n(S(\gamma);-\gamma) & \longrightarrow & \Omega_n(BSO(n-k);-\gamma) & \longrightarrow & \Omega_k(BSO(n-k);\text{trivial})\rightarrow\cdots \\ \uparrow{\scriptstyle St} & & \uparrow{\scriptstyle St} & & \\ \Omega_n(k+1,k+1) & & \Omega_n(k,k) & & \end{array}$$

involving the exact (normal bordism) Gysin sequence of the classifying bundle γ over $BSO(n-k)$ (see Salomonsen [57], 5.3). $\Omega_n(S(\gamma);-\gamma)$ can be interpreted as the bordism group of n-manifolds M, equipped with an oriented (n-k)-plane bundle η over M, an isomorphism $TM\oplus\mathbb{R} \cong \eta\oplus\mathbb{R}^{k+1}$ and a nonzero section of η (or, equivalently, a splitting $\eta = \eta' \oplus \mathbb{R}$). This is the same as the oriented bordism group of pairs $(M, \mathbb{R}^{k+2} \hookrightarrow TM\oplus\mathbb{R})$, which in turn is isomorphic to $\Omega_n(k+1,k+1)$ via the stabilizing map St of proposition 7.17. Similarly, also the second vertical arrow in the diagram above is bijective. Making suitable substitutions, we obtain the required exact sequence in the oriented setting. The unoriented version is deduced in precisely the same way. ∎

Next we compare singularity data of a morphism on a compact manifold and of its restriction to the boundary.

Given vectorbundles α and β of dimensions a and b $(0<a\leq b)$ over a compact n-manifold M, and a monomorphism $u^\partial : \alpha|\partial M \hookrightarrow \beta|\partial M$ over the boundary ∂M of M, let ξ denote a complement of the image of u^∂ in $\beta|\partial M$.

Consider the following diagram

$$\begin{array}{ccc} & & \omega_1(\partial M\times\mathbb{R},\xi) \in \Omega_{n-b+a-1}(\partial M;\xi-T(\partial M)) \\ \omega_a(\alpha,\beta,u^\partial) & & \downarrow i_* \\ \Omega_{n-b+a-1}(RP(\alpha);\gamma\otimes\beta+\mathbb{R}-\gamma\otimes\alpha-TM) & \xrightarrow{\ d\ } & \Omega_{n-b+a-1}(M;\beta+\mathbb{R}-\alpha-TM). \end{array}$$

Here $\omega_a(\alpha,\beta,u^\partial)$ is our singularity obstruction to extending u^∂ to a monomorphism $\alpha \hookrightarrow \beta$ over all of M, while $\omega_1(\partial M \times \mathbb{R},\xi)$ is the singularity obstruction to finding a nonzero section of ξ, or, equivalently, to extending u^∂ to a monomorphism $(\alpha|\partial M) \oplus \underset{\sim}{\mathbb{R}} \hookrightarrow \beta|\partial M$ over ∂M (see theorem 3.7). Also, i_* is induced by the inclusion $\partial M \subset M$ and by an isomorphism $T(\partial M) \oplus \underset{\sim}{\mathbb{R}} \cong TM|\partial M$, where $(0,1)$ always corresponds to an outward pointing tangent vector of M. Finally, define d as follows. Given $x = [L, h : L \longrightarrow RP(\alpha), \bar{h}]$ in the domain of d, $d(x)$ is represented by the double cover manifold $\tilde{L} = S(h^*(\gamma))$ (associated to the pullback of the canonical line bundle γ over $RP(\alpha)$), together with the obvious map

$$\tilde{h} : \tilde{L} \xrightarrow{\ \pi\ } L \xrightarrow{\ h\ } RP(\alpha) \longrightarrow M$$

and the isomorphism

$$T\tilde{L} \oplus \tilde{h}^*(\beta) \cong \pi^*(TL \oplus h^*(\gamma\otimes\beta)) \xrightarrow{\ \pi^*(\bar{h})\ } \pi^*(h^*(\gamma\otimes\alpha\oplus TM)) \cong \tilde{h}^*(\alpha\oplus TM)$$

(here we omit trivial factors; also, we use the obvious trivialization of the line bundle $\pi^*(h^*(\gamma))$ over $\tilde{L}$).

<u>Theorem 19.6.</u> <u>We have</u>

$$i_*(\omega_1(\partial M \times \mathbb{R},\xi)) = -d(\omega_a(\alpha,\beta,u^\partial)) .$$

<u>Proof.</u> According to proposition 2.15, $\omega_a(\alpha,\beta,u^\partial)$ is given by the zero set Z_o of a suitable generic section s_o of the bundle $\underset{\sim}{\mathrm{Hom}}(\gamma,\beta)$ over $RP(\alpha)$ (and by some additional singularity data). Similarly, $i_*(\omega_1(\partial M \times \mathbb{R},\xi))$ may be represented by the zero set $Z_1 \subset RP(\{0\}\oplus\underset{\sim}{\mathbb{R}})|\partial M$ of a suitable section s_1 of $\underset{\sim}{\mathrm{Hom}}(\gamma,\beta)$ over $RP(\alpha\oplus\mathbb{R})|\partial M$. Then the zero set Z of a suitable extension of s_o and s_1 to a section of $\underset{\sim}{\mathrm{Hom}}(\gamma,\beta)$ over all of $RP(\alpha\oplus\underset{\sim}{\mathbb{R}})$ gives rise to the bordism required to establish the identity above.

Indeed, the normal bundle of $RP(\alpha)$ in $RP(\alpha\oplus\underset{\sim}{\mathbb{R}})$ is isomorphic to γ. Hence also a tubular neighborhood T of Z_o in Z is diffeomorphic to the open disk bundle of $\gamma|Z_o$. Thus $Z - T$ is a bordism from $\partial\bar{T} \cong S(\gamma|Z_o) = \tilde{Z}_o$ to Z_1. Moreover, $Z - T$ lies entirely in $RP(\alpha\oplus\mathbb{R}) - RP(\alpha)$ which retracts to $RP(\{0\}\oplus\mathbb{R}) \cong M$; note that the canonical line bundle γ is trivial here. It is now not hard to add the necessary data to make $Z - T$ into a fullfledged normal bordism.∎

Given a vector bundle η over a topological space Y, define an involution ι_η on $\Omega_q(Y;\text{trivial})$ by

(19.7) $$\iota_\eta[N \xrightarrow{g} Y, is] = [N \xrightarrow{g} Y, (TN\oplus\mathbb{R}) \oplus g^*(\eta) \xrightarrow{is\oplus-Id} \mathbb{R}^{q+1} \oplus g^*(\eta)].$$

Corollary 19.8. Let M be a compact n-manifold, $n>k$, with a k-framefield $v : \mathbb{R}^k \hookrightarrow TM$ which restricts to a k-framefield $v| : \mathbb{R}^k|\partial M \hookrightarrow T(\partial M)$ on the boundary. Then we have in $\Omega_k(M;\text{trivial})$

$$i_*(\chi(\partial M, v|)) = (Id - (-1)^k \iota_{TM})(\chi(M,v))$$

and in $\Omega_k(B(S)O(n-k);\text{trivial})$

$$i_*(\chi'(\partial M, v|)) = (Id - \iota_{\gamma_{n-k}})(\chi'(M,v))$$

where i_* is induced by the inclusion $\partial M \subset M$ or $B(S)O(n-k-1) \subset B(S)O(n-k)$.

Proof. Apply theorem 19.6 to $\alpha = \mathbb{R}$, $\beta = \zeta$ (the complement of $v(\mathbb{R}^k)$ in TM) and to the monomorphism $u^\partial : \partial M\times\mathbb{R} \hookrightarrow \beta|\partial M$ given by an outward pointing vector field. Note that then $-i_*(\omega_1(\partial M\times\mathbb{R}, \xi)) = i_*(\chi(\partial M, v|))$, $\omega_a(\alpha,\beta,u^\partial) = \chi(M,v)$, and on $\Omega_{n-b+a-1}(RP(\alpha);\gamma\otimes\beta-\gamma\otimes\alpha-TM) = \Omega_k(M;\text{trivial})$ we have

$$d = Id - \iota_\zeta = Id - (-1)^k \iota_{TM}.$$

The first formula in 19.8 follows, and so does the second equation when we apply the homomorphism induced by a classifying map of ζ. ∎

As an example, consider the case when $k = 0$ (and M is connected). Then the groups in 19.1 can be identified canonically with $\mathbb{Z}$, and $\chi(M,v) = \chi'(M,v) = \chi''(M,v)$ equals the Euler number $\chi(M)$. In this situation corollary 19.8 amounts just to the well-known relation

$$\chi(\partial M) = (1 - (-1)^n)\, \chi(M) .$$

There is no hope, however, that for general k $\chi''(M,v)$ will always be as strong an invariant as e.g. $\chi'(M,v)$. Actually, we claim that already in the case $k = 1$ no analogon of corollary 19.8 can hold for χ''. To see this, we first need to establish the following auxiliary result.

Lemma 19.9. Let η be a t-planebundle over a pathconnected paracompact space Y, and let s be any integer. Identify $\Omega_1(Y;\text{trivial})$ with $\mathbb{Z}_2 \oplus H_1(Y;\mathbb{Z})$ via the augmentation in framed bordism, and the Hurewicz homomorphism. Then we have for all elements $x = (x_1,x_2)$ in this group

$$(\mathrm{Id} + (-1)^s \iota_\eta)(x_1,x_2) = (w_1(\eta)(r(x_2)),\ (1 + (-1)^{s+t})x_2)$$

(where r stands for mod 2 reduction of the coefficients).

Proof. Assume x is represented by a framed circle, together with a map $g : S^1 \longrightarrow Y$. If $g^*(\eta)$ is trivial, then $\iota_\eta(x) = (-1)^t x$, and our claim follows. If $g^*(\eta) = \text{Möbius} \oplus \mathbb{R}^{t-1}$, then we can use the punctured cylinder $(S^1 \times I - \text{point})$ to construct a bordism which shows that $x + (-1)^{t-1}\iota_\eta(x)$ is given by the invariantly framed Lie group S^1 with a constant map, i.e.

$$x + (-1)^{t-1}\iota_\eta(x) = (1,\ 0) = (w_1(\eta)\ (r(x_2)),\ 0)$$

(go back to 19.7). Since the first component of $(\mathrm{Id}+(-1)^s\iota_\eta)(x)$

needs to be computed only modulo 2, and since for the second component we have $x_2 = \mu(x) = (-1)^t \mu(\iota_\eta(x))$, the lemma holds in general.

Example 19.10. If γ denotes the universal bundle over $BO(t)$, $t > 0$, we have for all

$$x = (x_1, x_2) \in \Omega_1(BO(t);\text{trivial}) = \mathbb{Z}_2 \oplus H_1(BO(t);\mathbb{Z}) = \mathbb{Z}_2 \oplus \mathbb{Z}_2$$

that

$$(Id \pm \iota_\gamma)(x_1, x_2) = (x_2, 0) .$$

Now let ζ be the vectorbundle Möbius $\oplus \underset{\sim}{\mathbb{R}}^{n-2} \amalg \underset{\sim}{\mathbb{R}}^{n-1}$, $n \geq 2$, over the disjoint union $S^1 \amalg S^1$, and consider the n-manifold $M = D(\zeta)$ (= disc bundle of ζ), together with the "horizontal" vectorfield v (projecting down to the obvious nontrivial vectorfield on $S^1 \amalg S^1$). The outward pointing vectorfield at $\partial M = S(\zeta)$ can be extended in the obvious way to a vectorfield on all of M which is "vertical" (= tangential to the fiber discs) and whose zero set is the zero section of $D(\zeta)$. Thus clearly

$$\chi'(M,v) = [S^1 \amalg S^1, \zeta, is \amalg is] = (0,1) \in \Omega_1(BO(n-1);\text{trivial}),$$

and therefore we know from corollary 19.8 that

$$\chi'(\partial M, v|) = (1,0) \in \Omega_1(BO(n-2);\text{trivial}) .$$

In particular, $\chi''(M,v) = 0$, but $\chi''(\partial M, v|) = 1$, so there cannot be a formula analoguous to 19.8 holding for the invariant χ''.

Note, by the way, that the calculation above gives a new proof of lemma 12.7. ∎

We want to draw a few more conclusions from theorem 19.6.

Corollary 19.11. For $n > k$ the diagram

$$\begin{array}{ccccc}
\Omega_k(BSO(n-k);\text{trivial}) & \xrightarrow{\partial} & \Omega_{n-1}(k,k) & \xrightarrow{\chi'} & \Omega_k(BSO(n-k-1);\text{trivial}) \\
& \searrow^{\mathrm{Id}-\iota_{\gamma_{n-k}}} & & & \downarrow i_* \\
& & & & \Omega_k(BSO(n-k);\text{trivial}) \,,
\end{array}$$

and its unoriented analogue, commutes. (Here ∂ *and* χ' *come from two different exact sequences in theorem 19.5, and* ι_γ *and* i_* *are defined as in 19.7 and 19.8).*

Proof. Given $x = [L,\xi,\overline{h}]$ in the domain of ∂ (as in the definition preceding theorem 19.5), we obtain its value under $i_* \circ \chi' \circ \partial$ by first forming the spherebundle $S(\xi)$ and equipping it with the "horizontal" stable framefield $\overline{v}$ which projects back to $\overline{h}^{-1}$, then destabilizing the pair $(S(\xi),\overline{v})$ after a suitable fine bordism, and finally applying χ' and the obvious homomorphism i_*. Since χ' is compatible with bordisms and stabilizations, we see that $i_* \circ \chi' \circ \partial(x)$ is just $\omega_1(S(\xi) \times \mathbb{R}, TF(S(\xi))) \in \Omega_k(S(\xi);\text{trivial})$ when mapped to $B(S)O(n-k)$ by a classifying map of $TF(S(\xi)) \oplus \underset{\sim}{\mathbb{R}}$. Here $TF(S(\xi))$ denotes the "vertical" $(n-k-1)$-plane bundle tangential to the fibers of the fibration $\pi : S(\xi) \longrightarrow L$.

Now apply theorem 19.6 to the disc bundle $M=D(\xi)$, to $\alpha = M \times \mathbb{R}$ and $\beta = \pi^*(\xi)$ and to the obvious inclusion

$$u^\partial : S(\xi) \times \mathbb{R} \subset \pi^*(\xi)|S(\xi) = TF(S(\xi)) \oplus \underset{\sim}{\mathbb{R}}$$

given by the ("vertical") outward pointing vector field. We obtain

$$\begin{aligned}
-i_*(\omega_1(S(\xi) \times \mathbb{R}, TF(S(\xi)))) &= d(\omega_1(M \times \mathbb{R}, \pi^*(\xi), u^\partial)) \\
&= (\mathrm{Id}-\iota_{\pi^*(\xi)})[L \subset D(\xi), \overline{h}] \,.
\end{aligned}$$

If we apply a classifying map of $\pi^*(\xi)$, the lefthand term becomes $i_* \circ \chi' \circ \partial(x)$, while we get the expression $(\mathrm{Id}-\iota_{\gamma_{n-k}})(x)$ on the right hand side.■

As a last application of theorem 19.6 we fill a gap which was left in the proof of theorem 13.11.

Corollary 19.12. Let $1 \leq k < n$ and adopt the notation of 13.11. Then for all q the diagram

$$\begin{array}{ccc} \pi_q(V_{n,k}) & \xrightarrow{\partial} & \pi_{q-1}(S^{n-k-1}) \\ \downarrow \sigma & & \downarrow \sigma \\ \Omega_{q-n+k}(P^{k-1};(n-k)\lambda) & \xrightarrow{d} & \Omega_{q-n+k}(\text{point};\text{trivial}) \end{array}$$

commutes up to sign. Furthermore, for any two k-framefields v_o and v_1 on a closed smooth n-manifold M we have

$$(-1)^{k+1} d(d_k(v_o,v_1)) = \chi(M,v_1) - \chi(M,v_o) .$$

Proof. Given an element $x \in \pi_q(V_{n,k})$, we may represent it by a monomorphism

$$u^\partial : S^q \times \mathbb{R}^k \hookrightarrow S^q \times (\mathbb{R}^k \times \mathbb{R}^{n-k})$$

which restricts to the obvious inclusion over the upper halfsphere $S^q_+ = \{(x_1,..x_{q+1}) \in S^q | x_{q+1} \geq 0\}$. Let ξ denote the orthogonal complement of the image of u^∂ so that e.g. $\xi | S^q_+ = S^q_+ \times (\{0\} \times \mathbb{R}^{n-k})$. Pick a generic section of ξ whose singularity lies entirely in the interior of S^q_+; its singularity data yield $\pm\, \sigma \circ \partial(x)$ (see also the first argument in the proof of theorem 16.9).

Applying theorem 19.6 to $M=D^{q+1}$, $\alpha=D^{q+1} \times \mathbb{R}^k$, $\beta=D^{q+1} \times \mathbb{R}^n$ and u^∂, we obtain

$$\begin{aligned} \pm\, d \circ \sigma(x) &= d(\omega_k(D^{q+1} \times \mathbb{R}^k,\, D^{q+1} \times \mathbb{R}^n,\, u^\partial)) \\ &= -i_*(\omega_1(S^q \times \mathbb{R},\, \xi)) \\ &= \pm\, \sigma \circ \partial(x) . \end{aligned}$$

This proves the first part of our corollary.

If v_o and v_1 are k-framefields on any closed manifold M, we can apply theorem 19.6 also to the bundles $\underset{\sim}{\alpha}=\mathbb{R}^k$ and $\beta=\pi_1^*(TM)$ over $M\times I$ and to the monomorphism u^∂ given by v_j on $M \times \{j\}$, $j = 0,1$. We get

$$\begin{aligned} -d\,(d_k(v_o,v_1)) &= -\pi_{1*}d\,(\omega_k(\underset{\sim}{\mathbb{R}^k}, \pi_1^*(TM), u^\partial)) \\ &= \pi_{1*}i_*\,(\omega_1(\partial(M\times I)\times\mathbb{R}, (\mathrm{im}\,u^\partial)^\perp)) \\ &= (-1)^k\,(\chi(M,v_1) - \chi(M,v_o)) \end{aligned}$$

as desired.∎

Since we have now the full power of theorem 13.11 at our disposal, we can draw a few conclusions.

Theorem 19.13. Let M be a closed n-dimensional manifold admitting a k-framefield $v_o : M \times \mathbb{R}^k \hookrightarrow TM$.

If $n > 2k > 0$, we can use the notation of 13.11 to describe the set $\{\chi(M,v)\}$ of values of $\chi(M,-)$ (v runs through all k-framefields on M) as follows.

$$\{\chi(M,v)\} = \chi(M,v_o) + d\,(\Omega_k(P^{k-1}\times M;\phi_{M,k})) = \mathrm{incl}_*^{-1}((-1)^k\omega_{k+1}(M)),$$

and similarly

$$\{\chi'(M,v)\} = \chi'(M,v_o)+\tau'_*\circ d(\Omega_k(P^{k-1}\times M;\phi_{M,k}))\subset(\mathrm{incl}\circ i)_*^{-1}((-1)^k\omega'_{k+1}(M))$$

(where $\tau' : M \longrightarrow B(S)O(n-k)$ denotes a classifying map of a complement of $v_o(M\times\mathbb{R}^k)$ in TM), and

$$\{\chi''(M,v)\}=\chi''(M,v_o)+\mathrm{const}_*\circ d(\Omega_k(P^{k-1}\times M;\phi_{M,k}))\supset\chi''(M,v_o)+d(\Omega_k(P^{k-1};(n-k)\lambda))$$

(these last two sets are actually equal if M is k-connected).

If $n > 2k+2 > 2$, the invariant $\chi''(M,-)$ detects at least as many different homotopy classes (actually: fine bordism classes) of k-framefields on M as there are elements in $\partial(\pi_n(V_{n,k}))$. For $n \equiv k(2)$

this group contains e.g. the odd torsion of $\pi_{n-1}(S^{n-k-1}) \cong \pi_k^S$.

In particular, if $n \geq 9$ and $(n,k) = (4q+1,2)$ or $(4q+2,2)$ or $(4q+2,3)$, $q \in \mathbb{Z}$, then χ'' detects at least two different k-framefields on M. If n is odd, $n \geq 9$, then M has at least twelve 3-framefields (if any), detected by χ''. If n is odd, and $n \geq 17$ (resp. $n \geq 25$), then M has at least fifteen 7-framefields (resp. at least sixty-three 11-framefields) if any, detected by χ''.

Proof. According to theorem 13.11, we have

$$\{\chi(M,v)\} \subset \chi(M,v_0)+d(\Omega_k(P^{k-1}\times M;\phi_{M,k})) \subset \mathrm{incl}_*^{-1}((-1)^k\omega_{k+1}(M)) .$$

On the other hand, given an element x in the set to the right, $x - \chi(M,v_0) = d(y)$ for some $y \in \Omega_k(P^{k-1}\times M;\phi_{M,k})$. If $n > 2k > 0$, there is a k-framefield v on M such that $d_k(v_0,v) = (-1)^{k+1}y$; indeed, represent $(-1)^{k+1}y$ by a triple $(\mathcal{S},\tilde{G},\overline{G})$ and apply the proof of theorem 3.1 to the monomorphism v_0 and to the zero bordism $(\mathcal{S},\tilde{G},\overline{G})$ of its (empty) singularity data. We conclude that $d(y) = \chi(M,v) - \chi(M,v_0)$, and hence $x = \chi(M,v)$. The whole first claim of our theorem follows; just note that $i : B(S)O(n-k) \longrightarrow B(S)O(n)$ induces an isomorphism on k-dimensional bordism (since we assume $n - k > k$), and that $\Theta_M(\Omega_k(P^{k-1};(n-k)\lambda)) = \Omega_k(P^{k-1}\times M;\phi_{M,k})$ if M is k-connected.

If $n > 2k+2 > 2$, then all four homomorphisms σ in 13.11 are bijective. Hence

$$\{\chi''(M,v)\} \supset \chi''(M,v_0) + \sigma(\partial(\pi_n(V_{n,k})))$$

contains at least as many elements as $\partial(\pi_n(V_{n,k}))$. The remark about odd torsion follows from corollary 17.12 (put $\ell = 1$).

Finally, consider the following version of the top exact sequence in 13.11:

$$\pi_n(V_{n,k}) \xrightarrow{\partial} \pi_{n-1}(S^{n-k-1}) \xrightarrow{\mathrm{incl}_*} \pi_{n-1}(V_{n,k+1}) \xrightarrow{\mathrm{proj}_*} \mathrm{proj}_*(\pi_{n-1}(V_{n,k+1})) \rightarrow 0.$$

For low k, the values for the two groups to the right are given in 16.15. We deduce the following table for the group $\partial(\pi_n(V_{n,k})) = \ker \mathrm{incl}_*$.

(19.14)

$n > 2k+2$	$k = 1$ (or 4 or 5)	$k = 2$	$k = 3$
$n \equiv 1(4)$	0	$\mathbb{Z}_2$	$\mathbb{Z}_{12}$
$n \equiv 2(4)$	0	$\mathbb{Z}_2$	$\mathbb{Z}_2$
$n \equiv 3(4)$	0	0	$\mathbb{Z}_{12}$
$n \equiv 4(4)$	0	0	0

This and the wellknown results $\pi_7^S \cong \mathbb{Z}_{240}$ and $\pi_{11}^S \cong \mathbb{Z}_{504}$ (see the tables of Toda [66], pp. 186-188) imply the numerical illustrations in our theorem.

<u>Remark 19.15.</u> Starting from a k-framefield v on a closed n-manifold M, $n > k \geq 1$, we can produce many other k-framefields on M (taking many values under χ'') by the following construction (see Atiyah-Dupont [6], p.31).

Given an element $y = [w:(D^n, S^{n-1}) \longrightarrow (V_{n,k},*)]$ of $\pi_n(V_{n,k})$ and an embedded n-disk $D \subset M$, deform the k-framefield v until it restricts to the basepoint $* \in V_{n,k}$ everywhere in D, and replace $v|D$ by w to obtain a new k-framefield v^y. If M is connected and oriented (and the embedding $D \subset M$ is orientation preserving), and if $n > k+1$, then the homotopy class of v^y depends only on the homotopy class of v and on y.

Clearly we have always

$$d_k(v^y, v) = \Theta_M \bullet \sigma\,(y) \;,$$

and therefore

$$\chi\,(M,v^y) \;=\; \chi\,(M,v) + e\cdot\Theta_M \circ \sigma(\partial(y)),$$

$$i_*(\chi'(M,v^y)) = i_*(\chi'(M,v)) + e\cdot\Theta^{(or)}(\partial(y)), \quad \text{and}$$

$$\chi''(M,v^y) \;=\; \chi''(M,v) + e\cdot\sigma(\partial(y)),$$

where $e = \pm 1$ is a fixed sign.

In particular, if we fix a k-framefield v_o on M, we obtain the relation

$$\{\chi''(M,v)\} \;\supset\; \chi''(M,v_o) + \sigma(\partial(\pi_n(V_{n,k})))$$

valid under the very general assumption that $n > k \geq 1$. ■

Example 19.16. Let M be a connected closed n-manifold, $n \geq 3$, with vanishing Euler number $\chi(M)$, and identify $\Omega_1(M;\text{trivial})$ with $\mathbb{Z}_2 \oplus H_1(M;\mathbb{Z})$ (and $\Omega_1(B(S)O(n-1);\text{trivial})$ with $\mathbb{Z}_2 \oplus H_1(B(S)O(n-1);\mathbb{Z})$) via the isomorphism (const_*,μ) (cf. 19.9). For $k = 1$ the value sets of $\chi(M,-)$, $\chi'(M,-)$ and $\chi''(M,-)$ (see theorem 19.13) are given by the following table.

(19.17)

$n \geq 3$	$\{\chi(M,v)\}$ $\mathbb{Z}_2 \oplus H_1(M;\mathbb{Z})$	$\{\chi'(M,v)\}$ $\subset \mathbb{Z}_2 \oplus H_1(B(S)O(n-1);\mathbb{Z})$	$\{\chi''(M,v)\}$ $\subset \mathbb{Z}_2$
$w_1(M)=0$; $n \equiv 1(4)$	$(k(M), r^{-1}\{PD\,w_{n-1}(M)\})$	$(k(M), 0)$	$k(M)$
$w_1(M)=0$; $n \equiv 2(4)$	$(0, 0)$	$(0, 0)$	0
$w_1(M)=0$; $n \equiv 3(4)$	$(0, 2\cdot H_1(M;\mathbb{Z}))$	$(0, 0)$	0
$w_1(M)=0$; $n \equiv 4(4)$	$((\frac{\text{sign}(M)}{2})_2, 0)$	$((\frac{\text{sign}(M)}{2})_2, 0)$	$(\frac{\text{sign}(M)}{2})_2$
$w_1(M) \neq 0$; n even	$(\mathbb{Z}_2, -PD(W_{n-1}(M)))$		
$w_1(M) \neq 0$, but $w_1(M)^2=0$; n odd	A	$(\mathbb{Z}_2, 0)$	$\mathbb{Z}_2$
$w_1(M)^2 \neq 0$; n odd	$(\mathbb{Z}_2, r^{-1}\{PD\,W_{n-1}(M)\})$		

Here $\mathcal{k}(M)$ and $\mathrm{sign}(M)$ denote the real Kervaire semicharacteristic and the signature of M, respectively (see e.g. 12.1). $PD(W_{n-1}(M))$ is the Poincaré dual of the (integer, twisted) Stiefel-Whitney class of M. A is a subset of $\mathbb{Z}_2 \oplus H_1(M;\mathbb{Z})$ which projects bijectively onto the subset $r^{-1}\{PD(w_{n-1}(M))\}$ of $H_1(M;\mathbb{Z})$, where $r : H_1(M;\mathbb{Z}) \longrightarrow H_1(M;\mathbb{Z}_2)$ is reduction modulo 2. Therefore the sets $r^{-1}\{PD(w_{n-1}(M))\}$, $2\cdot H_1(M;\mathbb{Z}) = \ker r$ and A all have the same cardinality. Thus if n is odd and $w_1(M)^2=0$ (resp. $w_1(M)^2 \neq 0$) the invariant $\chi(M,-)$ takes precisely as many (resp. twice as many) different values as there are elements in $2\cdot H_1(M;\mathbb{Z})$.

For the proof of all these facts, consider the relevant exact sequence in 13.11 for $k=1$. We get the commuting diagram

$$(19.18)\qquad \begin{array}{ccccc} \ldots \longrightarrow \Omega_1(M;\text{trivial}) & \xrightarrow{d\,=\,\mathrm{Id}-\iota_{TM}} & \Omega_1(M;\text{trivial}) & \xrightarrow{\mathrm{incl}_*} & \Omega_1(P^1 \times M;\phi_{M,2}) \\ \| & & \| & & \\ \mathbb{Z}_2 \oplus H_1(M;\mathbb{Z}) & \xrightarrow{d} & \mathbb{Z}_2 \oplus H_1(M;\mathbb{Z}) & & \\ (x_1, x_2) & \longrightarrow & (w_1(M)(r(x_2)), (1-(-1)^n)x_2) & & \end{array}$$

(see also lemma 19.9). The homomorphism incl_* fits also into the following commuting diagram

$$(19.19)\qquad \begin{array}{ccccccc} 0 \longrightarrow \mathbb{Z}_2 & \overset{\mathrm{const}_*}{\rightleftarrows} & \Omega_1(M;\text{trivial}) & \xrightarrow{\mu} & H_1(M;\mathbb{Z}) & \longrightarrow & 0 \\ \| & & \downarrow \mathrm{incl}_* & & \downarrow j & & \\ \mathbb{Z}_2 & \xrightarrow{\delta_1} & \Omega_1(P^1 \times M;\phi_{M,2}) & \twoheadrightarrow & \begin{cases} \mathbb{Z} \oplus H_1(M;\mathbb{Z}) & n \text{ even} \\ H_1(M;\mathbb{Z}_2) & n \text{ odd}\,. \end{cases} & & \end{array}$$

Here the horizontal exact sequences are given by 9.3 and 14.2; in particular, the lower homomorphism δ_1 is injective iff $w_1(M) = 0$ (when n is even), resp. iff $w_1(M)^2 = 0$ (when n is odd). The vertical arrows are induced by $\mathrm{incl}=(\mathrm{const},\mathrm{id})$; note that $\ker j = (1-(-1)^n)\cdot H_1(M;\mathbb{Z})$ since $j = (0,\mathrm{Id})$ or $j = r$ according to whether

n is even or odd.

Given a vectorfield v_o on M, $\mathrm{incl}_*(\chi(M,v_o)) = -\omega_2(M)$, and hence j maps $-\mu(\chi(M,v_o))$ to the Poincaré-dual of the $(n-1)^{st}$ (integer or mod 2) Stiefel-Whitney class of M (see § 14). Conclude also from theorem 19.13 and diagram 19.18 that

$$\{\chi(M,v)\} = \chi(M,v_o) + d(\{0\} \oplus H_1(M;\mathbb{Z})) .$$

If n is even, then $\chi(M,v_o) = (\chi''(M,v_o),-PD(W_{n-1}(M)))$, and $d(\{0\}\oplus H_1(M;\mathbb{Z}))$ is 0 or $\mathbb{Z}_2\oplus\{0\}$ according to whether $w_1(M)$ vanishes or not. Moreover, if $w_1(M) = 0$, then it follows from work of Massey, Atiyah, Frank and Thomas (see e.g. [65], p. 650) that $W_{n-1}(M) = 0$ and $\chi''(M,v_o) = (\frac{\mathrm{sign}(M)}{2})_2$. To get the last equation for $n = 4q$ (and even when $n = 4$), observe that the obvious homomorphism

$$(\frac{\mathrm{sign}}{2})_2 \;:\; \Omega_n(1,\, 1) \longrightarrow \mathbb{Z}_2$$

vanishes on $\ker \chi'' = \ker \chi' = f(\Omega_n(2,2))$ (see Atiyah [4], theorem 1.1, and also our theorem 19.5). On the other hand, attaching enough handles to $\mathbb{C}P(2q)\#\mathbb{C}P(2q)$ we obtain an oriented n-manifold N such that $\mathrm{sign}(N) = 2$ and $\chi(N) = 0$. Thus the homomorphism above is non-trivial and agrees with χ''.

Next we consider the case when n is odd. If also $w_1(M) = 0$, then $d(\{0\} \oplus H_1(M;\mathbb{Z})) = \{0\} \oplus 2\cdot H_1(M;\mathbb{Z})$, and

$$\{\chi(M,v)\} = (\chi''(M,v_o),\, r^{-1}\{PD(w_{n-1}(M))\}) ;$$

our claims concerning $\chi''(M,v_o)$ and $w_{n-1}(M)$ follow again e.g. from [65], p. 650. E.g. if $n \equiv 1(4)$, conclude as above that the composed homomorphism

$$\Omega_n(1,\, 1) \xrightarrow{\;f\;} \Omega_n(0,\, 0) \xrightarrow{\;k\;} \mathbb{Z}_2$$

(cf. 12.1) vanishes on $\ker \chi'' = \ker \chi' = f(\Omega_n(2,2))$, is nontrivial

($\mathfrak{k}(S^n) = 1!$), and hence coincides with χ''.

Now consider the exact sequences

$$\begin{array}{ccccccc}
\cdots \longrightarrow H_2(M;\mathbb{Z}_2) & \xrightarrow{\ \delta\ } & H_1(M;\mathbb{Z}) & \xrightarrow{\ \cdot 2\ } & H_1(M;\mathbb{Z}) & \xrightarrow{\ r\ } & H_1(M;\mathbb{Z}_2) \\
\| & & \downarrow{\scriptstyle r} & & \downarrow & & \| \\
\longrightarrow H_2(M;\mathbb{Z}_2) & \xrightarrow{\ \delta\ } & H_1(M;\mathbb{Z}_2) & \longrightarrow & H_1(M;\mathbb{Z}_4) & \longrightarrow & H_1(M;\mathbb{Z}_2)
\end{array} ,$$

where δ denotes Bockstein homomorphisms. The lower one is dual to $H^2(M;\mathbb{Z}_2) \xleftarrow{\ Sq^1\ } H^1(M;\mathbb{Z}_2)$. Thus $w_1(M)^2 = Sq^1(w_1(M))$ is nonzero iff $w_1(M)$ is nontrivial on $\delta(H_2(M;\mathbb{Z}_2))$, i.e. if there exists an element $x_2 \in H_1(M;\mathbb{Z})$ such that $w_1(M)(r(x_2)) = 1$ and $2x_2 = 0$; in this case the image of d is $\mathbb{Z}_2 \oplus 2\cdot H_1(M;\mathbb{Z})$ since

$$d(x_1, x_2) = (w_1(M)\,(r(x_2)),\ 2\,x_2)$$

for odd n. If no such element exists, then the image of d maps bijectively onto $2\cdot H_1(M;\mathbb{Z})$ under the second projection on $\mathbb{Z}_2 \oplus H_1(M;\mathbb{Z})$. This completes the calculations for the first column in the table 19.17, provided we put

$$A = \{(\chi''(M,v_o) + w_1(M)(r(x_2)),\ \mu(\chi(M,v_o)) + 2x_2) \mid x_2 \in H_1(M;\mathbb{Z})\}.$$

Finally consider the second component $\mu(\chi'(M,v)) \in H_1(BO(n-1);\mathbb{Z}) = \mathbb{Z}_2$ of $\chi'(M,v)$. According to fact 9.11 and lemma 12.4 we have

$$\mu(\chi'(M,v)) = w_1(M)\,\dot{w}_{n-1}(M)[M] = 0$$

since the mod 2 Euler number $w_n(M)[M]$ vanishes. This finishes the proof of 19.17.∎

Before we have a glimpse at the case $k=2$, let us first establish a few auxiliary results.

<u>Lemma 19.20.</u> <u>Let X be a topological space as in 9.3. Then we have</u>

an isomorphism

$$\Omega_2(X;\text{trivial}) \xrightarrow{(\text{const}_*,\mu,h)} \Omega_2(\text{point};\text{trivial}) \oplus H_2(X;\mathbb{Z}) \oplus H_1(X;\mathbb{Z}_2) .$$

Here the homomorphism h *depends on the arbitrary choice of vector-bundles* $\eta_1,\ldots,\eta_r$ *on* X *such that* $w_1(\eta_1),\ldots,w_1(\eta_r)$ *forms a basis of* $H^1(X;\mathbb{Z}_2)$; *for* $j = 1,\ldots,r$ *the diagram*

$$\begin{array}{ccc} & \Omega_2(X;\ \text{trivial}) & \\ {}^{h}\swarrow & & \searrow^{\text{const}_*\circ(\mathrm{Id}\pm\iota_{\eta_j})} \\ H_1(X;\mathbb{Z}_2) & \xrightarrow{w_1(\eta_j)} & \mathbb{Z}_2 = \Omega_2(\text{point};\text{trivial}) \end{array}$$

(cf. 19.7) commutes.

Proof. In the diagram

(19.21)

$$\begin{array}{ccccccccc} & & 0 & & & & & & \\ & & \downarrow & & & & & & \\ & & \mathbb{Z}_2 & \overset{\text{const}_*}{\longleftarrow\!-\,-} & & & & & \\ & & \downarrow{\scriptstyle \delta_1'} & & & & & & \\ 0 & \longrightarrow & \Omega_1(X\times BO(2);\Gamma) & \xrightarrow{\delta_2} & \Omega_2(X;\text{trivial}) & \xrightarrow{\mu} & H_2(X;\mathbb{Z}) & \longrightarrow & 0 \\ & & \downarrow{\scriptstyle \text{proj}_*\cdot\mu} & \overset{h}{\swarrow\!-\,-} & \downarrow{\scriptstyle \text{const}_*\circ(\mathrm{Id}\pm\iota_\eta)} & & & & \\ & & H_1(X;\mathbb{Z}_2) & \xrightarrow{w_1(\eta)} & \mathbb{Z}_2 & & & & \\ & & \downarrow & & & & & & \\ & & 0 & & & & & & \end{array}$$

the cross of exact sequences is given by theorem 9.3 and fact 9.4. If η is any vector bundle on X, the square commutes; indeed, note that the obvious map

$$i_* : \Omega_1(X;\text{trivial}) \longrightarrow \Omega_1(X\times BO(2);\Gamma)$$

is onto and that for all $x\in\Omega_1(X;\text{trivial})$ we have

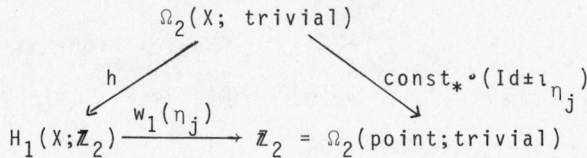

$$\mathrm{const}_*\circ(\mathrm{Id}\pm\iota_\eta)\circ\delta_2\circ i_*(x) = \mathrm{const}_*\circ(\mathrm{Id}\pm\iota_\eta)([S^1, \text{parallelization}]\cdot x)$$

$$(19.22) \qquad = \mathrm{const}_*([S^1, \text{parallelization}]\circ(\mathrm{Id}\pm\iota_\eta)(x))$$

$$= w_1(\eta)(\mathrm{proj}_*\circ\mu\circ i_*(x))$$

(use lemma 19.9 and the fact that multiplication with $[S^1$,parallelization$]$ gives an isomorphism from Ω_1(point;trivial) to Ω_2(point;trivial)).

If $\eta_1,\ldots\eta_r$ are vector bundles as above, their first Stiefel-Whitney classes allow us to identify $H_1(X;\mathbb{Z}_2)$ with $(\mathbb{Z}_2)^r$; define

$$h = (\mathrm{const}_*\circ(\mathrm{Id} + \iota_{\eta_1}),\ldots, \mathrm{const}_*\circ(\mathrm{Id} + \iota_{\eta_r})) .$$

Furthermore, observe that $\mathrm{const}_*\circ\delta_2\circ\delta_1' = \mathrm{Id}$, and therefore $\mathbb{Z}_2$ splits off both in $\Omega_1(X\times BO(2);\Gamma)$ and in $\Omega_2(X;\text{trivial})$. Thus (const_*,h) gives rise to a left inverse of δ_2.

Corollary 19.23. For $n \geq 2$ there are isomorphisms

$$(\mathrm{const}_*,\begin{Bmatrix} c_1 \\ w_2(\gamma_n)\end{Bmatrix}) : \Omega_2(BSO(n);\text{trivial}) \xrightarrow{\cong} \mathbb{Z}_2 \oplus \begin{cases}\mathbb{Z} & n = 2 \\ \mathbb{Z}_2 & n > 2\end{cases}$$

and

$$(\mathrm{const}_*,w_2(\gamma_n),\mathrm{const}_*\circ(\mathrm{Id}+\iota_{\gamma_n})) : \Omega_2(BO(n);\text{trivial})\xrightarrow{\cong}\mathbb{Z}_2 \oplus \mathbb{Z}_2 \oplus \mathbb{Z}_2 .$$

The corresponding generators of $\Omega_2(B(S)O(n);\text{trivial})$ are represented by (i) the invariantly parallelized torus $S^1 \times S^1$ with a constant map, (ii) $\mathbb{C}P(1) = S^2 = \partial D^3$ with the boundary framing and a classifying map of $\lambda_{\mathbb{C}} \oplus \underset{\sim}{\mathbb{R}}^{n-2}$, and, in the unoriented theory in addition, (iii) the torus $\partial(D^2\times S^1)$ with the boundary framing, and the first projection composed with the non-nulhomotopic map $S^1 \longrightarrow BO(n)$.

Furthermore, $\mathrm{Id} - \iota_{\gamma_n} \equiv 0$ on $\Omega_2(BSO(n);\text{trivial})$, while in the framed bordism of $BO(n)$ we have

$$(Id \pm \iota_{\gamma_n})(\mathbb{Z}_2 \oplus \mathbb{Z}_2 \oplus \{0\}) = \{0\} \quad \underline{\text{and}}$$

$$(Id \pm \iota_{\gamma_n})(0, 0, 1) = (1, 0, 0) .$$

Proof. Clearly, $H_2(BSO(2);\mathbb{Z}) = H_2(\mathbb{C}P(\infty);\mathbb{Z}) \cong \mathbb{Z}$. In all other cases $H_2(B(S)O(n);\mathbb{Z})$ consists only of elements of order 2 (see Borel-Hirzebruch [9], p. 373, or also [13], pp.23-24), and an easy application of the universal coefficient theorem leads to the composed isomorphism

$$H_2(B(S)O(n);\mathbb{Z}) \longrightarrow H_2(B(S)O(n);\mathbb{Z}_2) \xrightarrow{w_2(\gamma_n)} \mathbb{Z}_2 .$$

Our first claim follows now from lemma 19.20.

To complete the proof it suffices to note that $\iota_{\gamma_n}[S^2,\lambda_{\mathbb{C}} \oplus \underset{\sim}{\mathbb{R}}^{n-2}] = [S^2,\lambda_{\mathbb{C}} \oplus \mathbb{R}^{n-2}]$ since here ι_{γ_n} amounts to the action of an element in $\pi_2(SO(N)) = 0$; moreover, apply the last two equations in (19.22) to $X = BO(n)$, $\eta = \gamma_n$ and $x = [S^1 = \partial D^2, \text{Möbius} \oplus \underset{\sim}{\mathbb{R}}^{n-1}]$ in order to make the necessary calculations for the third generator. ∎

We are now prepared to study 2-fields.

Example 19.24. Let M be a closed connected manifold of dimension $n \geq 5$ such that $\mathrm{span}(M) \geq 2$. Using the isomorphisms of corollary 19.23 we can tabulate the value sets of $\chi'(M,-)$ and $\chi''(M,-)$ (see theorem 19.13) for $k=2$ as follows.

$n \geq 5$	$\{\chi'(M, v)\}$ $\subset \Omega_2(B(S)O(n-2);triv) = \mathbb{Z}_2 \oplus \mathbb{Z}_2((\oplus \mathbb{Z}_2))$	$\{\chi''(M, v)\}$ $\subset \Omega_2(point;triv) = \mathbb{Z}_2$
$w_1(M)=0$; $n \equiv 1(4)$	$(\mathbb{Z}_2, w_2(M)w_{n-2}(M)[M])$	$\mathbb{Z}_2$
$w_1(M)=0$; $n \equiv 2(4)$	$(\mathbb{Z}_2, 0)$	$\mathbb{Z}_2$
$w_1(M)=0$; $n \equiv 3(4)$	$(0, 0)$	0
$w_1(M)=0$; $n \equiv 4(4)$	$((\frac{\mathrm{sign}(M)}{4})_2, 0)$	$(\frac{\mathrm{sign}(M)}{4})_2$
$w_1(M) \neq 0$	$(\mathbb{Z}_2, w_2(M)w_{n-2}(M)[M], 0)$	$\mathbb{Z}_2$

Recall here the classical result that the signature of M is divisible by 4, so $(\frac{\text{sign}(M)}{4})_2 \in \mathbb{Z}_2$ is welldefined.

For the proof of the tabulated facts, compare the diagram (see 9.3 or 19.21),

$$
(19.25)\qquad
\begin{array}{ccccccccc}
 & & 0 & & & & & & \\
 & & \downarrow & & & & & & \\
 & & \mathbb{Z}_2 & \xleftarrow{\quad\text{const}_*\quad} & & & & & \\
 & & \downarrow{\scriptstyle \delta_1'} & & \nwarrow & & & & \\
0 & \longrightarrow & \Omega_1(M\times BO(2);\Gamma) & \xrightarrow{\delta_2} & \Omega_2(M;\text{trivial}) & \xrightarrow{\mu} & H_2(M;\mathbb{Z}) & \longrightarrow & 0 \\
 & & \downarrow & & & & & & \\
 & & H_1(M;\mathbb{Z}_2) & & & & & & \\
 & & \downarrow & & & & & & \\
 & & 0 & & & & & &
\end{array}
$$

with the diagram 14.7 (for $Y = M$) via the maps induced by the inclusion

$$\text{incl} \;:\; M \subset P^2 \times M \;.$$

If $n \equiv 1$ or $2(4)$ or $w_1(M) \neq 0$, then $\delta_2 \circ \delta_1'(1) = 0$ in 14.7 and hence the element $\delta_2 \circ \delta_1'(1) \in \Omega_2(M;\text{trivial})$ lies in $\ker(\text{incl}_*) = d(\Omega_2(P^1 \times M;\phi_{M,2}))$; according to theorem 19.13

$$\{\chi''(M,v)\} = \chi''(M,v_o) + \text{const}_* \circ d(\Omega_2(P^1 \times M;\phi_{M,2})) = \mathbb{Z}_2$$

in this case.

If $n \equiv 3(4)$ and $w_1(M) = 0$, then $w_2(\phi_{M,3}) = 0$ in 14.7, and it turns out that incl_* is injective; hence we see that even the strong equation

$$(19.26)\qquad \{\chi(M,v)\} = \text{incl}_*^{-1}(\omega_3(M)) = \text{incl}_*^{-1}(0) = 0$$

holds here (use also the work of Atiyah and Dupont, see e.g. remark

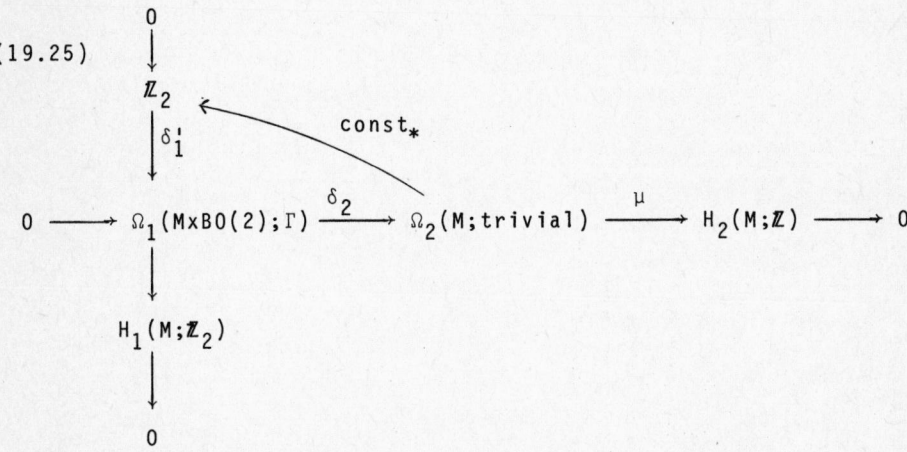

14.18).

Next consider the other components of $\chi'(M,v)$. It follows easily from fact 9.11 that the second component is given by

$$\begin{aligned} w_2(\gamma_{n-2})(\chi'(M,v)) &= w_2(\zeta)|S[S] \\ &= w_2(M)w_{n-2}(M)[M] \end{aligned}$$

(see also 19.1); if M is orientable and $n \equiv 0(2)$, then this characteristic number is a multiple of the mod 2 Euler number $w_n(M)[M]$ which vanishes here (see 16.7). Furthermore recall from 19.5 and 19.11 that

$$\begin{aligned} \mathrm{const}_* \circ (\mathrm{Id} + \iota_{\gamma_{n-2}})(\chi'(M,v)) &= \mathrm{const}_* \circ \chi' \circ \partial(\chi'(M,v)) \\ &= \mathrm{const}_* \circ \chi'(0) \\ &= 0 , \end{aligned}$$

so that in the unoriented theory the third component of $\chi'(M,v)$ vanishes always.

It remains to determine $\{\chi''(M,v)\}$ when $w_1(M) = 0$ and $n = 4q$. To this purpose consider the homomorphism

$$\left(\frac{\mathrm{sign}}{4}\right)_2 \ : \ \Omega_{4q}(2,2) \longrightarrow \mathbb{Z}_2$$

which is nontrivial (attach enough handles to $\mathbb{C}P(2q) \# \mathbb{C}P(2q) \# \mathbb{C}P(2q) \# \# \mathbb{C}P(2q)$ to kill the Euler number, and apply 19.17 to obtain a 2-frame-field) and which vanishes on

$$f(\Omega_{4q}(3,3)) = \ker \chi' = \ker(\chi'' : \Omega_{4q}(2,2) \longrightarrow \mathbb{Z}_2)$$

(use the criteria of Mayer, Frank and Atiyah, see e.g. [4], theorem 3.3); clearly, $\left(\frac{\mathrm{sign}}{4}\right)_2$ and χ'' must coincide. ■

Instead of calculating also the sets $\{\chi(M,v)\}$ completely for $k = 2$, we content ourselves to do so in a few relevant cases.

Theorem 19.27. Let M be a closed connected orientable manifold of dimension $n \geq 6$.

If $n \equiv 2(4)$, every nowhere vanishing vectorfield on M extends to a 2-framefield.

If $n \equiv 3(4)$, every 2-framefield on M extends to a 3-framefield. Moreover, we have: already every nowhere vanishing vectorfield extends to a 3-framefield if and only if $2 \cdot H_1(M;\mathbb{Z}) = 0$.

If $n \equiv 4(4)$, $\operatorname{span}(M) \geq 3$ and $2 \cdot H_2(M;\mathbb{Z}) = 0$, then we have already that every nowhere vanishing vectorfield and also every 2-framefield on M can be extended to a 3-framefield.

Proof. This follows from theorem 19.3, from an inspection of table 19.17, from the result 19.26 and from its analogon for $n \equiv 0(4)$ (we use the condition $2 \cdot H_2(M;\mathbb{Z})$ to obtain the injectivity of incl_*).∎

Next we study bordism groups of manifolds with framefields in some more detail. In the oriented theory (and, analogously, in the unoriented theory) the long exact sequences of theorem 19.5 fit together into an exact triangle described by the following diagram.

(19.28)

$$
\begin{array}{ccccccccc}
 & & 0 & & & & & & \\
 & & \downarrow & & & & & & \\
\Omega_n^{fr}(BSO(1)) & \longrightarrow & \Omega_n(n,n) & (\longrightarrow & \Omega_n^{fr}(BSO(0))) & \longrightarrow & 0 & & \\
 & & \downarrow f & & & & \downarrow & & \\
 & & \Omega_n(n-1,n-1) & \xrightarrow{\chi'} & \Omega_{n-1}^{fr}(BSO(1)) & \longrightarrow & \Omega_{n-1}(n-1,n-1) & & \\
 & & \downarrow f & & & & \downarrow f & & \\
 & & \vdots & & & & \vdots & & \\
 & & \downarrow & & & & \downarrow f & & \\
\Omega_{k+1}^{fr}(BSO(n-k)) & \xrightarrow{\partial} & \Omega_n(k+1,k+1) & \xrightarrow{\chi'} & \Omega_{k+1}^{fr}(BSO(n-k-1)) & \longrightarrow & \Omega_{n-1}(k+1,k+1) & & \\
 & & \downarrow f & & & & \downarrow f & & \\
\Omega_k^{fr}(BSO(n-k+1)) & \xrightarrow{\partial} & \Omega_n(k,k) & \xrightarrow{\chi'} & \Omega_k^{fr}(BSO(n-k)) & \xrightarrow{\partial} & \Omega_{n-1}(k,k) & \xrightarrow{\chi'} & \Omega_k^{fr}(BSO(n-k-1)) \\
 & & \downarrow f & & & & \downarrow f & & \\
 & & \Omega_n(k-1,k-1) & & & \longrightarrow & \Omega_{n-1}(k-1,k-1) & \xrightarrow{\chi'} & \Omega_{k-1}^{fr}(BSO(n-k)) \\
 & & \downarrow & & & & \downarrow & & \\
 & & \vdots & & & & \vdots & & \\
 & & \downarrow & & & & & & \\
\Omega_3^{fr}(BSO(n-2)) & \xrightarrow{\partial} & \Omega_n(3,3) & \xrightarrow{\chi'} & \Omega_3^{fr}(BSO(n-3)) & & & & \\
 & & \downarrow f & & & & \downarrow & & \\
\Omega_2^{fr}(BSO(n-1)) & \xrightarrow{\partial} & \Omega_n(2,2) & \xrightarrow{\chi'} & \Omega_2^{fr}(BSO(n-2)) & \longrightarrow & \Omega_{n-1}(2,2) & \longrightarrow & \Omega_2^{fr}(BSO(n-3)) \\
 & & \downarrow f & & & & \downarrow f & & \\
\Omega_1^{fr}(BSO(n)) & \xrightarrow{\partial} & \Omega_n(1,1) & \xrightarrow{\chi'} & \Omega_1^{fr}(BSO(n-1)) & \longrightarrow & \Omega_{n-1}(1,1) & \longrightarrow & \Omega_1^{fr}(BSO(n-2)) \\
 & & \downarrow f & & & & \downarrow f & & \\
\Omega_0^{fr}(BSO(n+1)) & \xrightarrow{\partial} & \Omega_n(0,0) & \xrightarrow{\chi'} & \Omega_0^{fr}(BSO(n)) & \longrightarrow & \Omega_{n-1}(0,0) & \longrightarrow & \Omega_0^{fr}(BSO(n-1)) \\
 & & \downarrow f & & & & \downarrow f & & \\
0 & \longrightarrow & \Omega_n & \longrightarrow & 0 & \longrightarrow & \Omega_{n-1} & \longrightarrow & 0 .
\end{array}
$$

In order to include the case $n = k$ in the discussion, we put $BSO(0) = S^0$; so the framed bordism group $\Omega_i^{fr}(BSO(j)) = \Omega_i(BSO(j);\text{trivial})$ consists of two copies of $\Omega_i^{fr} = \pi_i^S$ when $j = 0$.

In the usual manner we get a spectral sequence where the typical E^r-term (at $\Omega_k^{fr}(BSO(n-k))$) is the quotient

$$\partial^{-1}(f^{r-1}(\Omega_{n-1}(k+r-1,k+r-1)))\Big/\chi'(\ker f^{r-1} : \Omega_n(k,k) \longrightarrow \Omega_n(k-r+1,k-r+1));$$

in particular, the E^1-term is just $\Omega_k^{fr}(BSO(n-k))$, and it is not hard to establish a canonical isomorphism between the E^∞-term and the quotient

$$(19.29) \qquad f^{k+1}(\Omega_n(k,k))\Big/f^{k+2}(\Omega_n(k+1,\ k+1))$$

of subgroups of Ω_n (for which the notion of Ω-pleasantness, see 16.27, is very relevant).

The first differential d^1 is just $\chi'\circ\partial$. So corollary 19.11 gives us the commuting diagram

$$(19.30) \qquad \begin{array}{ccccc} \Omega_k^{fr}(BSO(n-k+1)) & \xrightarrow{d^1} & \Omega_k^{fr}(BSO(n-k)) & \xrightarrow{d^1} & \Omega_k^{fr}(BSO(n-k-1)) \\ & \searrow^{Id-\iota_{\gamma_{n-k+1}}} & \downarrow i_* & \searrow^{Id-\iota_{\gamma_{n-k}}} & \downarrow i_* \\ & & \Omega_k^{fr}(BSO(n-k+1)) & & \Omega_k^{fr}(BSO(n-k)) \end{array} .$$

If $n>2k\geq 0$, the left hand vertical arrow i_* is bijective, and the E^2-term at $\Omega_k^{fr}(BSO(n-k))$ consists only of elements of order 2; indeed, given $x\in\Omega_k^{fr}(BSO(n-k))$ such that $d^1(x) = 0$, we have $x = \iota_{\gamma_{n-k}}(x)$ and

$$\begin{aligned} i_*(2x) &= i_*(x) + i_*(\iota_{\gamma_{n-k}}(x)) \\ &= i_*(x) - \iota_{\gamma_{n-k+1}}(i_*(x)) \\ &= i_*(d^1(i_*(x))) \ ; \end{aligned}$$

hence $2x$ lies in $d^1(\Omega_k^{fr}(BSO(n-k+1)))$. As a consequence, in the metastable range all E^r-terms, $r \geq 2$, and in particular the E^∞-term 19.29, have only elements of order 2. This sheds some interesting light on the result of Mayer and Frank (see [65], theorem 14) that the signature is divisible by $2a_k$ on the image of

$$f^{k+1} \quad : \quad \Omega_{4q}(k,\ k) \longrightarrow \Omega_{4q}$$

for certain numbers $a_1, a_2, .., a_k, ..$ such that $a_{k+1} = a_k$ or $a_{k+1} = 2a_k$.

There are various multiplicative structures involved in diagram 19.28. E.g. both $\sum_{n,k} \Omega_n(k,k)$ and the total E^1-term $\sum_{i,j} \Omega_i^{fr}(BSO(j))$ are bigraded rings with respect to the product of manifolds (and the oriented Whitney sum of vectorbundles; however, in the product $z_1 \cdot z_2$, where $z_\ell = [Z_\ell, \eta_\ell] \in \Omega_{i_\ell}^{fr}(BSO(j_\ell))$, $\ell = 1,2$, reverse the orientation of $\eta_1 \times \eta_2$ if both j_1 and j_2 are odd); given this convention, χ' is a ring homomorphism. Furthermore, all the groups in 19.28 allow a multiplication by parallelized manifolds, and all the homomorphisms in 19.28 respect this multiplication. It is left to the interested reader to expand on these few remarks and find out e.g. to what extend d^1 has the properties of an antiderivation.

One aspect of this whole spectral sequence set-up will be very useful in practical computations: we obtain numerous new ("higher order") homomorphisms. E.g. if we denote the "first order" homomorphism

$$(\chi'_k =) \quad \chi' \quad : \quad \Omega_n(k,\ k) \longrightarrow \Omega_k^{fr}(BSO(n-k))$$

by χ_k' as indicated, there is a welldefined second order homomorphism (cf. diagram 19.28)

(19.31) $$\mathrm{proj} \circ \chi_{k+1}' \circ f^{-1} \ : \ \ker \chi_k' \longrightarrow \Omega_{k+1}^{fr}(BSO(n-k-1)) \big/ \chi' \circ \partial(\Omega_k^{fr}(BSO(n-k))$$

which we will also denote sometimes by χ_{k+1}''; composed with $\partial | (\partial^{-1}(\ker \chi_k'))$ it is essentially a differential d^2. As an example, consider the case $n \equiv 1(4)$, $n > 1$ and $k = 0$. Both χ_o' = Euler number and $\chi_1' \circ \partial$ vanish (see 19.9 and 19.11), so that our second order homomorphism takes the form

$$\chi_1'' \circ f^{-1} \quad : \quad \Omega_n(0,\ 0) \longrightarrow \Omega_1^{fr}(BSO(n-1)) \quad = \quad \mathbb{Z}_2\ .$$

It follows from table 19.17 that this is just the Kervaire semi-characteristic homomorphism k which plays such an important rôle in the calculation of $\Omega_n(0,0)$ (see theorem 12.1).

Now we want to apply the whole previous discussion and study the groups $\Omega_n(k,k)$ for $k = 1,2$ and 3.

<u>Lemma 19.32.</u> <u>If</u> $n > k$, <u>the composite homomorphism</u> $d^1 = \chi' \circ \partial$ <u>is given by the following table for small values of</u> k.

		$\Omega_k^{fr}(BSO(n-k+1)) \xrightarrow{\partial} \Omega_n(k,k) \xrightarrow{\chi'}$	$\Omega_k^{fr}(BSO(n-k))$
$k = 0$		$\mathbb{Z} \xrightarrow{(1+(-1)^n)\,\mathrm{Id}}$	$\mathbb{Z}$
$k = 1$		$\mathbb{Z}_2 \xrightarrow{0}$	$\mathbb{Z}_2$
$k = 2$	$n \geq 5$	$\mathbb{Z}_2 \oplus \mathbb{Z}_2 \xrightarrow{0}$	$\mathbb{Z}_2 \oplus \mathbb{Z}_2$
	$n = 4$	$\mathbb{Z}_2 \oplus \mathbb{Z}_2 \xrightarrow{0}$	$\mathbb{Z}_2 \oplus \mathbb{Z}$
	$n = 3$	$\mathbb{Z}_2 \oplus \mathbb{Z} \xrightarrow{0}$	$\mathbb{Z}_2$
$k = 3$		$\mathbb{Z}_{24} \xrightarrow{(1-(-1)^n)\,\mathrm{Id}}$	$\mathbb{Z}_{24}$

<u>Proof.</u> This follows directly from corollary 19.11 and from 19.9 and 19.23, as soon as we know that for $q \geq 1$ the homomorphisms

$$\mathbb{Z}_{24} \cong \Omega_3^{fr}(\text{point}) \underset{\text{const}_*}{\overset{i_*}{\rightleftarrows}} \Omega_3^{fr}(BSO(q)) , \tag{19.33}$$

induced by constant maps, are isomorphisms (inverse of one another). To see this, compare the diagram

$$
\begin{array}{ccccccc}
 & & 0 & & & & \\
 & & \downarrow & & & & \\
 & & \mathbb{Z}_4 & & & & \\
 & & \downarrow & & & & \\
\Omega_4(BSO(q)) & \xrightarrow{\sigma\circ j_4} & \Omega_2(BSO(q)\times BO(2);\Gamma) & \xrightarrow{\delta_3} & \Omega_3^{fr}(BSO(q)) & \xrightarrow{f_3} & \Omega_3(BSO(q)) \\
 & & \downarrow{\scriptstyle f_2'} & & & & \\
 & & H_2(BSO(q);\mathbb{Z}_2)\oplus\mathbb{Z}_2\oplus\mathbb{Z} & & & & \\
 & & \downarrow & & & & \\
 & & 0 & & & &
\end{array}
$$

(cf. theorem 9.3) with the corresponding diagram for $\Omega_3^{fr}(\text{point})$. For $j = 2,3$ combine the Künneth sequence

$$0 \longrightarrow H_j(BSO(q);\mathbb{Z}) \otimes \mathbb{Z}_2 \longrightarrow H_j(BSO(q);\mathbb{Z}_2) \longrightarrow Tor(H_{j-1}(BSO(q);\mathbb{Z}),\mathbb{Z}_2) \to 0$$

with the fact (see [9], p. 373, or also [13], p. 24) that all elements in $H_j(BSO(q);\mathbb{Z})$ have order 2 unless $j = q = 2$, to show that

$$H_2(BSO(q);\mathbb{Z}) \cong \begin{cases} \mathbb{Z}_2 & \text{for } q > 2 \\ \mathbb{Z} & \text{for } q = 2 \end{cases}$$

and hence $\Omega_3(BSO(q)) = H_3(BSO(q);\mathbb{Z}) = 0$ (see also fact 9.4); thus δ_3 in the diagram above is onto. Furthermore note that for $x = [\mathbb{C}P(2), \lambda_{\mathbb{C}} \oplus \underset{\sim}{\mathbb{R}}^{q-2}] \in \Omega_4(BSO(q))$ the first component of $f_2' \circ \sigma \circ j_4(x)$ (in $H_2(BSO(q);\mathbb{Z}_2) \xrightarrow[\cong]{w_2(\gamma)} \mathbb{Z}_2$) is given by $w_2(\lambda_{\mathbb{C}})^2[\mathbb{C}P(2)] = 1$ (unless $q = 1$). It follows immediately that the homomorphism i_* in 19.33 is onto. Since $const_* \circ i_* = Id$, both i_* and $const_*$ are bijective.

Example 19.34. The last lemma implies in particular that for $k = 1$ or 2 $E^1 = E^2$ at $\Omega_k^{fr}(BSO(n-k))$. Thus the second order differential d^2 takes the form

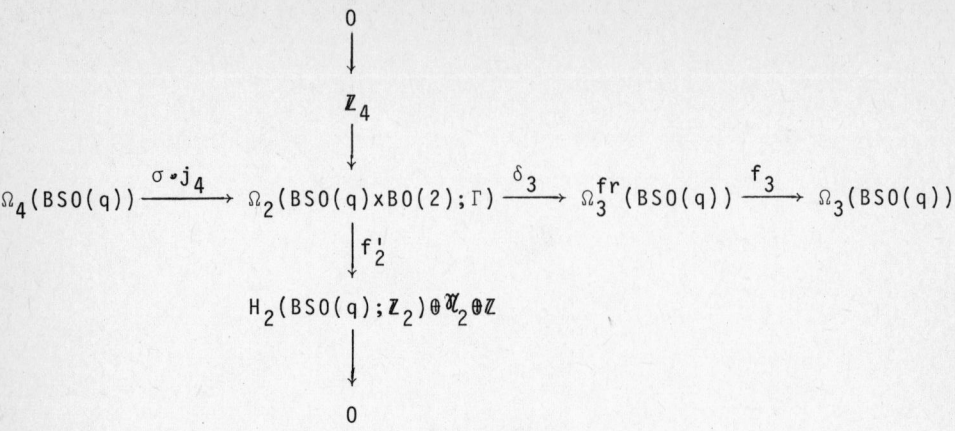

$$\chi' f^{-1}\partial \; : \; \mathbb{Z}_2 \; = \; \Omega_1^{fr}(BSO(n)) \longrightarrow \Omega_2^{fr}(BSO(n-2))$$

(see also diagram 19.28). We claim that d^2 is the inclusion of the (first) $\mathbb{Z}_2$-factor when $n \equiv 1$ or $2(4)$, and $d^2 = 0$ otherwise. Indeed, $\partial(1)$ is represented by $S^1 \times S^{n-1}$ and the vectorfield u along S^1. If $n \equiv 0(4)$, S^{n-1} has at least 3 vectorfields in its own right, so $(\chi' f^{-1})(\partial(1)) = 0$. If $n \equiv 2(4)$, S^{n-1} has only one vectorfield v, and since $\chi''(S^{n-1}, v)$ is given by the invariantly framed circle (see example 13.17), we have

$$\begin{aligned} \chi' f^{-1}\partial(1) &= \chi'(S^1 \times S^{n-1}, (u, v)) \\ &= ([S^1 \times S^1, \text{invariant framing}], 0) \; . \end{aligned}$$

So let us turn to the more complicated case when n is odd. Let v be a suitable vectorfield on S^{n-1} with zeroes at the poles $\pm (0,..,0,1)$, each having index 1. If we want to form $f^{-1}\partial(1)$, we should kill the singularity S of the 2-field (u,v) on $S^1 \times S^{n-1}$ (S consists of the two circles $S^1 \times \{\pm (0,..,0,1)\}$) and at the same time construct an $\Omega_n(1,1)$-bordism. So start from $S^1 \times S^{n-1} \times I$, identify a tubular neighborhood of S in the top $S^1 \times S^{n-1}$ $(\times \{1\})$ in the obvious way with $S^1 \times D^{n-1} \amalg S^1 \times D^{n-1}$, and glue a "handle" $S^1 \times D^{n-1} \times [1,2]$ on $S^1 \times S^{n-1} \times I$ by an attaching map

$$g \; : \; S^1 \times D^{n-1} \times \{1, 2\} \longrightarrow S^1 \times D^{n-1} \amalg S^1 \times D^{n-1}$$

satisfying $g|S^1 \times D^{n-1} \times \{+1\} = \mathrm{Id}$ and

$$g((x_1,x_2), (y_1,..,y_{n-1}), 2) = ((-x_1,x_2), (-y_1,y_2,..y_{n-1}))$$

for $x \in S^1$, $y \in D^{n-1}$. After smoothing corners we obtain an oriented bordism from the bottom copy of $S^1 \times S^{n-1}$ to the "top" part N of the boundary. N is the union of

$$\begin{aligned} N_1 \; &:= \; S^1 \times S^{n-1}(\times \{1\}) - (S^1 \times \mathring{D}^{n-1} \amalg S^1 \times \mathring{D}^{n-1}) \\ &= \; S^1 \times (S^{n-1} - (\mathring{D}^{n-1} \amalg \mathring{D}^{n-1})) \\ &\cong \; S^1 \times S^{n-2} \times [0, 1] \end{aligned}$$

and of

$$N_2 = S^1 \times S^{n-2} \times [1, 2]$$

with the identification

$$((x_1,x_2), (y_1,\dots,y_{n-1}), 2) = ((-x_1,x_2), (-y_1,y_2,\dots,y_{n-1}), 0) .$$

Clearly N can be fibered over the circle

$$\mathcal{C} = [0, 2] \big/ 0 \sim 2 ;$$

more precisely, N is the total space of the obvious subbundle (with fiber $S^1 \times S^{n-2}$) of the vectorbundle (Möbius $\oplus \underset{\sim}{\mathbb{R}}$) $\oplus$ (Möbius $\oplus \underset{\sim}{\mathbb{R}}^{n-2}$) over $\mathcal{C}$.

Let v_1 be the vectorfield on N which points always in the S^1-direction of the fiber except around the "twisting locus" $S^1 \times S^{n-2} \times [0]$ where a rotation through the vectorfield u_o on S^{n-2} (see example 4.16) makes up for the identification $(x_1,x_2) \sim (-x_1,x_2)$. Also,let v_2 be the "horizontal" vectorfield on N which projects onto the standard vectorfield on the circle $\mathcal{C}$. The bordism constructed above shows that

$$[N, v_1] = [S^1 \times S^{n-1}, u] = \partial(1) \in \Omega_n(1,1) ,$$

and hence $\chi' f^{-1}\partial(1) = \chi'(N,v_1,v_2)$ is determined by the singularity of a suitable third vectorfield v_3 on N. In view of the identifications built into N and v_1 we may choose v_3 in such a way that

(i) v_1, v_2, v_3 are linearly independent except maybe over $(\frac{1}{2}, \frac{3}{2}) \subset \mathcal{C}$, and

(ii) over $\frac{1}{2} \in \mathcal{C}$ (resp. $\frac{3}{2} \in \mathcal{C}$) v_3 is given by the vectorfield u_1 (resp. $-u_o$) along S^{n-2} (see example 4.16).

Applying the main argument of 4.16 repeatedly, we see that $\chi''(N,v_1,v_2)$ may be represented by $\frac{n-3}{2}$ parallelized circles, multiplied by the parallelized fibre S^1. Since the complement of $\mathbb{R}\, v_1 \oplus \mathbb{R}\, v_2$ is stably trivial over $(\frac{1}{2}, \frac{3}{2})$, we conclude that

(19.35) $$\chi' f^{-1}\partial(1) = ((\tfrac{n-3}{2})_2, 0)$$

for odd n. ∎

<u>Example 19.36.</u> For $n = 4q+1$, $q \geq 1$, consider the second order differential

$$\begin{array}{ccc} \mathrm{proj}\circ\chi'\circ f^{-1}\circ\partial : \Omega_2^{fr}(BSO(n-1)) & \longrightarrow & \Omega_3^{fr}(BSO(n-3))/\mathrm{Im}\ \chi'\circ\partial = 2\cdot\mathbb{Z}_{24} \\ \cong \big\downarrow (\mathrm{const}_*, w_2(\gamma)) & & \cong \\ \mathbb{Z}_2 \oplus \mathbb{Z}_2 & & \mathbb{Z}_2 \end{array}$$

(see 19.28, 19.32 and 19.23). We claim that it is just the projection onto the second factor.

Indeed, note first that our homomorphism vanishes on $\mathbb{Z}_2 \oplus \{0\}$ (since this is the image of the differential

$$d^2 : \Omega_1^{fr}(BSO(n+1)) \xrightarrow{\partial} \Omega_{n+1}(1,1) \xleftarrow{f} \Omega_{n+1}(2,2) \xrightarrow{\chi'} \Omega_2^{fr}(BSO(n-1)),$$

see the last example). Moreover, recall from corollary 19.23 that $\partial(0,1)$ is given, up to stabilization, by a sphere bundle space $N = S(\lambda_{\mathbb{C}} \oplus \underset{\sim}{\mathbb{C}}^{2q-1})$ over $\mathbb{C}P(1) = S^2$, together with a stable "horizontal" 2-framefield u. There is also a "vertical" vectorfield v_i along the fibers, defined by complex multiplication with i. Since χ' is compatible with stabilization, it remains only to show that the value $\chi''(N,(u,v_i)) \in \mathbb{Z}_{24}$ is odd.

Interpret S^2 as the union of two hemispheres S^2_+ and S^2_-, glued together along their common boundary $\partial S^2_+ = \partial S^2_- = S^1$. Then the total space of $\lambda_{\mathbb{C}} \oplus \underset{\sim}{\mathbb{C}}^{2q-1}$ is obtained from $S^2_+ \times \mathbb{H}^q \amalg S^2_- \times \mathbb{H}^q$ via the attaching map

$$\begin{array}{rccc} g : & S^1 \times \mathbb{H} \times \mathbb{H}^{q-1} & \longrightarrow & S^1 \times \mathbb{H} \times \mathbb{H}^{q-1} \\ & (z;\ a + jb,\ c) & \longmapsto & (z;\ za+jb,\ c) \end{array}$$

where $z, a, b \in \mathbb{C}$. On each of the resulting two parts of N quaternionic multiplication with j defines a second "vertical" vectorfield v_j. In order to obtain a welldefined global vectorfield over N we have to deform, near the equator S^1,

$$v_j\,(z;\, a + jb,\, c) \;=\; (-b + ja,\, jc)$$

into

$$Tg^{-1} \circ v_j \circ g\,(z;\, a + jb,\, c) \;=\; (-\bar{z}b + jza,\, jc)\ .$$

The linear homotopy between these two sections (of the complement of (u,v_i)) has its singularity at the locus $z = -1$, $c = 0$, and it is not hard to see that this 3-sphere with its invariant (quaternionic) framing represents $\pm\, \chi''(N,(u,v_i))$ (which thus generates Ω_3^{fr}). ∎

Theorem 19.37. <u>All pairs (n,k), $n > 0$, $k \le 4$, except the pairs $(4,3)$ and $(4,4)$, are Ω-pleasant.</u>
(See definition 16.27; cf. also theorem 12.13 for the unoriented analogue).

Theorem 19.38. <u>For $n \ge 2k$ the kernel of the homomorphism</u>

$$(f,\, \chi') : \Omega_n(k,k) \longrightarrow \Omega_n \oplus \Omega_k^{fr}(BSO(n-k))$$

<u>takes the following values. (The expressions in brackets are higher order homomorphisms - see e.g. 19.31 - detecting $\ker(f,\chi')$).</u>

	k = 0	k = 1	k = 2	k = 3
$n \equiv 1(4)$	$\mathbb{Z}_2$ (χ_1'')	$\mathbb{Z}_2$ (χ_2'')	$\mathbb{Z}_2$ (χ_3'')	0
$n \equiv 2(4)$	0	$\mathbb{Z}_2$ (χ_2'')	$\mathbb{Z}_2 \oplus \mathbb{Z}_2$ $(\chi_3'',\ ?)$	$\mathbb{Z}_2$
$n \equiv 3(4)$	0	0	$\mathbb{Z}_2$ (χ_3'')	0
$n \equiv 4(4)$	0	0	0	0

(See theorem 16.9 for the relevance of $\ker(f,\chi')$; at the end of this section we will use the table above to complete the calculations in example 16.15).

Theorems 19.37 and 19.38 will be byproducts when we prove the following results which give a nearly complete description of the groups $\Omega_n(k,k)$ themselves for $k = 1,2,3$.

<u>Theorem 19.39. Let $n \geq 2$. We have isomorphisms</u>

$$\Omega_n(1,1) \xrightarrow{(f,\ \chi_1'',\ \chi_2'')} \Omega_n \oplus \mathbb{Z}_2 \oplus \mathbb{Z}_2 \qquad \text{if } n \equiv 1(4)$$

<u>(where f stands for the obvious forgetful map, and where</u> χ_2'', <u>originally defined on</u> $\ker \chi_1'$, <u>see 19.31, is suitably extended to all of</u> $\Omega_n(1,1)$);

$$\Omega_n(1,1) \xrightarrow{(f,\ \chi_2'')} \Omega_n \oplus \mathbb{Z}_2 \qquad \text{if } n \equiv 2(4);$$

$$\Omega_n(1,1) \xrightarrow{f} \Omega_n \qquad \text{if } n \equiv 3(4);$$

$$\Omega_n(1,1) \xrightarrow{f} \ker(w_n : \Omega_n \longrightarrow \mathbb{Z}_2) \qquad \text{if } n \equiv 4(4).$$

<u>Proof.</u> We could easily establish this result without using the work of Atiyah, Frank and Thomas on 2-fields (and just using examples 4.16, 13.17 and 19.34 instead), but we will not bother to be so elementary.

The relevant piece of the exact sequence 19.5 is

$$\Omega_{n+1}(1,1) \xrightarrow{\chi''=\chi_1''} \mathbb{Z}_2 \xrightarrow{\partial} \Omega_n(1,1) \xrightarrow{f} \Omega_n(0,0) \xrightarrow{\chi'=\chi} \mathbb{Z}$$

where the left hand homomorphism χ'' is onto (and hence f is injective) if and only if $n + 1 \equiv 0$ or $1(4)$ (see 19.17), and the right hand χ is just the Euler number homomorphism. Recall from 12.1 the isomorphisms

$$\left.\begin{matrix} (f, \frac{1}{2}(\chi - \tau)) \\ f \\ (f, k) \end{matrix}\right\} : \Omega_n(0,0) \longrightarrow \Omega_n \oplus \begin{cases} \mathbb{Z} & \text{if } n \equiv 0(2) \\ 0 & \text{if } n \equiv 3(4) \\ \mathbb{Z}_2 & \text{if } n \equiv 1(4)\ . \end{cases}$$

For $n \equiv 3$ or $4(4)$ our claim concerning $\Omega_n(1,1)$ follows immediately. If $n \equiv 2(4)$ we get the diagram

$$\begin{array}{ccccccccc} 0 & \longrightarrow & \mathbb{Z}_2 & \xrightarrow{\partial} & \Omega_n(1,1) & \xrightarrow{f} & \Omega_n & \longrightarrow & 0 \\ & & \nwarrow^{\chi''_2} & & \parallel & & & & \\ & & & & \ker \chi'_1 & & & & \end{array}$$

(see 19.31); since $\chi''_2 \circ \partial$ is the first component of the differential $\chi' f^{-1} \partial$ calculated in example 19.34, the short exact sequence above splits as desired.

Finally, assume $n \equiv 1(4)$. First note that (S^n, u_o) (cf. 4.16) represents an element of order 2 in $\Omega_n(1,1)$; indeed, on the cylinder $S^n \times I$ rotate u_o through the I-direction into $-u_o$ to obtain a fine oriented zero bordism connecting (S^n, u_o) with the pair $(-S^n, -u_o)$ which is diffeomorphic, via reflection, to $(S^n, -u_1)$; in turn, $-u_1$ is homotopic to u_o (apply the main argument in 4.16 an even number of times). Thus we obtain the exact sequences

$$\begin{array}{ccccccccc} 0 & \longrightarrow & \mathbb{Z}_2 & \xrightarrow{\partial} & \Omega_n(1,1) & \xrightarrow{(f,\chi''_1)} & \Omega_n \oplus \mathbb{Z}_2 & \longrightarrow & 0 \\ & & & & \parallel & & & & \\ & & & & \ker \chi''_1 \oplus \mathbb{Z}_2\,[S^n, u_o] & & & & \end{array}$$

and

$$\begin{array}{ccccccccc} 0 & \longrightarrow & \mathbb{Z}_2 & \underset{\chi''_2}{\overset{\partial}{\rightleftarrows}} & \ker \chi''_1 & \xrightarrow{f} & \Omega_n & \longrightarrow & 0 \end{array}$$

(cf. 19.31 and 19.32). According to example 19.34 $\chi''_2 \circ \partial = \mathrm{Id}$, so $\Omega_n(1,1)$ splits as claimed. ∎

<u>Theorem 19.40.</u> <u>Let $n \geq 4$.</u>
<u>If $n \equiv 1(4)$, then</u> $\Omega_n(2,2)$ <u>is an extension of</u> $\Omega_n \oplus \mathbb{Z}_2$ <u>by</u> $\mathbb{Z}_2$.
<u>If $n \equiv 2(4)$, then</u> $\Omega_n(2,2)$ <u>is an extension of</u> $\Omega_n \oplus \mathbb{Z}_2 \oplus \mathbb{Z}_2$ <u>by</u> $\mathbb{Z}_2$.

If $n \equiv 3(4)$, then we have the isomorphism

$$\Omega_n(2,2) \xrightarrow{(f,\chi_3'')} \Omega_n \oplus \mathbb{Z}_2 .$$

If $n \equiv 4(4)$, then we have the isomorphism

$$\Omega_n(2,2) \xrightarrow{f} \ker(\text{mod } 4 \text{ signature} : \Omega_n \longrightarrow \mathbb{Z}_4) .$$

The proof will contain more details on the cases $n \equiv 1$ or $2(4)$.

Theorem 19.41. Let $n \geq 5$. We have isomorphisms

$$\Omega_n(3,3) \xrightarrow{(f,\,\chi'')} \ker(w_2w_{n-2} : \Omega_n \longrightarrow \mathbb{Z}_2) \oplus \mathbb{Z}_{24} \qquad \text{if } n \equiv 1(4);$$
$$\Omega_n(3,3) \xrightarrow{f} \ker(\chi'' : \Omega_n(2,2) \longrightarrow \mathbb{Z}_2) \qquad \text{if } n \equiv 2(4);$$
$$\Omega_n(3,3) \xrightarrow{(f,\,\chi'')} \Omega_n \oplus \mathbb{Z}_{24} \qquad \text{if } n \equiv 3(4);$$
$$\Omega_n(3,3) \xrightarrow{f} \ker(\text{mod } 8 \text{ signature} : \Omega_n \rightarrow \mathbb{Z}_8) \qquad \text{if } n \equiv 4(4).$$

Proof of theorems 19.40 and 19.41. Most information on $\Omega_n(2,2)$ is obtained from the exact sequence (cf. 19.5)

(19.42)

$$\Omega_{n+1}(2,2) \xrightarrow[(\chi_2'',w_2w_{n-1})]{\chi_2' =} \mathbb{Z}_2 \oplus \mathbb{Z}_2 \xrightarrow{\partial} \Omega_n(2,2) \xrightarrow{f} \ker(\chi_1'' : \Omega_n(1,1) \longrightarrow \mathbb{Z}_2).$$

Here we have identified $\Omega_2^{fr}(BSO(n-1))$ with $\mathbb{Z}_2 \oplus \mathbb{Z}_2$ via the isomorphism $(\text{const}_*, w_2(\gamma))$ (cf. 19.23 and the calculations in example 19.24). The table in 19.24 (and in 19.16) implies that χ_2'' is nontrivial (on $\Omega_{n+1}(2,2)$!) precisely when $n \not\equiv 2(4)$, and that w_2w_{n-1} is nontrivial on $\Omega_{n+1}(2,2)$ precisely if $n \equiv 0(4)$; in the latter case, (χ_2'', w_2w_{n-1}) is actually onto. Indeed, first note that the characteristic number w_2w_{4q-1} is nontrivial on Ω_{4q+1} for $q \geq 1$: just calculate it on $M = N^5 \times \mathbb{C}P(2q-2)$, where $[N^5]$ generates $\Omega_5 \cong \mathbb{Z}_2$ and therefore must be detected by w_2w_3; according to theorem 19.39 there is an element $x \in \Omega_{4q+1}(1,1)$ satisfying $f(x) = [M]$, $\chi_1''(x) = 0$ (so that $x = f(y)$ for some $y \in \Omega_{4q+1}(2,2)$)

and $\chi_2''(x) = 0$; clearly,

$$(\chi_2'', w_2 w_{4q-1})(y) = (0, 1) .$$

On the other hand, recall from the table 19.17 that χ_1'' is onto precisely if $n \equiv 0$ or $1(4)$, and $\ker \chi_1''$ can be calculated from theorem 19.39.

E.g. if $n \equiv 0(4)$ then $\chi_1'' = (\frac{1}{2}$ signature, reduced mod 2) on

$$\Omega_n(1,1) \cong \ker (\text{mod 2 signature} : \Omega_n \longrightarrow \mathbb{Z}_2) ;$$

our claim on $\Omega_n(2,2)$ follows.

If $n \equiv 1(4)$, we have the commuting diagram of horizontal and vertical exact sequences

$$(19.43)\qquad \begin{array}{ccccccccc}
0 & \longrightarrow & (0\oplus)\mathbb{Z}_2 & \underset{\partial|}{\overset{\chi_3'}{\rightleftarrows}} & \ker \chi_2' & \xrightarrow{f} & \ker(w_2 w_{n-2} : \Omega_n \longrightarrow \mathbb{Z}_2) & \longrightarrow & 0 \\
 & & & & \cap & & \cap & & \\
0 & \rightarrow & (0\oplus)\mathbb{Z}_2 & \xrightarrow{\partial|} & \Omega_n(2,2) & \xrightarrow{(f,\chi_2'')} & \Omega_n \oplus \mathbb{Z}_2 & \longrightarrow & 0 \\
 & & & & \downarrow{\scriptstyle \chi_2' = (\chi_2'', w_2 w_{n-2})} & & & & \\
 & & & & \mathbb{Z}_2 \oplus \mathbb{Z}_2 & & & &
\end{array}$$

According to example 19.36 $\chi_3' \circ \partial| = \mathrm{Id}$, so the upper sequence splits and

$$\ker \chi_2' \cong \ker w_2 w_{n-2} \oplus \mathbb{Z}_2 .$$

Note also that we have $\chi_2'[N, v_1, v_2] = (1,0)$ for the element $[N, v_1, v_2] \in \Omega_n(2,2)$ discussed in example 19.34.

If $n \equiv 2(4)$, (19.42) gives again a diagram

$$
\begin{array}{lccccccc}
 & & & \mathbb{Z}_{24} \supset \mathbb{Z}_2[12] \xleftarrow{\chi_3'} \ker\chi_2' & & & & \\
 & 0 \longrightarrow \mathbb{Z}_2 \oplus \mathbb{Z}_2 & \xrightarrow{\partial} & \ker \chi_2' & \xrightarrow{f} & \Omega_n & \longrightarrow & 0 \\
(19.44) & \| & & \cap & & \cap & & \\
 & 0 \longrightarrow \mathbb{Z}_2 \oplus \mathbb{Z}_2 & \xrightarrow{\partial} & \Omega_n(2,2) & \xrightarrow{(f,\chi_2'')} & \Omega_n \oplus \mathbb{Z}_2 & \longrightarrow & 0 .
\end{array}
$$

(In order to see that the second order homomorphism χ_3' maps $\ker \chi_2'$ indeed into the elements of order 2 in $\mathbb{Z}_{24}$, just contemplate the commuting diagram

$$
\begin{array}{ccccccc}
\Omega_n(3,3) & \xrightarrow{\chi_3'} & \mathbb{Z}_{24} & \xrightarrow{\partial} & \Omega_{n-1}(3,3) & \xrightarrow{\chi_3'} & \mathbb{Z}_{24} , \\
\big\downarrow f & & & & & & \\
\Omega_n(2,2) & & & & & &
\end{array}
$$

with $\partial \circ \chi_3' = 0$ and $\chi_3' \circ \partial = \cdot\, 2$, cf. 19.28 and 19.32). It follows from example 19.34 (- just multiply the whole situation there by a parallelized circle -) that $\chi_3' \circ \partial(1,0)$ can be represented by the invariantly framed torus $S^1 \times S^1 \times S^1$ and hence is the nontrivial element of order 2 in $\Omega_3^{fr} \cong \mathbb{Z}_{24}$. Note also that the product manifold $S^1 \times S^{n-1}$, together with the obvious product 2-field $u \times u_o$, represents an element of order 2 in $\Omega_n(2,2)$ such that

$$\chi_2'' \, [S^1 \times S^{n-1}, u \times u_o] = 1$$

(cf. the calculation of $\Omega_{n-1}(1,1)$ in the last proof, and example 13.17). Thus

$$(19.45) \qquad \Omega_n(2,2) = \ker \chi_2' \oplus \mathbb{Z}_2 \, [S^1 \times S^{n-1}, u \times u_o] .$$

In particular, χ_3' can be extended to all of $\Omega_n(2,2)$, and we obtain the exact sequence

$$0 \longrightarrow \mathbb{Z}_2 \longrightarrow \Omega_n(2,2) \xrightarrow{(f,\chi_2'',\chi_3'')} \Omega_n \oplus \mathbb{Z}_2 \oplus \mathbb{Z}_2 \longrightarrow 0$$

alluded to in the statement of our theorem.

Finally assume $n \equiv 3(4)$. Then 19.42 gives rise to the following short exact sequence

$$(19.46) \qquad 0 \longrightarrow (0\oplus)\mathbb{Z}_2 \underset{\partial|}{\overset{\chi_3'}{\rightleftarrows}} \Omega_n(2,2) \xrightarrow{\ f\ } \Omega_n \longrightarrow 0 \ .$$

In the discussion of $\Omega_n(3,3)$ we will see that the second order homomorphism χ_3' (cf. 19.31) is a left inverse of $\partial|$; therefore, the sequence splits as claimed in the statement of theorem 19.40.

So let us now turn to the bordism of 3-framefields. If n is odd, the relevant piece of the exact sequence 19.5 appears as the horizontal line in the following commuting diagram.

(19.47)

$$\begin{array}{ccccccccccc}
 & & & & & & & & 0 & & \\
 & & & & & & & & \downarrow & & \\
 & & & & \mathbb{Z}_{24} & & & & (0\oplus)\mathbb{Z}_2 & & \\
 & & & \cdot 2 \nearrow & & \nwarrow \chi_3' & & & \partial| \downarrow \uparrow \chi_3' & & \\
\Omega_{n+1}(4,4) & \xrightarrow{f} & \Omega_{n+1}(3,3) & \xrightarrow{\chi_3'} & \mathbb{Z}_{24} & \xrightarrow{\partial} & \Omega_n(3,3) & \xrightarrow{f} & \ker \chi_2' & \longrightarrow & 0 \\
 & & & & & & & & \downarrow f & & \\
 & & & & & & & & \ker(w_2w_{n-2}:\Omega_n \to \mathbb{Z}_2) & & \\
 & & & & & & & & \downarrow & & \\
 & & & & & & & & 0 & &
\end{array}$$

The commutativity of the triangle was established in lemma 19.32. The vertical short exact sequence comes from 19.43 (when $n \equiv 1(4)$) and from 19.46 (when $n \equiv 3(4)$ and hence $\ker \chi_2' = \Omega_n(2,2)$ and $\ker w_2w_{n-2} = \Omega_n$). Observe that $\chi_3'(\Omega_{n+1}(3,3))$ is nontrivial: if $n \equiv 1(4)$, a nontrivial element was exhibited above; if $n \equiv 3(4)$, $f(\Omega_{n+1}(4,4)) \neq \Omega_{n+1}(3,3)$ since there are (n+1)-manifolds with signature 8 admitting 3-framefields (use 19.40 and 19.24), but (according to Mayer and Frank (cf. [65], p. 653) certainly no 4-framefields. It follows that

(19.48) $\chi_3'(\Omega_{n+1}(3,3)) = \ker \partial = \ker(\chi_3' \circ \partial) = \{0,12\}$

and that

$$\ker (f : \Omega_n(3,3) \longrightarrow \ker \chi_2') \cong \mathbb{Z}_{12} .$$

Next we claim the existence of an element $x \in \Omega_n(3,3)$ such that $f(x) = \partial(0,1)$ and that $\chi_3'(x) \in \mathbb{Z}_{24}$ is odd. For $n \equiv 1(4)$ this was established in example 19.36. If $n = 4q+3$, let x be the class of $S^3 \times \mathbb{C}P(2q)$, together with the standard invariant 3-framefield along S^3; then $\chi_3'(x)$ can be represented by $2q+1$ copies of the invariantly framed 3-sphere (which is a generator of $\Omega_3^{fr} \cong \mathbb{Z}_{24}$, see [7]) and does not lie in $\chi_3' \circ \partial(\mathbb{Z}_{24})$; hence $x \notin \partial(\mathbb{Z}_{24}) = \ker f$, and $f(x)$ must be the nonzero element in $\partial|(\mathbb{Z}_2)$ since $S^3 \times \mathbb{C}P(2q)$ is trivial in oriented bordism.

We conclude that the vertical sequence at the right hand side in 19.47 splits as indicated. Also the claims in theorem 19.41 on $\Omega_n(3,3)$ for odd n follow immediately. Furthermore, if n is even, then $\chi_3'(\Omega_{n+1}(3,3))$ is all of $\mathbb{Z}_{24}$, and hence

$$f \;:\; \Omega_n(3,3) \longrightarrow \ker \chi_2' \qquad (\subset \Omega_n(2,2))$$

is bijective. In view of 19.24 and 19.40 this completes the proof of theorem 19.41. ∎

Theorems 19.37 and 19.38 can now be deduced easily from theorems 19.39 - 19.41 and their proofs. To see e.g. that the pair $(n,4)$ is Ω-pleasant for $n \geq 5$, combine the left hand part of the exact sequence 19.47 with 19.41 (and 19.44), and conclude that the forgetful map

$$f \;:\; \Omega_n(4,4) \longrightarrow \Omega_n$$

is onto when $n \equiv 2$ or $3(4)$ and has $\ker w_2 w_{n-2}$ as its image when

$n \equiv 1(4)$. If $n \equiv 0(4)$, recall the exact sequence

$$\begin{array}{ccccccc} \Omega_n(4,4) & \xrightarrow{f} & \Omega_n(3,3) & \xrightarrow{\chi_3'} & \mathbb{Z}_2[12] & \longrightarrow & 0 \\ & & \Big\downarrow{\scriptstyle (\frac{1}{8}\text{ sign, mod } 2)} & \swarrow{\scriptstyle \cong} & & & \\ & & \mathbb{Z}_2 & & & & \end{array}$$

and the vertical epimorphism from the last proof. Since the signature is divisible by 16 on $\Omega_n(4,4)$, we get

$$(19.49) \qquad \chi_3' \equiv (\tfrac{1}{8}\text{ signature, mod } 2) .$$

We conclude from theorem 19.41 that

$$f \ : \ \Omega_n(4,4) \longrightarrow \ker(\text{mod } 16 \text{ signature} : \Omega_n \longrightarrow \mathbb{Z}_{16})$$

is onto. ∎

<u>Remark 19.50.</u> We can use theorem 19.38 to show that

$$\tilde{\pi}^{or}(n,3) \cong \mathbb{Z}_4 \qquad \text{and} \qquad \tilde{\pi}^{or}(n,4) \cong \mathbb{Z}_8$$

when $n \equiv 4(8)$, $n \geq 12$ (thus completing the calculations of example 16.15). Indeed, consider the homomorphism

$$(f, \chi') \ : \ \Omega_{n-1}(k-1,k-1) \longrightarrow \Omega_{n-1} \oplus \Omega^{fr}_{k-1}(BSO(n-k)) \ ;$$

the arguments in the proof of theorem 16.9 (see especially the first diagram) lead to the exact sequence

$$(19.51) \qquad \pi^{or}(n,k) \subset \operatorname{proj}_*(\pi_{n-1}(V_{n,k+1})) \xrightarrow{St^{-1} \circ\, \varepsilon^{or}} \ker(f,\chi')$$

valid for all n,k at least if $n > 2k+1 > 1$. If n is as above and $k = 3$ (resp. 4), then $\ker(f,\chi')$ ($\subset \Omega_{n-1}(k-1,k-1)$) is $\mathbb{Z}_2$ (resp. 0), see 19.38. In the first case, we have

$$\begin{array}{rcll} \pi^{or}(n,\ 3) & = & \tilde{\pi}^{or}(n,\ 3) \oplus \mathbb{Z}z_2 , & \text{and} \\ \operatorname{proj}_*(\pi_{n-1}(V_{n,4})) & = & \mathbb{Z}_4 \oplus \mathbb{Z}z_1 , & \end{array}$$

where $pr_*(z_i) = i$ (cf. 16.2 and 16.4), and therefore $St^{-1} \circ \varepsilon^{or}$ is just pr_*, taken modulo 2; thus $\tilde{\pi}^{or}(n,3) = \pi^{or}(n,k) \cap \ker pr_* \cong \mathbb{Z}_4$. Our claim on $\tilde{\pi}^{or}(n,4)$ follows similarly from an inspection of the last table in 16.15. ∎

<u>Exercise 19.52.</u> Compute the E^3-term in our spectral sequence (cf. 19.28) at $\Omega_k^{fr}(BSO(n-k))$, $k \leq 2$.

<u>Exercise 19.53.</u> For $k = 0$, $n > 1$ and $1 \leq r \leq \infty$ consider the differential d^r defined on the E^r-term at $\Omega_0^{fr}(BSO(n+1))$. Show that d^r is nontrivial precisely when $n \equiv 0(2)$, $r = 1$ and hence $d^r = (\text{Euler number}) \circ \partial$, or when $n \equiv 1(4)$, $r = 2$ and hence $d^r = k \circ \partial$ (k = real Kervaire semicharacteristic).

<u>Exercise 19.54.</u> Use the unoriented version of 19.5 and 19.11, as well as 19.10, 19.23 and the exact sequence

$$0 \longrightarrow \Omega_1^{fr}(BO(n-1)) \xrightarrow{\cdot[S^1,\ \text{inv.framing}]} \Omega_2^{fr}(BO(n-1)) \xrightarrow{w_2(\gamma)} \mathbb{Z}_2 \longrightarrow 0,$$

to get a very simple way to compute $\mathcal{N}_n(k,k)$, $k = 0$, 1 or 2 (compare this approach with the proofs of theorems 12.1 and 12.8).

<u>Exercise 19.55.</u> Consider the unoriented analogue of diagram 19.28. Prove the following claims.

(i) If a pair $(n,k+1)$ is $\mathcal{N}$-pleasant (cf. definition 12.12) then the kernel of the homomorphism

$$\Omega_k^{fr}(BO(n-k)) \xrightarrow{(\mu,\ \partial)} H_k(BO(n-k);\mathbb{Z}_2) \oplus \mathcal{N}_{n-1}(k,k)$$

equals the group $\chi'(\ker f : \mathcal{N}_n(k,k) \longrightarrow \mathcal{N}_n)$ (which is a "B^∞-term" of our spectral sequence).

(ii) Given a pair (n,k), we have:
(n,ℓ) is $\mathcal{N}$-pleasant for all $\ell \leq k$ if and only if the sequence

$$\ker(f\colon \mathcal{N}_n(\ell,\ell) \longrightarrow \mathcal{N}_n) \xrightarrow{\chi'} \Omega_\ell^{fr}(BO(n-\ell)) \xrightarrow{(\mu,\partial)} H_\ell(BO(n-\ell);\mathbb{Z}_2) \oplus \mathcal{N}_{n-1}(\ell,\ell)$$

is exact for all $\ell < k$.

It would be interesting to spell out a similar characterization of Ω-pleasantness. This is closely connected with the problem of factorizing the signature, taken modulo suitable powers of 2, through χ'.

Exercise 19.56. Consider the unoriented version of the exact sequence 19.5. Show that $2f \equiv 0$ and deduce an estimate for the order of elements in $\mathcal{N}_n(k,k)$ in terms of the order of elements in $\Omega_k^{fr}(BO(n-k+1))$.

§20. Stable versus unstable framefields.

In this section we compare the span and the stable span of a nonempty closed smooth n-manifold M (see definitions 13.1 and 13.12), making use of some of the techniques in the proof of the stability theorem 6.6.

Theorem 20.1.*) Let M be a closed connected even-dimensional manifold. Then

$$\text{span}(M) = \begin{cases} 0 & \text{if } \chi(M) \neq 0, \\ \text{stable span}(M) & \text{otherwise}. \end{cases}$$

Here $\chi(M)$ denotes the Euler number of M.

Next consider the homomorphism

$$(20.2) \qquad i_{k*} : \Omega_{k-1}(P^{k-1} \times M; \phi_{M,k}) \longrightarrow \Omega_{k-1}(P^k \times M; \phi_{M,k})$$

induced by the inclusion $i_k : P^{k-1} \times M \subset P^k \times M$, where k is an integer, and $\phi_{M,k} = \lambda \otimes TM - k\lambda - TM$ as in § 13.

Definition 20.3. $s(M) := \max \{k \leq n \mid i_{k*} \text{ is bijective}\}$.

Clearly, $0 \leq s(M) \leq n$.

Theorem 20.4. Let M be a closed connected manifold of odd dimension n.

Then we have

$$\min \{\tfrac{n-1}{2}, s(M), \text{stable span}(M)\} \leq \text{span}(M) \leq \text{stable span}(M)$$

and $\min \{\text{span}(S^n), \text{stable span}(M)\} \leq \text{span}(M)$.

If stable span$(M) > s(M)$, then $\omega_{s(M)+1}(M)$ is a $\mathbb{Z}_2$-valued

---

*) This and some of the following results were also obtained by V. Eagle, see remark 20.21.

invariant which vanishes precisely if

$$\text{span}(M) = \text{stable span}(M) .$$

The second inequality recovers an old result of W. Sutherland (cf. [62], 5.1). Also, theorems 20.1 and 20.4 have, in particular, the following consequence.

Corollary 20.5. Let M be a closed connected n-manifold such that $2 \cdot s(M) \leq n$. Then

$$\text{span}(M) = \begin{cases} s(M) & \text{if stable span}(M) \geq s(M) \text{ and } \omega_{s(M)+1}(M) \neq 0 \\ \text{stable span}(M) & \text{otherwise} . \end{cases}$$

In view of these results (which will be proved below) it is important to get more information on $s(M)$.

Proposition 20.6. For every closed n-manifold M we have

$$s(M) \geq \text{span } S^n .$$

If in addition the tangent bundle TM is trivial over the $(\text{span}(S^n)+1)$-skeleton of M (with respect to some cell decomposition), and if $n \neq 15$, then

$$s(M) = \text{span}(S^n) .$$

For the values of $\text{span}(S^n) = h(n)$, known from the work of Adams and Hurwitz-Radon, see remark 16.5. It is not hard to show that

$$(20.7) \qquad 2 \cdot \text{span}(S^n) < n - 2$$

except for $n = 1,2,3,7$ or 15 (see also [10], p. 86). In particular, if M is any closed n-manifold, $n \neq 15$, such that TM is stably trivial over the $(\text{span}(S^n)+1)$-skeleton, then the conclusion of corollary 20.5 holds with $s(M) = \text{span}(S^n)$. This implies e.g. much of the work of Bredon and Kosinski [10] on the span of π-manifolds.

The assumption on TM in proposition 20.6 and its consequences is often unnecessarily strong. E.g. if $n \equiv 1(4)$, it requires TM to be trivial over the 2-skeleton, and hence at least orientable; even that is not absolutely needed in order to guarantee that $s(M) = \text{span } S^n$ (=1).

<u>Proposition 20.8.</u> <u>Let M be a closed connected n-manifold. When $n \equiv 1(4)$, we have:</u>

$s(M) = 1$ if and only if $w_1(M)^2 = 0$;

$s(M) = 2$ if and only if $w_1(M)^2 \neq 0$, but $w_1(M)^2 = yw_1(M) + y^2$ for some $y \in H^1(M);\mathbb{Z}_2)$;

$s(M) = 3$ if $w_1(M)^2 \neq yw_1(M) + y^2$ for all $y \in H^1(M;\mathbb{Z}_2)$, but there is an element $z \in H^2(M;\mathbb{Z}_2)$ such that

$$w_1(M)^4 = w_1(M)^2 z + w_1(M) Sq^1(z) + z^2$$

(this holds e.g. when $w_1(M)^4 = 0$) and if in addition either

$$n \equiv 1(8) \text{ and } w_2(M) = yw_1(M) + y^2 \text{ for some } y \in H^1(M;\mathbb{Z}_2)$$

or

$$n \equiv 5(8) \text{ and } w_2(M) = w_1(M)^2.$$

<u>When $n \equiv 3(8)$, we have:</u>

$s(M) = 3$ if and only if $w_1(M)^4 + w_1(M)w_3(M) + w_2(M)^2 = 0$.

<u>Corollary 20.9. If $n \equiv 1(4)$ and $w_1(M)^2 = 0$, then</u>

$$\text{span}(M) = \begin{cases} 1 & \text{if } R_L(M) \neq 0 ; \\ \text{stable span}(M) & \text{if } R_L(M) = 0 . \end{cases}$$

This is based on work of Atiyah-Dupont [6] and Dupont-Lusztig [19] identifying the "twisted" Kervaire semicharacteristic $R_L(M) \in \mathbb{Z}_2$ with the index of any 2-field with finite singularities on M.

Corollary 20.10. *If* $n \equiv 3(8)$ *and* $w_1(M) = w_2(M) = 0$, *then*

$$\text{span}(M) = \begin{cases} 3 & \text{if } \chi_2(M) \neq 0\,; \\ \text{stable span}(M) & \text{if } \chi_2(M) = 0\,. \end{cases}$$

Here we have used a result of Duane Randall relating the mod 2 Kervaire semicharacteristic $\chi_2(M)$ with the index of a 4-field on a spin manifold M (see also 15.13 and 15.14), as well as the theorem of Atiyah-Dupont to the effect that $\text{span}(M) \geq 3$. ∎

Now we turn to the proof of our claims. It suffices to consider closed connected n-manifolds M.

Lemma 20.11. *For* $0 < k \leq n$ *the diagram*

$$\begin{array}{ccccccccc}
 & & & \overset{(1+(-1)^n)\cdot}{\nearrow} & \pi_{n-1}(S^{n-1}) & & & & \\
 & & & & \uparrow{\scriptstyle pr_*} & & & & \\
\pi_n(V_{n+1,k+1}) & \xrightarrow{pr_*} & \pi_n(S^n) & \xrightarrow{\partial} & \pi_{n-1}(V_{n,k}) & \longrightarrow & \pi_{n-1}(V_{n+1,k+1}) & \longrightarrow & 0 \\
\downarrow{\scriptstyle \Theta_M \circ \sigma} & & \| \; \mathbb{Z} \; \| & & \downarrow{\scriptstyle \Theta_M \circ \sigma} & & & & \\
\Omega_k(P^k \times M;\phi_M) & \xrightarrow{\Delta^k} & \Omega_0(M;\text{triv}) & \xrightarrow{d} & \Omega_{k-1}(P^{k-1} \times M;\phi_M) & \xrightarrow{i_{k*}} & \Omega_{k-1}(P^k \times M;\phi_M) & \longrightarrow & 0 \\
 & & & & \cup & & \cup & & \\
 & & & & \omega_k(M) & \longrightarrow & \omega_{k+1}(\mathbb{R}^{k+1}, TM \oplus \mathbb{R}) & &
\end{array}$$

commutes (at least up to signs); here the top horizontal line is the exact homotopy sequence of the fibration $pr : V_{n+1,k+1} \longrightarrow S^n$; *the bottom line is the exact Thom-Gysin sequence (cf. [57], 5.2) of the pair* $(P^k \times M, (P^k - \{\ell\}) \times M)$, $\ell = \mathbb{R}\cdot(0,..,0,1)$, *with the coefficient bundle* $\phi_M = \phi_{M,k}$ *(as in 20.2); the vertical arrows are defined as in 13.7.*

Furthermore, i_{k*} *is an isomorphism for all* $k \leq s(M)$. *In particular,* $s(M) = 0$ *if and only if* n *is even.*

Proof. Clearly, $\partial[\text{Id}]$ is the index of a k-field on S^n. Thus $pr_* \circ \partial$ is multiplication by $\chi(S^n) = 1 + (-1)^n$. Also, $\sigma \circ \partial[\text{Id}]$ is given by the singularity data of a nondegenerate (k-1)-morphism over S^n such as

$$u \; : \; S^n \times \mathbb{R}^k \subset S^n \times \mathbb{R}^{n+1} \xrightarrow{\text{projection}} TS^n \;;$$

the singularity of u is the unit sphere S^{k-1} of $\mathbb{R} \subset \mathbb{R}^{n+1}$, the kernel map is just the double cover $S^{k-1} \longrightarrow P^{k-1}$, and an analysis of the relevant stable isomorphism $\overline{g}$ shows that, after obvious deformations, the singularity data of u coincides with the standard representative of $\pm d(\text{point})$. Thus $\Theta_M \circ \sigma \circ \partial = \pm d$.

The commutativity of the other square is established more easily. Note that pr_* and Δ^k are k-fold iterates of homomorphisms $proj_*$ and Δ as the ones in the commuting diagram 13.11. In particular, since Δ^k is onto (or, equivalently, i_{k*} is injective) for $k = s(M)$, the same must hold for all smaller k. Finally deduce from theorem 9.3 that i_{1*} is bijective iff n is odd.■

<u>Proof of proposition 20.6.</u> Consider the diagram in the last lemma. If $k = \text{span}(S^n)$, then the left hand homomorphisms pr_* and Δ^k are onto, and hence i_{k*} is bijective.

If $k = \text{span}(S^n) + 1 \leq n$, then pr_* is not onto, and therefore $\partial[Id] \neq 0$. If in addition TM is trivial over the k-skeleton M_k of M, then we can define a homomorphism

$$\rho \; : \; \Omega_{k-1}(P^{k-1} \times M;\, \phi_M) \;\cong\; \Omega_{k-1}(P^{k-1} \times M_k;\, \phi_M) \longrightarrow \Omega_{k-1}(P^{k-1};(n-k)\lambda)$$

such that $\rho \circ (\Theta_M \circ \sigma) = \sigma$; for $n \neq 15$, σ is bijective (cf. 13.7 and 20.7), and hence $\Theta_M \circ \sigma(\partial[Id]) = \pm d(\text{point})$ is a nontrivial element in the kernel of i_{k*}. Thus, by lemma 20.11, $k > s(M)$ in this case, and therefore $\text{span}(S^n) = s(M)$.■

<u>Lemma 20.12.</u> <u>There exists an element</u> $r((M)) \in \pi_n(S^n) = \mathbb{Z}$ <u>such that for all</u> $0 < k \leq$ stable span(M) <u>M carries a k-field w with finite singularities whose index is</u> $\partial(r((M)))$.

Proof. Since ∂ is compatible with the obvious projections $V_{n,k} \longrightarrow V_{n,j}$, we need to consider only the case when $k = $ stable span(M).

Given a "stable k-framefield" $u : \mathbb{R}^{k+1} \hookrightarrow TM \oplus \mathbb{R}$, interprete it as a (k+1)-framefield on $M \times I$, defined at the points $M = M \times \{0\}$. Let v be a nondegenerate vectorfield on $M \times I$ such that v points outward along $M \times \{1\}$ and v agrees with the first component vectorfield of u along $M \times \{0\}$ (i.e. $v(x,0) = u(x,(1,0,..,0))$ for all $x \in M$). Clearly, we may assume that v vanishes only at points of the form $(z_\ell, \frac{3}{4})$, $z_\ell \in M$, $\ell = 1,..,t$. By covering homotopy techniques we obtain a (k+1)-framefield $\tilde{u}$ on $M - \bigcup_\ell z_\ell \times [\frac{3}{4}, 1]$ whose first component vectorfield is v and such that $\tilde{u}|M \times \{0\} = u$ (the lower half of figure 6.7 provides a picture of this type of situation); $\tilde{u}|(M - \{z_1,...z_t\}) \times \{1\}$ is given by a k-framefield w tangent to $M \times \{1\}$ and by the outward normal vectorfield v. If we define $r((M)) = \text{index}(v)$, then the index of the k-field w is clearly just $\partial(r((M)))$. ■

Proof of theorem 20.1. Since $pr_*(\text{index } w)$ is the index of a 1-field on M, we have for even n

$$2 \cdot r((M)) = pr_* \circ \partial(r((M))) = pr_*(\text{index } w) = \chi(M) .$$

If $\chi(M)$ vanishes, then so does $r((M))$ and index(w) (for $k =$ stable span(M)), and M carries a k-framefield. If $\chi(M) \neq 0$, every vectorfield on M has zeroes. ■

Proof of theorem 20.4. Consider diagram 20.11 and note that $i_{k*}(\omega_k(M))$ equals the obstruction $\omega_{k+1}(\mathbb{R}^{k+1}, TM \oplus \mathbb{R})$ to the existence of a stable k-framefield on M. If $k = \min\{\frac{n-1}{2}, s(M), \text{stable span}(M)\}$, this obstruction vanishes, i_{k*} is bijective, and the vanishing of $\omega_k(M)$ implies that $\text{span}(M) \geq k$, by theorem 13.3. If

$k = \min\{\text{span}(S^n), \text{ stable span}(M)\}$, then in 20.12 $\text{index}(w) = \partial(r((M))) = 0$, and again $k \leq \text{span}(M)$.

Next apply lemma 20.12 when stable span$(M) \geq k > s(M)$ and n is odd. Consider the subgroups

$$2 \cdot \mathbb{Z} \;=\; pr_* \cdot \partial(\pi_{n+1}(S^{n+1})) \;\subset\; \text{im } pr_* \;\subset\; \text{im } \Delta^k$$

of $\pi_n(S^n) = \mathbb{Z}$; since i_{k*} is not injective and hence $d \not\equiv 0$, $\text{im } \Delta^k \neq \mathbb{Z}$, and all the subgroups above must coincide. Consequently, $\Theta_M \circ \sigma$ restricts to an isomorphism

$$\text{(20.13)} \qquad \left(\begin{array}{ccc} \mathbb{Z}_2 & & \\ \cup\!\!\!| & & \\ (r((M)),\text{mod } 2) & \longleftrightarrow & \end{array}\right) \overset{\cong}{} \begin{array}{ccc} \text{im}(\partial) & \xrightarrow{\;\cong\;} & \text{im}(d) \\ \cup\!\!\!| & & \cup\!\!\!| \\ \text{index}(w) & \longleftrightarrow & \omega_k(M) \end{array},$$

and $\omega_k(M) = 0$ if and only if $r((M))$ is even. Thus if $\omega_k(M)$ vanishes for $k = s(M) + 1$, then $\omega_k(M)$, and hence index(w), are also trivial for $k =$ stable span(M), and the last claim in theorem 20.4 follows. ∎

<u>Remark 20.14.</u> Fix $0 < k \leq n$, n odd. Suppose $\ell(M) \in \mathbb{Z}_2$ is an invariant, defined for closed connected n-manifolds M having stable span $\geq k$, such that

(i) $\ell(M) \neq \ell(M \# S^1 \times S^{n-1})$ for such M, and

(ii) $\ell(M) = 0$ if span$(M) \geq k$.

Then $\ell(M)$ coincides with the $\mathbb{Z}_2$-valued invariant $\omega_{s(M)+1}(M)$ of theorem 20.4 whenever stable span$(M) \geq k > s(M)$.

Indeed, if $\omega_{s(M)+1}(M)$ vanishes, then so does $\ell(M)$, since then span$(M) \geq k$. If $\omega_{s(M)+1}(M) \neq 0$, then by the last proof, the index $r((M))$ of the vector field v on $M \times I$ (constructed in 20.12) is odd. As in the proof of the stability theorem 6.6 (see also figure 6.7) we can attach a handle $D^n \times D^1$ and extend v so that its index becomes even and v is pointing outward everywhere on the new "top" boundary $M \# S^1 \times S^{n-1}$. Thus the extended k-field w has index zero there, and we deduce from properties (i) and (ii) above that $\ell(M) \neq \ell(M \# S^1 \times S^{n-1}) = 0$.

Exercise 20.15. Consider the case $k = n \equiv 1(2)$. Given a closed n-manifold M with a stable framing $u : \mathbb{R}^{n+q} \xrightarrow{\cong} TM \oplus \mathbb{R}^q$, $q \geq 2$, let $g : M \longrightarrow V_{n+q,q}$ correspond to the monomorphism $u^{-1} \mid \mathbb{R}^q$, and define the generalized curvatura integra by

$$c(M, u) \;=\; g_*([M]) \;\in\; H_n(V_{n+q,q}; \mathbb{Z}) \;=\; \mathbb{Z}_2 \;.$$

In his paper "Relative Characteristic Classes", Amer. J. of Math. 79 (1957), 517-588, M. Kervaire showed that $c(M,u) - \chi_2(M)$ is the stable Hopf invariant of $[M,u] \in \Omega_n^{fr} \cong \pi_n^S$, so that for $n \neq 1, 3, 7$, by the work of Adams, $c(M,u)$ coincides with the mod 2 Kervaire semicharacteristic $\chi_2(M)$ of M (cf. 15.14).

(i) Deduce the main result of Bredon-Kosinski [10]

$$\operatorname{span}(M) = \begin{cases} \operatorname{span}(S^n) & \text{if } \chi_2(M) \neq 0 \,, \\ n & \text{if } \chi_2(M) = 0 \,, \end{cases}$$

for connected π-manifolds M from 20.4, 20.6 and 20.14 in case $n \neq 15$.

(ii) Show by a direct geometric argument that $c(M,u)$ equals the index $r((M))$ (of the vector field v in 20.12), reduced mod 2. ∎

Proof of proposition 20.8. In sections 14 and 15 we have studied the obstruction groups $\Omega_{k-1}(P^{k-1} \times M; \phi_{M,k})$ for $k = 2, 3$ and 4. Now we have to compare them to the groups $\Omega_{k-1}(P^k \times M; \phi_{M,k})$ via homomorphisms such as i_{k*} (cf. 20.2) which are compatible with the diagrams 14.2, 14.7 and 15.2 (for $P^{k-1} \times M$) and the corresponding diagrams for $P^k \times M$.

First, let $k = 2$ and $n \equiv 1(4)$. Then i_{2*} fits into the commuting diagram of exact sequences

$$\begin{array}{ccccccccc}
\overline{\Omega}_2(P^1\times M;\lambda) & \xrightarrow{\sigma\circ j_2} & \mathbb{Z}_2 & \xrightarrow{\delta_1} & \Omega_1(P^1\times M;\phi_{M,2}) & \xrightarrow{f_1} & \overline{\Omega}_1(P^1\times M;\lambda) & \longrightarrow & 0 \\
\Big\downarrow & & \Big\| & & \Big\downarrow{\scriptstyle i_{2*}} & & \Big\downarrow\wr\!\| & & \\
 & & & & & & H_1(M;\mathbb{Z}_2) & & \\
 & & & & & & \Big\uparrow\wr\!\| & & \\
\overline{\Omega}_2(P^2\times M;\lambda) & \xrightarrow{\sigma\circ j_2} & \mathbb{Z}_2 & \xrightarrow{\delta_1} & \Omega_1(P^2\times M;\phi_{M,2}) & \xrightarrow{f_1} & \overline{\Omega}_1(P^2\times M;\lambda) & \longrightarrow & 0
\end{array}$$

(see also 14.2). For $x_2 = [P^2,(\mathrm{id},\mathrm{const}),\mathrm{or}] \in \overline{\Omega}_2(P^2\times M;\lambda)$ we have (cf. 9.3 and 9.10)

$$\begin{aligned}
\sigma\circ j_2(x_2) &= (\mathrm{id},\mathrm{const})^* \, (w_1(\lambda)^2 + w_1(\lambda)\, w_1(M))\, [P^2] \\
&= w_1(P^2)^2\, [P^2] \\
&= 1\,,
\end{aligned}$$

and therefore the lower homomorphism f_1 is bijective. By lemma 20.11 and the calculation preceding theorem 14.4, the following are equivalent:

(i) $s(M) \geq 2$;

(ii) i_{2*} is bijective;

(iii) the upper homomorphism f_1 is bijective;

(iv) $w_1(M)^2 \neq 0$.

We conclude that $s(M) = 1$ if and only if $w_1(M)^2 = 0$.

Similarly, for $k = 3$ and $n \equiv 1(4)$ we obtain the commuting diagram of exact sequences

$$\begin{array}{ccccccc}
\Omega_3(P^2\times M) & \xrightarrow{f_1'\circ\sigma\circ j_3} & H_1(P^2\times M;\mathbb{Z}_2) & \xrightarrow{\delta_2\circ f_1'^{-1}} & \Omega_2(P^2\times M;\phi_{M,3}) & \xrightarrow{f_2} & \Omega_2(P^2\times M) \\
\Big\downarrow & & \Big\downarrow\wr\!\| & & \Big\downarrow{\scriptstyle i_{3*}} & & \Big\downarrow\wr\!\| \\
\Omega_3(P^3\times M) & \xrightarrow{f_1'\circ\sigma\circ j_3} & H_1(P^3\times M;\mathbb{Z}_2) & \xrightarrow{\delta_2\circ f_1'^{-1}} & \Omega_2(P^3\times M;\phi_{M,3}) & \xrightarrow{f_2} & \Omega_2(P^3\times M)
\end{array}$$

(see also 14.7, 9.3 and 14.6); the bijectivity of the right hand vertical arrow follows from fact 9.4, the universal coefficient theorem and the fact that $H_2(P^2;\mathbb{Z}) = H_2(P^3;\mathbb{Z}) = 0$.

Given $x = [N,g] \in \Omega_3(P^i \times M)$, $i = 2$ or 3, we have as in 14.10 and 14.11 that

$$w_1(\lambda)(f_1^! \circ \sigma \circ j_3(x)) = (w_1(\lambda)^3 + w_1(\lambda)w_1(M)^2)[N]$$

and

$$y(f_1^! \circ \sigma \circ j_3(x)) = w_1(\lambda)(y^2 + y\,w_1(M))\;[N]$$

for $y \in H^1(M;\mathbb{Z}_2)$.

If $w_1(M)^2 = y^2 + yw_1(M)$ for some $y \in H^1(M;\mathbb{Z}_2)$ then $w_1(\lambda)+y$ is trivial on the image of $\Omega_3(P^2 x M)$ under $f_1^! \circ \sigma \circ j_3$, but not on the image $f_1^! \circ \sigma \circ j_3(x_3) = (1;0)$ of $x_3 = [P^3,(id,const)] \in \Omega_3(P^3 x M)$; hence $\delta_2 f_1^{!\,-1}(1;0)$ is a nontrivial element of $\ker(i_{3*})$. On the other hand, if $w_1(M)^2$ is not of the form above, then the examples in the proof of theorem 14.14 show that $\Omega_3(P^2 x M)$ and $\Omega_3(P^3 x M)$ have equal images under $f_1^! \circ \sigma \circ j_3$; therefore i_{3*} is now injective. We conclude that $s(M) \geq 3$ precisely in the second case.

Finally, assume $k = 4$ and n odd. As in § 15 we will consider the cases Y = point, M and $B(S)O(n)$ simultaneously, and the n-plane bundle η over Y is $\mathbb{R}^n$, TM or γ respectively. The relevant coefficient bundle is $\phi_Y = \lambda \otimes \eta - 4\lambda - \eta$ over $P^i \times Y$, where we study the cases $i = 3$ and 4. From theorem 9.3 and the facts following it we obtain the commuting diagram of exact sequences

(20.16)

$$\begin{array}{ccccccc}
 & & 0 & & & & \\
 & & \downarrow & & & & \\
 & & \ker f_2' & & & & \\
 & & \downarrow & & & & \\
\overline{\Omega}_4(P^i x Y;\lambda) & \xrightarrow{\sigma\circ j_4} & \Omega_2(P^i x Y x BO(2);\phi_Y+\Gamma) & \xrightarrow{\delta_3} & \Omega_3(P^i x Y;\phi_Y) & \longrightarrow & \mathfrak{N}_3(Y) \\
 & \searrow^{\overline{\sigma}=(\overline{\sigma}_1,\overline{\sigma}_2,\overline{\sigma}_3)} & \downarrow f_2' & & & & \\
 & & (\ker : H_2(P^i x Y;\mathbb{Z}_2) \xrightarrow{\binom{n}{2}w_1(\lambda)^2+w_1(\lambda)w_1(\eta)} \mathbb{Z}_2) \oplus \mathbb{Z}_2 \oplus \mathbb{Z}_2 & & & & \\
 & & \downarrow & & & & \\
 & & 0 & & & &
\end{array}$$

where for $x = [N^4 \xrightarrow{\ g\ } P^i \times Y; g^*(\lambda) \cong \xi_N] \in \overline{\Omega}_4(P^i \times Y;\lambda)$

$$w_1(\lambda)^2(\overline{\sigma}_1(x)) = ((1+\tbinom{n}{2}))w_1(N)^4 + w_1(N)^2 w_1(\eta)^2)[N];$$
$$w_1(\lambda)y(\overline{\sigma}_1(x)) = w_1(N)^2((1+\tbinom{n}{2}))y^2 + yw_1(\eta))[N], \qquad y\in H^1(Y;\mathbb{Z}_2);$$
$$z(\overline{\sigma}_1(x)) = (\tbinom{n}{2})w_1(N)^2 z + w_1(\eta)^2 z + w_1(\eta)Sq^1(z) + z^2)[N], \quad z\in H^2(Y;\mathbb{Z}_2);$$

and

$$\overline{\sigma}_2(x) = (w_4(N) + \tbinom{n}{4}w_1(N)^4 + w_1(N)^2(w_2(\eta) + \tbinom{n}{2}w_1(\eta)^2) + w_1(\eta)^4)[N]$$

and

$$\overline{\sigma}_3(x) = (w_4(N) + \tbinom{n}{2}w_1(N)^4 + w_1(N)^2 w_1(\eta)^2)[N]$$

(cf. also 15.2 and 15.4). E.g. for $x_4 = [P^4,(\mathrm{id},\mathrm{const}),\mathrm{or}] \in \overline{\Omega}_4(P^4 x Y;\lambda)$ we have

$$\overline{\sigma}\,(x_4) = ((1+\tbinom{n}{2});\ 0,\ 0);\ \ 1+\tbinom{n}{4};\ 1+\tbinom{n}{2})\ ;$$

on the other hand, note that $w_1(N)^4 = g^*(w_1(\lambda)^4) = 0$ when $i = 3$.

Observe that the second term in the horizontal sequence in 20.16 does not depend on whether we chose i to equal 3 or 4. If there is an element $z \in H^2(M;\mathbb{Z}_2)$ such that

$$w_1(M)^4 = w_1(M)^2 z + w_1(M)Sq^1(z) + z^2 ,$$

and if in addition either

$n \equiv 1(8)$ and $w_2(M) = y^2 + yw_1(M)$ for some $y \in H^1(M;\mathbb{Z}_2)$,

or

$n \equiv 5(8)$ and $w_2(M) = w_1(M)^2$,

then, for $Y = M$, $\bar{\sigma}(x_4)$ lies in $\bar{\sigma}(\bar{\Omega}_4(P^4 xM;\lambda))$, but not in $\bar{\sigma}(\bar{\Omega}_4(P^3 xM;\lambda))$, and hence δ_3 (taken for $i = 3$) maps $\sigma \circ j_4(x_4)$ to a nontrivial element of $\ker(i_{4*})$; consequently, $s(M) \leq 3$ in this situation.

It remains to discuss the case $n \equiv 3(8)$. If

$$w_1(M)^4 + w_1(M)w_3(M) + w_2(M)^2 = 0$$

then $w_2(M)\bar{\sigma}_1 + \bar{\sigma}_2 + \bar{\sigma}_3$ vanishes on $\bar{\Omega}_4(P^3 xM;\lambda)$, but not on $x_4 \in \bar{\Omega}_4(P^4 xM;\lambda)$, and again $\ker(i_{4*}) \neq 0$, and $3 \geq s(M)$ ($\geq$ span $S^n = 3$).

To prove the converse, note first that

$$i_{4*} : \Omega_3(P^3 \times BO(n); \phi_{BO(n)}) \longrightarrow \Omega_3(P^4 \times BO(n); \phi_{BO(n)})$$

is bijective; indeed, consider the diagram in 20.11 with M replaced by $BO(n)$; then $\ker i_{4*}$ is spanned by $d(1) = \Theta \circ \sigma(\partial[Id] = \omega_4'(S^n) = 0$ since $\partial[Id]$ is the index of a 4-field on S^n (cf. proof of 20.11) and $\omega_4'(S^n)$ depends only on the bordism class of S^n in $\mathfrak{N}_n(0,0) \cong \mathfrak{N}_n$ (cf. 13.4 and 12.1). Therefore it follows from proposition 15.6 that $\delta_3(\ker f_2')$ (cf. 20.16) is isomorphic to $\mathbb{Z}_2$ for all choices of Y and $i = 3$ and 4. In particular, when $Y = M$ then $\delta_3(\ker f_2')$ cannot contribute to the kernel of i_{4*}. If $w_1(M)^4 + w_1(M)w_3(M) + w_2(M)^2 \neq 0$, we can find enough examples of elements in $\bar{\Omega}_4(P^3 xM;\lambda)$ (e.g. $[\mathbb{C}P(2),\text{const}]$, $[P^2 \times L \xrightarrow{\pi_2} L \longrightarrow M$, or] etc., see the discussion preceding 15.4), to show that $\bar{\sigma}(\bar{\Omega}_4(P^4 xM;\lambda))$ is already all of $\bar{\sigma}(\bar{\Omega}_4(P^4 xM;\lambda))$; it follows then that i_{4*} is injective and $s(M) \geq 4$. This completes the proof of proposition 20.8. ∎

Exercise 20.17. Given a closed connected manifold N and an even number $q > 0$ show that

$$\operatorname{span}(N \times S^q) = \begin{cases} \text{stable span}(N) + q & \text{if } \chi(N) = 0 \\ 0 & \text{if } \chi(N) \neq 0 . \end{cases}$$

Compare this to the stable span of $N \times S^q$.

Example 20.18. Consider the n-manifold $M = N \times S^q$ where N is closed and connected, and $q \geq 1$ is odd.

If the Euler number $\chi(N)$ is even, then

$$\operatorname{span}(M) = \text{stable span}(M) = \text{stable span}(N) + q > q .$$

Indeed, since $\omega_2(\underset{\sim}{\mathbb{R}}^2, TN \oplus \underset{\sim}{\mathbb{R}}) = i_{1*}(\chi(N)) = 0$ (compare 20.11), we can write $TN \oplus \underset{\sim}{\mathbb{R}} \cong \rho \oplus \underset{\sim}{\mathbb{R}}^2$ and $TS^q \cong \underset{\sim}{\mathbb{R}} \oplus \tau$ and therefore

$$\begin{aligned} T(N \times S^q) &\cong \pi_1^*(TN) \oplus \underset{\sim}{\mathbb{R}} \oplus \pi_2^*(\tau) \\ &\cong \pi_1^*(\rho) \oplus \underset{\sim}{\mathbb{R}}^2 \oplus \pi_2^*(\tau) \\ &\cong \pi_1^*(\rho) \oplus \underset{\sim}{\mathbb{R}}^{q+1} \\ &\cong \pi_1^*(TN \oplus \underset{\sim}{\mathbb{R}}^q) . \end{aligned}$$

Now assume that $\chi(N)$ is odd. Then

$$\operatorname{span}(S^q) \leq \operatorname{span}(M) \leq \text{stable span}(M) = q$$

(since $w_{n-q}(M) = w_{n-q}(N) \times 1 \neq 0$) and

(20.19) $$\max \{\operatorname{span}(S^n), \operatorname{span}(M)\} \leq s(M) .$$

To see the last inequality, identify $S^q \times I$ with $\{x = (x_1, .., x_{q+1}) \in \mathbb{R}^{q+1} \mid 1 \leq \sum x_i^2 \leq 4\}$ and pick a nondegenerate vectorfield which points outwards everywhere along the outer boundary sphere (of radius 2) and which takes the constant value $(1,0,..,0)$ at all points of the inner boundary sphere (of radius 1). Combine this with a nondegenerate vectorfield on N to obtain a vectorfield v on $N \times S^q \times I$ which extends to a stable q-framefield along $N \times S^q \times \{0\}$

and which points outwards at $N \times S^q \times \{1\}$. If $s(M)$ were strictly smaller than span(M), then we could apply (20.13) to $k = s(M) + 1$ and $r((M)) \equiv \text{index } v \equiv \chi(N) \equiv 1(2)$, and deduce $\omega_k(M) \neq 0$ which contradicts the assumption $k \leq \text{span}(M)$.

If in particular $M = P^2 \times S^{n-2}$, $n \equiv 5(8)$, $n > 5$, then proposition 20.8 combines with the inequalities above to show that $\text{span}(M) = 3$ which is different both from $\text{span}(S^n) = 1$ and from stable $\text{span}(M) = n - 2 \geq 11$.

<u>Exercise 20.20.</u> Consider $M = P^2 \times S^{n-2}$ for $n \equiv 3(8)$, $n > 3$. Show that $\text{span}(S^{n-2}) = 1$, but $\text{span}(M) = s(M) = \text{span}(S^n) = 3$.∎

<u>Remark 20.21.</u> In his recent Rutgers University thesis (unpublished), V. Eagle has also obtained versions of theorem 20.1 and corollaries 20.9 and 20.10 by different methods. On the basis of this and other evidence he conjectured for all closed n-manifolds M that span(M) equals $\text{span}(S^n)$ or stable span(M). This is disproved by the counterexample in 20.18. Further counterexamples (among them orientable ones) will be given below.∎

Next we outline a general approach which sometimes gives an upper bound for $s(M)$. Let the dimension n of M be odd, and let k be a natural number. Put $\phi_M^+ = TM \otimes \lambda \oplus \underset{\sim}{\mathbb{R}}$, $\phi_M^- = \underset{\sim}{\mathbb{R}}^k \otimes \lambda \oplus TM$, and extract the following (horizontal) exact sequence from 7.12 and the beginning of § 9

$$\begin{array}{ccccc}
\Omega_k(P^\ell \times M, (\phi_M^+, \phi_M^-), \lambda^{\otimes k+1}, n+k-1) & \xrightarrow{\sigma \circ j} & \Omega_{k-2}(P^\ell \times M \times BO(2); \phi_M + \Gamma) & \xrightarrow{\delta} & \Omega_{k-1}(P^\ell \times M; \phi_M) \\
(20.22) & & \Big\downarrow f' & & \\
 & & \overline{\Omega}_{k-2}(P^\ell \times M \times BO(2); \lambda^{\otimes k+1} \otimes \xi_{\gamma_2}) & & .
\end{array}$$

Here we consider the two cases $\ell = k - 1$ and $\ell = k$; the resulting two diagrams are compatible with the obvious homomorphisms induced by

$P^{k-1} \subset P^k$, which can actually be used to identify corresponding terms in the middle of the diagrams. Clearly,

$$i_{k*} \ : \ \Omega_{k-1}(P^{k-1} \times M; \phi_M) \longrightarrow \Omega_{k-1}(P^k \times M; \phi_M)$$

is not injective (and hence $s(M) < k$) if the following two conditions hold;

(a) the charateristic number

$$\Omega_k(P^k \times M, (\phi_M^+, \phi_M^-), \lambda^{\otimes k+1}, n+k-1) \xrightarrow{\overline{f}} \overline{\Omega}_k(P^k \times M; \lambda^{\otimes k+1}) \xrightarrow{w_1(\lambda)^k \circ \mu} \mathbb{Z}_2$$

factors through the homomorphism $f' \circ \sigma \circ j$ (most of which can be calculated using fact 9.11); and

(b) the characteristic number $w_1(\lambda)^k \circ \mu \circ \overline{f}$ is nontrivial on some element x_k in its domain.

To get such an x_k, start from

$$\overline{x}_k \ = \ [P^k, (\mathrm{id}, \mathrm{const}), \mathrm{or}] \ \in \ \overline{\Omega}_k(P^k \times M; \lambda^{\otimes k+1}) \ ,$$

decompose $\overline{f}$ into a string of forgetful maps, each fitting into an exact sequence ã la 7.12 (for the calculation of the $\sigma \circ j$-terms see also proposition 5.3 and fact 9.7), and try step by step to construct an element $x_k \in \overline{f}^{-1}(\overline{x}_k + t)$, where the correction term t lies in the kernel of $w_1(\lambda)^k \circ \mu$; this recipe for finding the desired x_k seems to work at least for $k \leq 8$.

For an illustration of this whole approach, assume that $k = 5$. Then $w_1(\phi_M)$ vanishes, and we have

$$\begin{aligned}
w_2(\phi_M) &= \tbinom{n}{2} w_1(\lambda)^2 + w_1(\lambda) w_1(M) \\
w_3(\phi_M) &= w_1(\lambda)^2 w_1(M) + w_1(\lambda) w_1(M)^2 = Sq^1(w_1(\lambda) w_1(M)) \\
w_4(\phi_M) &= \tbinom{n+3}{4} w_1(\lambda)^4 + \tbinom{n+2}{2} w_1(\lambda)^3 w_1(M) + w_1(\lambda)^2 (w_1(M)^2 + w_2(M)) + \\
&\quad + w_1(\lambda)(w_1(M)^3 + w_1(M) w_2(M) + w_3(M))
\end{aligned}$$

(see in particular 9.10). Also we get the diagram

$$
\begin{array}{ccccc}
\Omega_5(P^\ell \times M, (\phi_M^+, \phi_M^-), \underset{\sim}{\mathbb{R}}, n+4) & \xrightarrow{\sigma \circ j} & \Omega_3(P^\ell \times M \times BO(2); \phi_M + \Gamma) & \xrightarrow{\delta} & \Omega_4(P^\ell \times M; \phi_M) \\
\Big\downarrow \overline{f} \text{ onto} & & \Big\downarrow f' & & \\
\Omega_5\ (P^\ell \times M) & & \overline{\Omega}_3(P^\ell \times M \times BO(2); \xi_{\gamma_2})\ . & &
\end{array}
\qquad (20.23)
$$

Given an element

$$x \ = \ [g : N \longrightarrow P^\ell \times M,\ \overline{g} : TN \oplus (TM \otimes \lambda \oplus \underset{\sim}{\mathbb{R}}) \xrightarrow{n+4} \underset{\sim}{\mathbb{R}}^5 \otimes \lambda \oplus TM,\ \text{or}]$$

in the domain of $\sigma \circ j$, note first that N is an oriented 5-manifold and hence

$$
\begin{array}{llll}
(Sq^1 u) \cdot v & = & u \cdot (Sq^1 v) & \text{for } u,\ v \in H^*(N;\mathbb{Z}_2),\ \dim(u \cdot v) = 4; \\
w_2(N) \cdot w & = & Sq^2(w) & \text{for } \quad w \in H^3(N;\mathbb{Z}_2) \quad ; \\
w_3(N) \cdot z & = & Sq^2 \circ Sq^1(z) & \text{for } \quad z \in H^2(N;\mathbb{Z}_2) \quad ; \text{ and} \\
w_4(N) & = & w_2(N)^2 & (= w_2(N)^2 + U_4)
\end{array}
$$

(use facts 9.5, 9.6 and the characterization of the Wu classes $U_1, U_2, \ldots$ of N by the equations $w_s(N) = \sum_{r=0}^{s} Sq^{s-r}(U_r)$, see e.g. [46], theorem 11.14). Combining this with fact 9.11, we obtain the following table.

(20.24)

	$\alpha^5 = w_1(\lambda)^5$	$\alpha^3 w_1(M)^2$	$\alpha^3 w_2(M)$	$\alpha\, w_1(M)^4$	$\alpha\, w_1^2 w_2(M)$	$\alpha\, w_1 w_3(M)$	$\alpha\, w_2(M)^2$	
$w_1(M)^3$				$\binom{n}{2}+1$				$+w_1(M)^5$
$w_1(\lambda)\ \overline{w}_2(\gamma_2)$	$\binom{n+3}{4}$	$\binom{n}{2}$	1	1				
$w_1(\lambda)^3$	$\binom{n}{2}+1$	1						
$w_1(M)\ \overline{w}_2(\gamma_2)$		$\binom{n-1}{4}+\binom{n}{2}$			1			
$w_1(M) w_2(M)$					1	$\binom{n}{2}$		$+w_1 w_2^2(M)$
$w_1(\lambda) w_2(M)$			$\binom{n}{2}$			1	1	

Here we put, for short, $\alpha = w_1(\lambda)$ and $\overline{w}_2(\gamma_2) = w_1(\gamma_2)^2 + w_2(\gamma_2)$;

the expressions in the left hand column, when evaluated on $f\circ\sigma\circ j(x)$, equal the sum of 5-dimensional cohomology classes given to the right, when pulled back to N and evaluated on its fundamental class $[N]$; e.g. $w_1(M)^3(f\circ\sigma\circ j(x)) = (((\binom{n}{2}+1)\, w_1(\lambda)w_1(M)^4 + w_1(M)^5)[N]$. Similarly, for every $y \in H^1(M;\mathbb{Z}_2)$ we have the following table.

(20.25)	α^3y^2	$\alpha^3yw_1(M)$	αy^4	$\alpha y^3w_1(M)$	$\alpha y^2w_1(M)^2$	$\alpha y^2w_2(M)$	$\alpha yw_1(M)^3$	
$y\bar{w}_2(\gamma_2)$	$\binom{n+3}{4}$	$\binom{n}{2}+1$			1	1	1	
$yw_1(\lambda)^2$	$\binom{n}{2}+1$	1						
$yw_1(\lambda)w_1(\gamma_2)$	1	1	1		1			
y^3			$\binom{n}{2}$	1				$+y^5$

Of course, these few calculations are far from giving a complete picture of $f'\circ\sigma\circ j$; nevertheless, they suffice sometimes to yield useful information about $s(M)$.

<u>Proposition 20.26. Let M be a closed n-manifold. Then</u> $s(M) \leq 4$ <u>in each of the following cases:</u>

<u>(i) $n \equiv 1(8)$ and $w_2(M) = 0$ and $w_1(M)^5 = 0$;</u>

<u>(ii) $n \equiv 5(8)$ and $w_1(M)w_2(M)^2 = 0$;</u>

<u>(iii) $n \equiv 3(8)$ and $w_1(M)^4 + w_1(M)w_3(M) + w_2(M)^2 = y^4 + y^2w_2(M) + yw_1(M)^3$ for some $y \in H^1(M;\mathbb{Z}_2)$;</u>

<u>(iv) $n \equiv 3(8)$ and there exists some $y \in H^1(M;\mathbb{Z}_2)$ such that $w_1(M) = 0$, $w_2(M) = y^2$ and $y^5 = 0$.</u>

<u>Proof.</u> We have only to check that the conditions (a) and (b) (following 20.22) hold. Since $\bar{f}$ is onto in (20.23), condition (b) presents no problem. It remains to express the characteristic number $w_1(\lambda)^5[N]$ of x in terms of characteristic numbers of $f\circ\sigma\circ j(x)$. In case (i) just add the first two lines in the table (20.24); in case

(ii) add lines 3, 4 and 5 there; in case (iii) add lines 2, 3 and 6 in (20.24) together with the first three lines of table (20.25); and finally, in case (iv) the sum of line 2 in (20.24) and of the last two lines in (20.25) will give the required factorization of $w_1(\lambda)^5 \circ \mu \circ \bar{f}$ through $f' \circ \sigma \circ j$. ∎

Example 20.27. If $n \equiv 1(4)$ and Q is a nonempty closed spin manifold of dimension $n - 4$, then $s(P^4 \times Q)$ equals 3 or 4 (use 20.8 and 20.26 (i), (ii)). In particular, if

$$M = P^4 \times \mathbb{H}P(2t) \times S^q, \quad t \geq 0, \quad q \equiv 1(4), \quad q > 1,$$

then, by 20.5 and 20.19, $s(M) = \mathrm{span}(M) = 3$ or 4, while $\mathrm{span}(S^n) = \mathrm{span}(S^q) = 1$ and stable $\mathrm{span}(M) = q \geq 5$.

Example 20.28. If $n \equiv 3(8)$, $r \equiv 2(4)$, $r > 2$, and $Q \neq \emptyset$ is a closed spin manifold of dimension $n - r$, then $s(P^r \times Q) = 4$; indeed, since $w_1(Q)$, $w_2(Q)$ and hence $w_3(Q) = Sq^1(w_2(Q))$ vanish, we have $w_i = w_i(P^r \times Q) = y^i$ for $i = 1, 2, 3$ and y the nontrivial element in $H^1(P^r;\mathbb{Z}_2)$; therefore $w_1^4 + w_1 w_3 + w_2^2$ has the form required in 20.26 (iii), but does not vanish (as required in proposition 20.8).

In particular, the n-manifold

$$M = P^r \times \mathbb{H}P(2t) \times S^q, \ r \equiv 2(4), \ r > 2, \ t \geq 0, \ q+r \equiv 3(8), \ q > 1,$$

satisfies the (in)equalities

$$\mathrm{span}(S^n) = 3 < \mathrm{span}(M) = s(M) = 4 < \text{stable span}(M) = q$$

(use again corollary 20.5 and the estimates in example 20.18); in contrast, note that $\mathrm{span}(S^q) = 1$. ∎

So far all the counterexamples (in 20.18, 20.27 and 20.28) to V. Eagle's conjecture (see remark 20.21) have been nonorientable. We can use proposition 20.26(iv) to exhibit a family of orientable counterexamples.

Example 20.29. Consider the n-manifold

$$M = (((P^5 \# -P^5) \times S^{8t-5}) \# \mathbb{H}P(2t)) \times S^q$$

for $t > 0$ and $q \equiv 3(8)$, $q > 3$. In view of 20.5, 20.8, 20.19 and 20.26(iv) we obtain the (in)equalities

$$\operatorname{span}(S^q) = \operatorname{span}(S^n) = 3 < \operatorname{span}(M) = s(M) = 4 < \text{stable span}(M) = q. \blacksquare$$

Finally, let us note that $s(M)$ can take fairly high values for suitable M.

Proposition 20.30. *If $n \equiv 1(4)$, $n > 1$, (resp. if $n \equiv 3(4)$), there exists a nonorientable (resp. an orientable) closed connected n-dimensional manifold M such that*

$$s(M) \geq \frac{n-1}{2} .$$

Proof. Put $k = (n-1)/2$. We may assume that $k > \operatorname{span}(S^n)$ (otherwise our claim follows from 20.6). Then the index z of a suitable k-field on S^n is a nontrivial element in the kernel of

$$\Theta^{(or)} = \Theta \circ \sigma : \pi_{n-1}(V_{n,k}) \longrightarrow \Omega_{k-1}(P^{k-1} \times B(S)O(n); \phi)$$

since $\Theta^{(or)}(z) = \omega_k'(S^n)$ depends only on the (trivial) bordism class of S^n (see 13.8, 13.4 and 12.1). According to theorem 16.11 there exists an arbitrary (resp. oriented) closed connected n-manifold M satisfying $\omega_k(M) = 0$ which carries also a k-field with finite singularities with index z. Now apply lemma 20.11 and the first observation in its proof. Thus the kernel of i_{k*} is generated by

$$\pm\, d(\text{point}) = \Theta_M \circ \sigma\, (\partial[Id]) = \Theta_M \circ \sigma\, (z) = \omega_k(M) = 0 ,$$

and therefore $s(M) \geq k$. If $n \equiv 1(4)$, $n > 1$, then M cannot be orientable, by proposition 20.8. ∎

Exercise 20.31. Given a natural number n, let $s(B(S)O(n))$ denote the maximum number $k \leq n$ for which the obvious homomorphism

$$i_{k*} \ : \ \Omega_{k-1}(P^{k-1} \times B(S)O(n); \phi) \longrightarrow \Omega_{k-1}(P^{k} \times B(S)O(n); \phi)$$

(see also 13.2) is bijective.

(i) Using the techniques of lemma 20.11 and of § 13, show that for $1 \leq k \leq n$ the kernel of i_{k*} is generated by $\omega_k'(S^n)$ and hence

$$s(BO(n)) = \begin{cases} 0 & \text{if } n \equiv 0(2) , \\ n & \text{if } n \equiv 1(2) , \end{cases}$$

and

$$s(BSO(n)) = \begin{cases} 0 & \text{if } n \equiv 0(2) , \\ 1 & \text{if } n \equiv 1(4) , \\ n & \text{if } n \equiv 3(4) . \end{cases}$$

(ii) Deduce again that $s(M) = 0$ for all even-dimensional closed manifolds M, and that $s(M) \leq 1$ for orientable closed manifolds M of dimension $n \equiv 1(4)$. ∎

References

[1] J.F. Adams, Vector fields on spheres, Ann. Math. 75 (1962), 603-632.

[2] ——————, On the groups J(X). IV. Topology 5 (1966), 21-71.

[3] M.F. Atiyah, Thom complexes, Proc. London Math. Soc. 11 (1961), 291-310.

[4] ——————, Vector fields on manifolds, Arbeitsgemeinschaft für Forschung des Landes Nordrhein-Westfalen, Heft 200, Düsseldorf, 1969.

[5] ——————, R. Bott and A. Shapiro, Clifford modules, Topology 3 (1964), suppl. 1, 3-38.

[6] —————— and J. Dupont, Vectorfields with finite singularities, Acta math. 128 (1972), 1-40.

[7] —————— and Ł. Smith, Compact Lie groups and the stable homotopy of spheres, Topology 13, (1974), 135-142.

[8] J.C. Becker, Vector fields on quotient manifolds of spheres, Indiana Univ. Math. J. 22 (1973), 859-871.

[9] A. Borel and F. Hirzebruch, On characteristic classes of homogeneous spaces II, Amer. J. Math. 81 (1959), 315-382.

[10] G. Bredon and A. Kosinski, Vector fields on π-manifolds, Ann. Math. 84 (1966), 85-90.

[11] R. Brown, Embeddings, immersions and cobordism of differentiable manifolds, Bull. AMS 76 (1970), 763-766.

[12] O. Burlet, Cobordismes de plongements et produits homotopiques, Comm. Math. Helv. 46 (1971), 277-288.

[13] P.E. Conner and E.E. Floyd, Differentiable periodic maps, Springer Verlag, 1964.

[14] M. Crabb and B. Steer, Vector bundle monomorphisms with finite singularities, Proc. London Math. Soc. (3) 30 (1975), 1-39.

[15] J.P. Dax, Etude homotopique des espaces de plongements, Ann. Sc. Ec. Norm. Sup. 5 (1972), 303-377.

[16] ——————, Classes de Stiefel-Whitney en cobordisme normal, C.R. Acad. Sc. Paris 279 (1974), 545-548.

[17] A. Douady, Le problème des modules pour les sous-espaces analytiques compacts d'un espace analytique donné, Ann. Inst. Fourier 1 (1966), 1-95.

[18] J. Dupont, K-theory obstructions to the existence of vector fields, Acta math. 133 (1974), 67-80.

[19] ________ and G. Lusztig, On manifolds satisfying $w_1^2 = 0$, Topology 10 (1971), 81-92.

[20] J. Eells, Alexander-Pontrjagin duality in function spaces, Proc. Symp. Pure Math AMS 3 (1961), 109-129.

[21] S.D. Feit, k-mersions of manifolds, Acta mathematica 122 (1969), 173-195.

[22] D. Frank, On the index of a tangent 2-field, Topology 11 (1972), 245-252.

[23] ________, Vector fields and plane fields on non-orientable manifolds, old preprint, (probably appeared somewhere).

[24] A. Haefliger and M. Hirsch, Immersions in the stable range, Ann. Math. 75 (1962), 231-241.

[25] A. Hatcher and F. Quinn, Bordism invariants of intersections of submanifolds, Trans. AMS 200 (1974), 327-344.

[26] P. Hilton, Generalizations of the Hopf invariant, Colloque de topologie algêbrique, Louvain, 1956, 9-27.

[27] F. Hirzebruch, Neue topologische Methoden in der algebraischen Geometrie, Springer-Verlag, 1962.

[28] H. Hopf, Vektorfelder in n-dimensionalen Mannigfaltigkeiten, Math. Annalen 96 (1926), 225-250.

[29] I. James, Spaces associated with Stiefel manifolds, Proc. London Math. Soc. (3) 9 (1959), 115-140.

[30] U. Karras, M. Kreck, W. Neumann and E. Ossa, Cutting and pasting of manifolds; SK-groups, Publish or Perish 1973, VII, Math. Lecture Series 1.

[31] U. Koschorke, Infinite dimensional K-theory and characteristic classes of Fredholm bundle maps, Proc. Symp. Pure Math., vol 15, AMS 1970, pp. 95-133.

[32] ________, Concordance and bordism of line fields, Inventiones math. 24 (1974), 241-268.

[33] ________, Singularities and bordism of q-plane fields and of foliations, Bull. AMS 80 (1974) 760-765.

[34] U. Koschorke, Bordism of framefields, immersions and k-mersions, Notices AMS 21 (1974), A-17 and A-405.

[35] ____________, Framefields and non-degenerate singularities, Bull. AMS 81 (1975), 157-160.

[36] ____________, Framefields with finite singularities, Notices AMS 21, 6 (1974), A-563.

[37] ____________, Indices of framefields with finite singularities, Notices AMS 22, 1 (1975), A-230.

[38] ____________, Planefields and 2-codimensional foliations, Notices AMS 25, 2 (1978), A-259.

[39] ____________, Multiple points of immersions, and the Kahn-Priddy theorem, Math. Z 169 (1979), 223-236.

[40] N. Kuiper, The homotopy type of the unitary group of Hilbert space, Topology 3 (1965), 19-30.

[41] S. Lang, Introduction to differentiable manifolds, Interscience Publishers, New York, 1962.

[42] G. Lusztig, J. Milnor and F.P.Peterson, Semi-characteristics and cobordism, Topology 8 (1969), 357-359.

[43] J. Maffei, A cobordism theory for k-mersions, Berkeley dissertation, 1973.

[44] W. Massey, On the Stiefel-Whitney classes of a manifold Amer. J. Math. 82 (1960), 92-102.

[45] E.Y. Miller, The homotopy Euler characteristic of manifolds with vectorfields, (open) letter to J.C. Becker, May 1974.

[46] J. Milnor, Lectures on characteristic classes, Annals of Mathematics Studies Nr. 76, (1974).

[47] __________, Differential topology, mimeographed lecture notes, Princeton, 1958.

[48] __________, Morse theory, Ann. of Math. Studies, Princeton Univ. Press, 1963.

[49] __________, Topology from the differentiable viewpoint, University Press of Virginia, 1965.

[50] C. Olk, Immersionen von Mannigfaltigkeiten in euklidische Räume, Dissertation, Siegen, 1980.

[51] G.F. Paechter, The groups $\pi_r(V_{n,m})(I)$, Quart. J. Math. Oxford (2), 7 (1956), 249-268.

[52] R. Palais, Foundations of global non-linear analysis, Benjamin, New York, 1968.

[53] D. Randall, Tangent frame fields on spin manifolds preprint, PUC, Rio de Janeiro

[54] __________, private communication.

[55] B. Reinhart, Cobordism and the Euler number, Topology 2 (1963), 173-177.

[56] G. Ruget, A propos des cycles analytiques de dimension infinie, Inventiones math. 8 (1969), 267-312.

[57] H.A. Salomonsen, Bordism and geometric dimension, Math. Scand. 32 (1973), 87-111.

[58] J.-P. Serre, Homologie singulière des espaces fibrés, Applications, Ann. Math. 54 (1951), 425-505.

[59] N. Steenrod, The topology of fiber bundles, Princeton University Press, 1951.

[60] __________, (with D. Epstein), Cohomology operations, Ann. of Math. Study 50, Princeton 1962.

[61] R. Stong, On fibering of cobordism classes, Trans. Am. Math. Soc. 178 (1973), 431-447.

[62] W. Sutherland, Fibre homotopy equivalence and vector fields, Proc. London Math. Soc. (3) 15 (1965), 543-556.

[63] R. Switzer, Algebraic topology - homotopy and homology. Springer Verlag, 1975.

[64] R. Thom, Quelques propriétés globales des varietés différentiables, Comm. Math. Helv. 28 (1954), 17-86.

[65] E. Thomas, Vector fields on manifolds, Bull. Am. Math. Soc. 75 (1969), 643-683.

[66] H. Toda, Composition methods in homotopy groups of spheres, Ann. of Math. Studies 49, Princeton University Press, 1962.

[67] C.T.C. Wall, Surgery on compact manifolds, Academic Press, London and New York, 1970.

[68] R. Wells, Cobordism groups of immersions, Topology 5 (1966), 281-294.

[69] G.W. Whitehead, On the Freudenthal theorems, Ann. Math. (2) 57 (1953), 209-228.

[70] J.H.C. Whitehead, Combinatorial homotopy I and II, Bull. Am. Math. Soc. 55 (1949), 213-245 and 453-496.

[71] H. Whitney, Singularities of a smooth n-manifold in (2n-1)-space, Ann. Math. 45 (1944), 247-293.

[72] W. Wu, Classes characteristiques et i-carrés d'une variété, C.R. Acad. Sci. Paris 230 (1950), 508-511.

[73] ______, Les i-carrés dans une variété grassmannienne, C.R. Acad. Sci. Paris 230 (1950), 918-920.

Index